The IoT Physical Layer

Ibrahim (Abe) M. Elfadel
Mohammed Ismail
Editors

The IoT Physical Layer

Design and Implementation

Springer

Editors
Ibrahim (Abe) M. Elfadel
Masdar Institute at Khalifa University
Abu Dhabi, United Arab Emirates

Mohammed Ismail
Wayne State University
Detroit, MI, USA

ISBN 978-3-319-93099-2 ISBN 978-3-319-93100-5 (eBook)
https://doi.org/10.1007/978-3-319-93100-5

Library of Congress Control Number: 2018945070

Printed on acid-free paper

This Springer imprint is published by the registered company Springer International Publishing AG part of Springer Nature
The registered company address is: Gewerbestrasse 11, 6330 Cham, Switzerland

On reaching the saloon thou shalt there find a Lamp hanging from its ceiling; so mount the ladder and take that Lamp and place it in thy breast-pocket after pouring out its contents; nor fear evil from it for thy clothes because its contents are not common oil. And on return thou art allowed to pluck from the trees what so thou pleasest, for all is thine so long as the Lamp is in thy hand.
Aladdin and the Magic Lamp
Tales of One Thousand and One Night

To Our Beloved Families

Foreword I

From its founding in 1982, Semiconductor Research Corporation (SRC) has partnered with industry, government, and academia to fund university research to address industry relevant challenges. By focusing the academic research programs in the pre-competitive space, it has led to industrial breakthroughs in the design, fabrication, and modeling of integrated circuits that have been a key contributor in maintaining the pace of Moore's law. For more than three decades, the SRC research programs have been at the forefront of semiconductor research advancing the state of the art by investing over $ 2B and producing over 10,000 Ph.D. students, many of whom have become R&D leaders in the global semiconductor industry. As recognition for these achievements, SRC was awarded the US National Medal of Technology in 2005.

Of course, the role of SRC extends far beyond that of being a funding source and a research sponsor to include the following functions:

1. Ensuring that the ongoing research remains well aligned with the goals and the research needs of the industry.
2. Monitoring the research deliverables to ensure that they remain on track and provide value to SRC consortium members.
3. Evaluating university submissions for IP filings and assisting universities in filing their worthy ideas.
4. Holding review and organizing feedback to academic research teams.
5. Receiving all research publication submissions and posting them to the SRC Web site for early access by the members.
6. Assisting the principal investigators with collaboration efforts.
7. Managing the financial aspects of the program and making timely payments to the contracting universities for all invoices.
8. Assisting students in making connections to industry.

All of these SRC functions have been successfully applied to the Abu Dhabi SRC Center of Excellence on Energy-Efficient Electronic Systems (ACE^{4}S). The same criteria and metrics SRC use to monitor all US centers have been the basis for recognizing the success of the ACE^{4}S center. These metrics include number of students graduated under the program, number of publications at top venues, number of publication citations, number of patents issued and IP disclosures in progress, member importance and satisfaction ratings of each research project, and potential for industrial commercialization. The case of ACE^4S is particularly interesting to SRC for three main reasons:

1. While SRC has extended its reach to the global semiconductor community, ACE^4S has been, to date, its only full-fledged multi-university research center operating outside the USA.
2. As a corollary, SRC and ACE^4S have faced challenges that are not normally encountered in a US context. I am happy to observe that by and large, all these challenges have been overcome.
3. ACE^4S was one of the first centers worldwide to address an integrated, holistic program, namely the integrated circuit design aspects of the Internet of Things.

This edited book on the research work conducted under ACE^4S is indeed a very good illustration of the ACE^4S success story not only in terms of research output but also in terms of being an international pilot experiment for SRC. Most importantly, a solid foundation has been established for the future of semiconductor R&D in the United Arab Emirates.

Durham, NC
March 2018

Ken Hansen
CEO, Semiconductor Research Corporation

Foreword II

This book on the IoT physical layer is an up-to-date account on a significant R&D milestone in the drive of a young, ambitious country, the United Arab Emirates, to establish solid foundations for its knowledge-based economy. The milestone was achieved through a visionary partnership, involving local and international stakeholders from both academia and industry, which culminated in the development of a very special research center. The Abu Dhabi SRC Center of Excellence on Energy-Efficient Electronic Systems (ACE^{4}S) was meant to create a hub of excellence in semiconductor education and research to train and develop globally competitive homegrown talent in one of the world's most knowledge-intensive industries.

This book is a compilation of the success stories of ACE^{4}S in producing knowledge in an area that is foundational for all ICT-related technologies that society benefits from today and for the foreseeable future. From the transportation industry, to the communication industry, to health care, the ACE^{4}S has produced tangible contributions in the form of novel low-power semiconductor technology and products.

I have had the pleasure to watch ACE^{4}S blossom from an idea into an engine that produced award-winning research and highly qualified graduates. The success of ACE^{4}S can be directly attributed to the talent of the students and faculty in the UAE, as well as the triple helix support model that brings together academia with government and industry. Congratulations to all those who contributed to this success, students, faculty, and members of the SRC technical advisory board.

This book is not just a historical account of ACE^{4}S, but indeed a celebration of its achievements, and an important resource for current students and faculty, and future generations to come.

Abu Dhabi, UAE
March 2018

Dr. Rafic Makki
Executive Fellow, Globalfoundries
Senior Advisor, Mubadala Investment Company

Foreword III

As members of the ACE^{4}S External Advisory Board, it is heartening to see the work of this unique research center being documented in this comprehensive book. We have had the privilege of following ACE^{4}S since its inception, and we have been keenly aware of the opportunities, strengths, weaknesses, and threats that ACE^{4}S has had to deal with to get to where it is now. The value of this book is therefore twofold. First, it does provide a representative cross section of the scope and depth of the IoT hardware research that has been conducted under the center. Second, and perhaps more importantly, it shows that ACE^{4}S has lived up to its initial promises and has overcome many of the challenges it has faced.

One particular threat to the center's existence that we highlighted very early on was its relative isolation from the rest of the global semiconductor research community in Asia, Europe, and USA. Yet, we have the pleasure to note that several of the chapters included in this volume describe work that has already achieved international exposure and recognition, including best paper awards at such top global venues as IEEE International Symposium of Circuits and Systems (ISCAS) and IEEE Transactions on Very Large Scale Integration.

The center itself has also acted as a magnet to attract and host excellent conferences in Abu Dhabi such as the 2013 IEEE Conference on Electronic Circuits and Systems, the 2016 IEEE Midwest Symposium on Circuits and Systems, and the 2017 IFIP/IEEE International Conference on Very Large Scale Integration. This very book will further enhance the global exposure of ACE^{4}S and encourage a more sustained collaboration between the semiconductor R&D community in the UAE and their international colleagues in both industry and academia.

Minneapolis—Twin City, MN
Montreal, Quebec
Ann Arbor, MI
March 2018

Prof. Ramesh Harjani
Prof. Mohamed Sawan
Prof. Dennis Sylvester
ACE^{4}S External Advisory Board

Preface

Whoever sitteth on this carpet and willeth in thought to be taken up and set down upon other site will, in the twinkling of an eye, be borne thither, be that place nearhand or distant many a day's journey and difficult to reach.

Prince Ahmad and the Fairy Peri-Banu

Tales of One Thousand and One Night

The story of this book started almost exactly five years ago when we submitted a proposal "to establish a world-class Center of Excellence for Energy-Efficient Electronics Systems in the United Arab Emirates." The submitted proposal itself was the culmination of almost one year of work with our academic colleagues, university administrators, industry sponsors, and local stakeholders. This preparatory work involved a few UAE-wide workshops, a competitive solicitation for proposals, and many long meetings to get all the stars aligned on an initiative that was first-of-a-kind in the UAE context. Little did we know back then that the Center of Excellence we were proposing would be one of the world's pioneering multi-university research centers in the now red-hot field of the Internet of Things (IoT). Indeed, our proposal had been submitted and the Center of Excellence established a few years before the leading consulting companies and top corporate strategists of the world started publishing their exponential predictions about the *market and business opportunities* of the Internet of Things: the billions upon billions of connected devices and trillions upon trillions of US dollars in monetary opportunity.

Yet back in 2012, it was not difficult for us to recognize the *innovation opportunity* that was presenting itself to young talents all over the world in the technical areas that became core disciplines for the IoT physical layer. Indeed, in the decade that preceded our proposal, the cost of bandwidth had decreased 40X, the cost of storage had decreased 50X, the cost of processing had decreased 60X, and sensors had become ubiquitous through the smartphone platform. The time was therefore right in the context of the semiconductor industry to start exploring the far-reaching implications of energy-autonomous, application-specific integration of sensors, storage, processing, and communication on the same silicon chip while satisfying stringent constraints on its footprint and energy efficiency.

The time was also right in the UAE context where several important factors converged to make such explorations possible under the framework of a Research Center of Excellence. These factors included:

1. The presence of a large body of faculty across UAE universities with world-class expertise in semiconductor manufacturing and design. Such body needed a solid platform where all areas of expertise could be leveraged toward overarching research goals and achievements.
2. The significant capital investment that the UAE had already made in creating the infrastructure of a local semiconductor R&D ecosystem. The investment was in support of the establishment of state-of-the-art programs, laboratories, and facilities aimed at the education of a world-class local workforce in the various areas of semiconductors.
3. The interest of the academic leadership in initiating research programs that contribute to the establishment of an academic research culture in the UAE at par with the standards of the best research universities in the world. The dynamic, knowledge-intensive field of semiconductors was well positioned to lead the way in shaping this local culture in harmony with the UAE vision of the post-hydrocarbon knowledge economy.
4. The visionary leadership of the UAE Mubadala Investment Company, which through its creation and expansion of GLOBALFOUNDRIES has put the UAE at the forefront of the global semiconductor industry. Such bold financial and industrial initiative resonated deeply with the local semiconductor research community and called for equally bold action in establishing ambitious research programs that are compatible with the technology roadmap of GLOBALFOUNDRIES.

Another important factor is that Mubadala had also reached out to the US Semiconductor Research Corporation (SRC) to help manage academic semiconductor research in the UAE. SRC has had decades of experience in managing industry-supported, pre-competitive semiconductor research at US universities. The SRC consortium membership has included top semiconductor companies such as Texas Instruments, Intel, IBM, and GLOBALFOUNDRIES, and its portfolio of supported US universities has included world-leading research universities such as MIT, UC Berkeley, CMU, Stanford, UT Austin, Purdue, and Princeton. The role of SRC has been decisive in making sure that the UAE center's work is not only industrially relevant but also at par with the research quality of similar multi-university SRC research centers in the USA. The UAE center, which has been known with its ACE^{4}S acronym, has also been a unique experience for SRC itself as its only research center outside the USA. Realizing the relevance of ACE^{4}S work to the Internet of Things, SRC has placed it under their umbrella program on the Intelligent Internet of Things.

While ACE^{4}S has stood for the obvious "*Center of Excellence for Energy-Efficient Electronic Systems*," the opening letter A has metamorphosed from ATIC-SRC (ATIC for Advanced Technology Investment Company now reintegrated within Mubadala) to the very obvious Abu Dhabi SRC. We are of course proud of both associations of the letter A in ACE^{4}S.

ACE^{4}S sponsored a total of 18 projects, all concerned with IoT sensor node design with focus on healthcare applications. About half of these projects were integrated so as to result in two system demonstrators illustrating the successful completion of all their deliverables. Furthermore, half of the projects involved VLSI circuit design with the GLOBALFOUNDRIES 65 nm technology node being chosen as the common technology platform across all ACE^{4}S design projects.

This book is devoted to providing an up-to-date account of the research work that has been conducted under the ACE^{4}S program. The content of this book comprises 20 chapters organized in five parts that are meant to address the various aspects of the IoT physical layer from novel materials and sensors to system demonstrators along with their hardware security. These parts are:

1. Part I: Advanced Materials and Sensors, which covers Chaps. 1 through 6.
2. Part II: Architectures and Circuits, which covers Chaps. 7 through 10.
3. Part III: Algorithms and Protocols, which covers Chaps. 11 through 14.
4. Part IV: Power Management, which covers Chaps. 15 though 17.
5. Part V: Systems and Security, which covers Chaps. 18 through 20.

To help the reader navigate through these parts, we now provide a short description of the content of each chapter.

Part I on Advanced Materials and Sensors opens with Chap. 1 that deals with the synthesis of a novel dry EKG electrode using reduced graphene oxide (rGO_x). Such dry electrodes are essential for making EKG sensors wearable. The interesting observation of this chapter is that the conductive properties of the electrode fabric can be tightly controlled by incrementally applying thin coatings of graphene in order to provide good signal acquisition with relatively low-motion artifact.

In Chap. 2, a similar issue of sensor wearability is addressed but at the level of shoe-integrated pressure sensors. Such sensors are meant to be used for continuous weight measurements, and the focus of this chapter is on the suitability of flexible piezoresistive sensors for such measurements. The flexible sensors that have been evaluated in this study have been found quite sensitive to motion artifact. They have also been found to suffer from memory and drift effects.

The issues of flexible sensors are taken up in a series of Chaps. 3 through 5, devoted to the *ZnO* material system which is shown to be a versatile system for both sensors and transistors on flexible substrates. Chapter 3 introduces several thin-film processes for producing *Al*-doped *ZnO* thin film using atomic layer deposition (ALD). Chapter 4 is devoted to the electrical, mechanical, and electromechanical characterization of such thin films, while Chap. 5 addresses the *Al*-doped *ZnO* thin film as active channel for flexible thin-film transistors. In particular, it deals with several of the reliability and thermal challenges of these transistors when deposited on flexible substrates. One important observation about this series of three chapters on the *ZnO* material system is that all the experimental and characterization work has been conducted in the micro-fabrication and microscopy facilities on the Masdar Campus of Khalifa University.

Part I closes with a chapter on photonic sensing for lab-on-chip applications. The sensor is made of two rings in a Vernier configuration integrated with a planar echelle grating demultiplexer. The sensor is capable of achieving a limit of detection as low as 80×10^6 RIU. This design has been the very first silicon photonics chip ever designed and fabricated out of a UAE University.

Part II opens with Chap. 7 where the design of an orthodontic wireless sensor node is presented. The sensor is used to monitor the orthodontic braces and warn of any bond failure between the braces and teeth. One of the challenges of this orthodontic sensor system is related to wireless power transfer and the need to overcome the changing impedance due to mouth motion dynamics.

Human body impedance change is also one of the major themes addressed in Chap. 8 which is concerned with the design of energy-efficient transceivers for body-coupled communication where the human body itself is used as a communication medium. One important outcome of the research presented in this chapter is that the use of orthogonal frequency division multiplexing (OFDM) can mitigate the issues inherent to the human body channel, including variable path losses, multiple signal return paths, and variable impedance between skin and electrode.

In Chap. 9, a system-on-chip (SoC) for ECG signal processing is presented. The purpose of this chip is to predict ventricular arrhythmia well ahead of onset to improve cardiac care outcomes. All the components of the SoC are described, including the analog front end, the ECG feature extractor, and the prediction classifier. This research shows that it is possible to predict arrhythmia episodes up to three hours before onset. The signal processing and prediction algorithms used in the SoC design are fully described in Chap. 13.

One important aspect of IoT wireless communication, especially for such applications as autonomous driving and urban transformation, is the use of a massive number of antennas to improve transmitted signal power and directionality while increasing the signal-to-noise ratio of received signals. Beamforming is the technique used to achieve such improvements. Chapter 10 takes a close comparative look at beamforming in antenna phase arrays and antenna delay arrays. In particular, it discusses the impact of the number of antennas and fractional bandwidth on beamforming performance.

Interestingly, the opening chapter of Part III on Algorithms and Protocols is also concerned with multiple antennas but in a different IoT context. The problem addressed in Chap. 11 is that of the optimal topology design of an indoor distributed antenna system as may be needed in high-rise buildings with many floors requiring the installation of access antennas. The chapter proposes the use of particle swarm optimization for solving this problem and shows the viability of such solution, especially in terms of its scalability to a large number of building floors.

Chapter 12 illustrates one of the innovation opportunities of the IoT era, namely the need to take a fresh look at single-channel communication protocols to propose novel signal encoding and transmission strategies that are more in line with the requirements of constrained sensor nodes. This chapter provides a comprehensive survey of one such proposal called pulsed-index communication which has the distinguishing feature that it does not require any circuits for clock and data recovery at the receiver, thus significantly contributing to footprint reduction and power savings.

Chapter 13 delves into the algorithmic details of the ECG signal processor presented in Chap. 9. It explains the timing and waveforms features that are extracted from the ECG signal to enable ventricular arrhythmia prediction. It also describes a hardware-friendly linear discriminant analyzer to identify the ECG signals that are most susceptible to arrhythmia. When tested on standard ECG databases, the prediction algorithm achieves 99.1% accuracy for a prediction window of three hours.

Part III closes with Chap. 14 which addresses the IoT context of video sensing with applications ranging from object detection and recognition to mobile object tracking. The image processing algorithm presented in this chapter is called maximally stable extremal regions (MSER) and is used to extract local image features that are invariant under an affine transformation. The interest of MSER is that it is amenable to very economical hardware implementation in terms of both area and power. Aside from describing the theoretical foundations of MSER, this chapter describes one such hardware implementation.

Part IV deals with the important IoT topic of power management. Chapters 15 and 16 address two alternative DC–DC power conversion architectures and circuits for energy-harvesting applications. In Chap. 15, the proposed design is that of a high-gain, inductor-based DC–DC conversion with zero-current switching control. The design achieves 81% peak efficiency at 55 °C μW output power and 50 mV input voltage.

On the other hand, Chap. 16 proposes a reconfigurable, multiple voltage gain, switched-capacitor buck convertor. The design achieves a measured peak efficiency of 80% at a load current of 800 μA and regulated load voltage of 1V.

Part IV on power management closes with Chap. 17 on micro-batteries for IoT micro-power source integration. Such sources will play the role of energy buffers between energy harvesters and sensor electronics, and their tight integration with CMOS will enable power management solutions that can deal with various usage cases, including ones in which the energy harvester is inactive. The chapter focuses on the macromodeling of such batteries in view of their integration in the circuit and sensor design framework.

Finally, Part V addresses system-level and hardware security issues in IoT. As mentioned earlier in this preface, ACE^{4}S has developed two system-level IoT platforms for healthcare monitoring. The first platform is described in Chap. 18 and is concerned with the cardiac arrhythmia predictor. This chapter should be read as the culmination of Chaps. 1, 9, 13, and 15.

The second mobile health platform is described in Chap. 19 and proposes a shoe-integrated, dual-path system for continuous weight and walk-cycle monitoring. While the original motivation of this IoT healthcare platform is weight-driven congestive heart failure prediction, the platform itself can be applied in many other healthcare conditions, including Parkinson's disease, physical therapy, and sports medicine. This chapter should also be read as the culmination of three preceding chapters, namely 2, 8, and 12.

The final chapter in Part V and in the book addresses the very important and pressing issue of hardware security in IoT sensors. In IoT healthcare applications where accuracy and privacy are of utmost concern, the very first thing the IoT application user should be able to trust is the IoT hardware itself. Chapter 20 gives an up-to-date survey of the methodology of logic locking for producing trustworthy digital design solutions.

As co-directors of ACE^4S, we are happy with and proud of the path-breaking work that the center has accomplished since its inception in 2013. This assessment is not just ours but also that of the larger international professional community as is evidenced by the prestigious awards that the research of our students has already garnered. They include the 2016 Best Paper Award from IEEE Transactions on Very Large Scale Integration (Chaps. 9, 13, 15, and 18) and the Best Student Paper Award in Biomedical Circuits and Systems from IEEE International Symposium on Circuits and Systems, Lisbon, Portugal, May 2015 (Chap. 8). Several of our ACE^4S students have also been active participants in the competitive, worldwide TECHCON that SRC holds every September for graduate students working on SRC-sponsored projects. We are particularly proud that ACE^4S has sponsored the research of the very first UAE national student to attend TECHCON.

Many excellent monographs devoted to the topic of IoT hardware have already been published, including some by Springer. The one that is closest in spirit to the present book is the 2017 Springer volume edited by our esteemed colleague, Prof. Massimo Alioto, National University of Singapore, under the title: "Enabling the Internet of Things: From Integrated Circuits to Integrated Systems." However, the readers of the present volume will notice several important differences with Prof. Alioto's book, including our emphasis on sensors, protocols, and algorithms and our highlighting of application-driven architectures and designs. Furthermore, the present volume is meant to give an up-to-date account of a specific research program on the IoT physical layer rather than a broad overview of this exciting IoT topic.

To make the most out of the chapters, the reader should have some familiarity with the elements of a wireless sensor node, at the architecture and component design levels. Our targeted audience are faculty and graduate students in EECS and ECE programs, engineers and technologists in the semiconductor industry, and R&D managers and leaders interested in keeping apace with the latest in academic research on the IoT physical layer.

When ACE^4S started, we belonged to two different, independent Abu Dhabi institutions: Khalifa University of Science, Technology and Research and the Masdar Institute of Science and Technology. These two universities along with the

Petroleum Institute have now become one. Of course, we would like to believe that the collaborative, collegial, and nurturing research environment that ACE^4S has created across the various university campuses has contributed, in no small part, to making the university merger idea a compelling proposition for all academic stakeholders. As we are certain the merger will succeed, we just hope that ACE^4S will get some credit for being ahead of everybody else by about 4 years!

The compilation of this book would not have been possible without the dedication, diligence, hard work, and commitment of all the contributing authors. They are the ones who have designed and implemented the enchanted IoT objects described in this book. To them go our deepest gratitude and warmest thanks!

Abu Dhabi, UAE
Detroit, MI, USA
March 2018

Ibrahim (Abe) M. Elfadel
Mohammed Ismail

Acknowledgements

The IoT research work described in this book has been conducted under the Abu Dhabi SRC *Center of Excellence on Energy-Efficient Electronic Systems* (ACE^{4}S). We are proud and grateful that the Semiconductor Research Corporation (SRC) has selected Abu Dhabi, capital city of the United Arab Emirates, to host the only SRC Center outside the USA. ACE^{4}S would not have been possible without the generous funding of the Mubadala Investment Corporation, Abu Dhabi, UAE, and the matching grants that ACE^{4}S has received from the five participating UAE universities: Khalifa University of Science, Technology and Research, the Masdar Institute of Science and Technology, the United Arab Emirates University, New York University—Abu Dhabi, and the American University of Sharjah.

As in all SRC centers, the projects under ACE^{4}S have been subject to the rigorous SRC review and evaluation process through both annual on-site reviews in Abu Dhabi and topical reviews in the USA and India. We would like to acknowledge all the members of the SRC Technical Advisory Board (TAB) who have attended such reviews and shared with our faculty, researchers, and students, their pertinent comments and candid feedback. The SRC TAB members are technical leaders in major semiconductor companies such as GLOBALFOUNDRIES, Intel, IBM, Texas Instruments, Freescale Semiconductor, and Mentor Graphics.

We also acknowledge the help of all the SRC Industrial Liaisons who have actively contributed to the academia–industry interlock that is a hallmark of the SRC *modus operandi.*

To balance the industrial feedback of the SRC Industrial Liaisons and TAB members, ACE^{4}S has included, in its organizational structure, an External Advisory Board (EAB) comprised of renowned academics with expertise directly relevant to the technical scope of the center. We would like to acknowledge our distinguished EAB members: Prof. Ramesh Harjani from the University of Minnesota, Prof. Mohamad Sawan from Polytechnique Montréal, and Prof. Dennis Sylvester from the University of Michigan, Ann Arbor, for attending the ACE^{4}S annual reviews and providing us with their expert comments.

We are fortunate to have had dedicated colleagues from the SRC staff to manage the complex web of academic, industrial, and corporate relationships in which ACE^{4}S has been embedded. In particular, we would like to thank Mr. Dale Edwards and Dr. Kwok Ng for being such patient managers.

ACE^{4}S has been a unique, first-of-a-kind, experience in graduate education that has contributed in no small way to the introduction of a culture of excellence in externally funded multi-university research to the UAE. Its success has depended on the dedication and hard work of many of our academic support staff in finance, procurement, contracts, legal, and intellectual property. In particular, we would like to single out and acknowledge the constant contributions of Ms. Marianne Hassan from Khalifa University of Science and Technology to the ACE^{4}S grant administration.

The inception of ACE^{4}S is due in large part to the visionary leadership of the Mubadala Investment Company. Our thanks go more particularly to Ibrahim Ajami, Sami Issa, and Rafic Makki for all the support they have provided to ACE^{4}S.

This book was finalized as the merger of Khalifa University, the Masdar Institute, and the Petroleum Institute, into a single, integrated academic entity was well under way. We would like to claim that ACE^{4}S, with the collaborative, collegial, multi-university research environment it has created, has been the merger's "feasibility" proof and has significantly contributed to its facilitation. The major credit though should go to our senior university administrators who have supported ACE^{4}S since it was a mere idea. In particular, we would like to thank Profs. Tod Laursen, Fred Moavenzadeh, Steve Griffiths, and Mohammed Al-Mualla for supporting the hosting of ACE^{4}S at Khalifa University and the Masdar Institute, now two campuses of one university: Khalifa University of Science and Technology.

Finally, we acknowledge the LaTeX technical support we have received from Shahzad Muzaffar during the composition of this book as well as the advice and guidance provided to us by Charles Glaser from Springer.

Contents

Editors and Contributors

About the Editors

Dr. Ibrahim (Abe) M. Elfadel is Professor of Electrical and Computer Engineering at the Masdar Institute, a part of Khalifa University of Science and Technology, Abu Dhabi, UAE. Since May 2013, he has been Founding Co-director of the Abu Dhabi SRC Center of Excellence on Energy-Efficient Electronic Systems (ACE^{4}S), and since May 2014, he has been Program Manager of TwinLab MEMS, a joint collaboration with GLOBALFOUNDRIES and the Singapore Institute of Microelectronics on microelectromechanical systems. Between November 2012 and October 2015, he was Founding Co-director of Mubadala's TwinLab 3DSC, a joint research center on 3D integrated circuits with the Technical University of Dresden, Germany. He also headed the Masdar Institute Center for Microsystems (iMicro) from November 2013 until March 2016. Between 1996 and 2010, he was with the corporate CAD organizations at IBM Research and the IBM Systems and Technology Group, Yorktown Heights, NY, where he was involved in the research, development, and deployment of CAD tools and methodologies for IBM's high-end microprocessors. In addition to IoT platform prototyping, his current research interests include energy-efficient IoT communications; energy-efficient edge and cloud computing; power and thermal management of multi-core processors; low-power, embedded digital signal processing; 3D integration; and CAD for VLSI, MEMS, and silicon photonics. He is the recipient of *six Invention*

Achievement Awards, one Outstanding Technical Achievement Award and one Research Division Award, all from IBM, for his contributions in the area of VLSI CAD. He is the inventor or co-inventor of 50 issued US patents. In 2014, he was the co-recipient of the *D. O. Pederson Best Paper Award from the IEEE Transactions on Computer-Aided Design Automation for Integrated Circuits and Systems.* He is also the co-editor (with Prof. Gerhard Fettweis) of *3D Stacked Chips: From Emerging Processes to Heterogeneous Systems,* Springer, 2016, and the co-editor (with Duane Boning and Xin Li) of the upcoming book: *"Machine Learning in VLSI Computer-Aided Design,"* Springer, 2018. Between 2009 and 2013, he served as Associate Editor of IEEE Transactions on Computer-Aided Design. He is currently serving as Associate Editor of IEEE Transactions on VLSI Systems and on the Editorial Board of the Microelectronics Journal (Elsevier). He has also served on the Technical Program Committees of several leading conferences, including DAC, ICCAD, ASPDAC, DATE, ICCD, ICECS, and MWSCAS. Most recently, he was General Co-chair of IFIP/IEEE 25th International Conference on Very Large Scale Integration (VLSI-SoC 2017), Abu Dhabi, UAE, October 23–25, 2017. He received his Ph.D. from MIT in 1993.

Mohammed Ismail Ph.D. Fellow IEEE Founding Director, the Analog VLSI Lab, Ohio State University (www.ece.osu.edu/VLSI) **Founding Director, The RaMSiS Research Group, KTH, Sweden Mubadala Technology Professor and Founding Chair, ECE, KUSTAR, UAE, Founding Director, KSRC. Co-Director, the ACE4S ATIC-SRC Center of Excellence** (http://www.src.org/program/grc/ace4s/) **ECE Professor and Chair, Wayne State University, Detroit, MI, USA.** He is a prolific author and entrepreneur in the field of chip design/test and nanotechnology, spent over 25 years in academia and industry in the USA and Europe. He obtained his BS and MS from Cairo University, Egypt, and his Ph.D. from the University of Manitoba, Canada, in 1983, all in electrical engineering. He is Founder of the Ohio State University's (OSU) Analog VLSI Lab, one of the foremost research entities in the field of nano-electronics, analog, mixed-signal and RF integrated circuits and served as its Director. He also served on the Faculty of OSU's ElectroScience Lab. He held Research Chair on Nano-electronics Circuits at the Swedish Royal Institute of Technology (KTH) where he founded the RaMSiS (Radio and Mixed Signal Integrated Systems) Research Group there. He had visiting appointments in Finland (Aalto University), Norway (NTH and University of Oslo), the Netherlands (Twente University), and Japan (Tokyo Institute of Technology). He joined Khalifa University of Science, Technology, and Research, Abu Dhabi, UAE in 2011, where he holds the ATIC (now Mubadala Technology) Professor Chair and is Founding Chair of the ECE Department. He is Founding Director of the Khalifa Semiconductor Research Center (KSRC) and Co-director of the ATIC-SRC Center of Excellence on Energy-Efficient Electronic systems (ACE4S) targeting self-powered nanoscale chipsets for wireless sensing and monitoring, biochips, and power management solutions. His current research focuses on CMOS RF and mm-wave ICs, energy harvesting and power management, wearable biochips and systems-on-chip (SoCs) for the Internet of Things (IoTs) and smart self-driving vehicles. He joined Wayne State University in Detroit, MI, as Professor and Chair of ECE while maintaining his position at KUSTAR as Adjunct. He lead a research team that developed the first CMOS combo 802.11a/b/g Wi-fi radio chip which was commercialized. More recently, he developed with his colleagues the world first self-powered wearable device that predicts the onset of a heart attack long before it happens and is currently in the process of commercializing it in partnership with the Cleveland Clinic, a world-leading hospital in cardiology. He served as a corporate consultant to over 30 companies worldwide and is Co-founder of Micrys Inc., Columbus, Ohio; Spirea AB, Stockholm; Firstpass Technologies Inc., Dublin, Ohio; and ANACAD-Egypt (now part of Mentor Graphics and Siemens). He advised the work of over 54 Ph.D. students and of over 100 MS students. He authored or co-authored over 20 books and over 170 journal publications, 300 conference papers, and has 15 US patents granted and several pending. He is Founding Editor of the Springer Journal of Analog

Integrated Circuits and Signal Processing and serves as the journal's editor in chief. He served IEEE in many editorial and administrative capacities. He is Founder of IEEE International Conference on Electronics, Circuits and Systems (ICECS), the flagship Region 8 Conference of IEEE Circuits and Systems Society and Co-founder of IEEE International Symposium on Quality Electronic Design (ISQED). He received the US Presidential Young Investigator Award from the White House, the Ohio State Lumley Research Award four times, in 1992, 1997, 2002, and 2007, and the US Semiconductor Research Corporation's Inventor Recognition Award twice as well as several best paper awards. He is Fellow of IEEE.

Contributors

S. A. Al Junaibi Nanyang Technological University, Singapore, Singapore

Osama Al Mrayat Masdar Institute at Khalifa University, Abu Dhabi, UAE

Mahmoud Al-Qutayri Khalifa University, Abu Dhabi, UAE

Lutfi Albasha American University of Sharjah, Sharjah, UAE

Mohammad Alhawari Khalifa University, Abu Dhabi, UAE

D. Y. Atia Khalifa University, Abu Dhabi, UAE

Nourhan Bayasi Khalifa University, Abu Dhabi, UAE

Ahmed Elian Department of Electrical and Computer Engineering, University of Toronto, Toronto, Canada

Ismail Elnaggar Khalifa University, Abu Dhabi, UAE

Temesghen Habte Khalifa University, Abu Dhabi, UAE

Nicholas Halfors Khalifa University, Abu Dhabi, UAE

N. G. Hallfors Khalifa University, Abu Dhabi, UAE

Basil Hatahet Denton Clinic, Denton, UK

H. Hildmann Universidad Carlos III de Madrid (UC3M) Dep. de Ingenieria de Sistemas y Automatica - Av. Universidad, 30, Leganes, Spain

A. F. Isakovic Khalifa University, Abu Dhabi, UAE

M. Ismail Wayne State University, Detroit, MI, USA

Yonatan Kifle Khalifa University, Abu Dhabi, UAE

Dima Kilani Khalifa University, Abu Dhabi, UAE

K. Liao Khalifa University, Abu Dhabi, UAE

Bodhisatwa Mazumdar New York University, Abu Dhabi, UAE

Baker Mohammad Khalifa University, Abu Dhabi, UAE

Shahzad Muzaffar Masdar Institute at Khalifa University, Abu Dhabi, UAE

Mohammed Shemso Nesro Masdar Institute at Khalifa University, Abu Dhabi, UAE

K. Poon EBTIC, Khalifa University, Abu Dhabi, UAE

Nasser Qaddoumi American University of Sharjah, Sharjah, UAE

Nasir Quadir American University of Sharjah, Sharjah, UAE

Jeyavijayan Rajendran University of Texas at Dallas, Richardson, TX, USA

Mahmoud S. Rasras Engineering Division, New York University Abu Dhabi, Abu Dhabi, UAE

Ayman Rezk Masdar Institute at Khalifa University, Abu Dhabi, UAE

D. Ruta EBTIC, Khalifa University, Abu Dhabi, UAE

Irfan Saadat Masdar Institute at Khalifa University, Abu Dhabi, UAE

Nisrein Sada Masdar Institute at Khalifa University, Abu Dhabi, UAE

Mohammed Saeed University of Michigan, Ann Arbor, MI, USA

Hani Saleh Khalifa University, Abu Dhabi, UAE

Ayman Shabra MediaTek, Woburn, MA, USA

Ozgur Sinanoglu New York University, Abu Dhabi, UAE

Andrzej Sluzek Khalifa University, Abu Dhabi, UAE; Warsaw University of Life Sciences-SGGW, Warsaw, Poland

Mansour Taghadosi American University of Sharjah, Sharjah, UAE

Temesghen Tekeste Khalifa University, Abu Dhabi, UAE

Owais Talaat Waheed Masdar Institute at Khalifa University, Abu Dhabi, UAE

Muhammad Yasin New York University, Brooklyn, NY, USA

Jerald Yoo National University of Singapore, Singapore, Singapore

Part I
Advanced Materials and Sensors

"The Silver Shoes," said the Good Witch, "have wonderful powers. And one of the most curious things about them is that they can carry you to any place in the world in three steps, and each step will be made in the wink of an eye. All you have to do is to knock the heels together three times and command the shoes to carry you wherever you wish to go." L. Frank Baum, The Wonderful Wizard of Oz

Chapter 1
Reduced Graphene Oxide for the Design of Electrocardiogram Sensors: Current Status and Perspectives

N. G. Hallfors, S. A. Al Junaibi, K. Liao, M. Ismail and A. F. Isakovic

1.1 Introduction

Cardiovascular disease (CVD) is the leading cause of mortality worldwide. As much as 31 percent of global deaths, estimated at 17.5 million deaths in 2012, are caused by CVD. Cardiac diseases account for deaths more than any other cause, and it is primarily driven by aging populations and unhealthy lifestyle including tobacco, diet and low levels of physical activity [9]. Early detection of heart problems, especially myocardial infarction (MI), is essential for successful treatment [12]; as such, there is a great interest in the field of cardiac health monitoring, including long-term, wearable, and wireless monitoring systems (Fig. 1.1).

The electrocardiogram (ECG) has long been the standard method for detailed measurement of heart activity [10]. ECG data contains useful information about heart activity, such as arrhythmia, ventricular dysfunction, and murmur, and can even give early warning of impending stroke and MI [8]. ECG measures current produced by the depolarization of the cardiac muscle during the cardiac cycle through electrodes placed on the skin. Quality of signal received can vary widely, based on sensor position and sensor type, as well as physiological differences among subjects.

The standard clinical ECG sensor is the gel-type Ag/AgCl sensor [1]. Ag/AgCl sensors rely on a gel electrolyte and adhesive patches, providing both excellent elec-

N. G. Hallfors · K. Liao · A. F. Isakovic (✉)
Khalifa University, Abu Dhabi, UAE
e-mail: abdel.isakovic@kustar.ac.ae; iregx137@gmail.com

N. G. Hallfors
e-mail: nicholas.hallfors@kustar.ac.ae

K. Liao
e-mail: kin.liao@kustar.ac.ae

S. A. Al Junaibi
Nanyang Technological University, Singapore, Singapore
e-mail: saif001@e.ntu.edu.sg

M. Ismail
Wayne State University, Detroit, MI, USA
e-mail: ismail@wayne.edu

Ibrahim (Abe) M. Elfadel and M. Ismail (eds.), *The IoT Physical Layer*,
https://doi.org/10.1007/978-3-319-93100-5_1

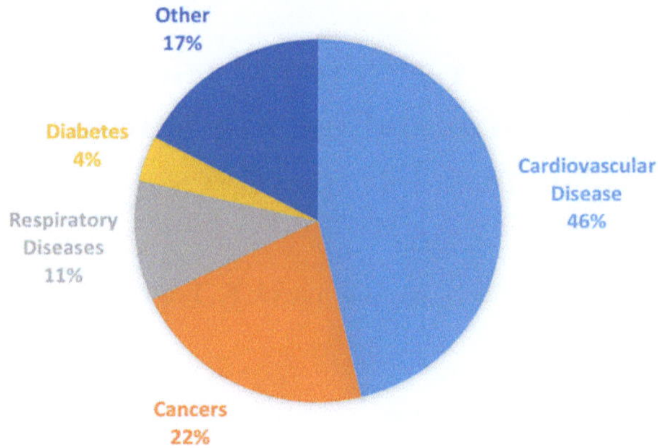

Fig. 1.1 Main mortality contributors of noncommunicable diseases [9]

trical conductivity and protection from motion artifacts. The primary drawbacks associated with Ag/AgCl electrodes are related to convenience and patient comfort. Also, gel-type electrodes are not practical for continuous monitoring systems, as electrolyte gel needs to be regularly reapplied, and may cause irritation over long periods of use. Likewise, adhesives need to be reapplied and may cause discomfort. Dry electrodes are therefore preferable when designing a wearable monitoring system.

Dry electrodes address the main problems associated with gel-type electrodes, as they can be incorporated into clothing and accessories for long-term, comfortable use. The main drawbacks associated with dry electrodes are the lower conductivity across the dry interface, and sensitivity to motion-induced noise in the signal [3, 6]. The primary goal in the study of dry electrodes is therefore to address these two issues. Dry electrodes can be categorized into surface contact, penetrating contact and capacitive non-contact electrodes [13]. The majority of electrical resistance comes from the outer layer of skin, especially if it is dry [4]. This resistance can be dramatically reduced by abrasion or bypassed with a penetrating electrode, but both of these methods add discomfort and inconvenience not suitable for a constant monitoring system.

Taking a broader view, there is a need for a nonstandard, post-Ag/AgCl sensors in ECG signal probing and collection, driven in part by the requirements of wearable electronics in IoT era. These requirements include, but are not limited to:

1. Complete insensitivity to motion artifacts.
2. Long-term, uninterrupted usability, two or more orders of magnitude longer application than the standard clinical sensors.
3. Time independence of the skin-to-sensor conductivity, preferably without any need for the periodic re-attachment or contacts improvements.
4. Wireless connectivity.

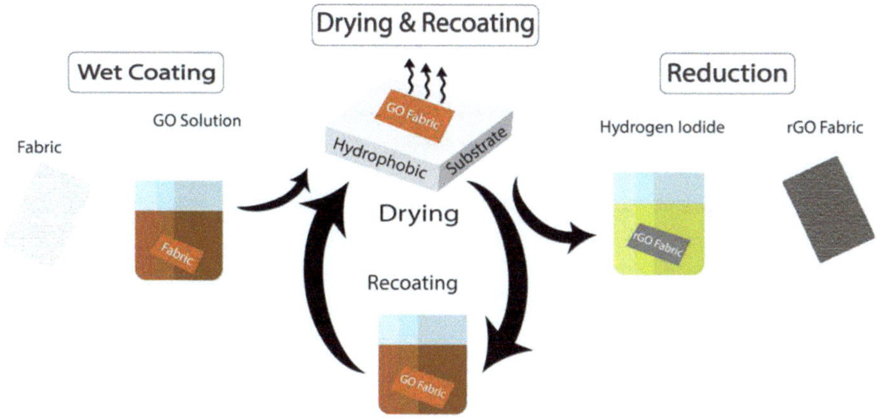

Fig. 1.2 Schematic of fabrication process for graphene-coated fabric. The drying/re-coating steps can be repeated to produce additional layers of graphene

1.2 Fabrication Process

Prior to the process of fabricating the rGO_x fabric, a diluted GO solution is prepared using the modified Hummer's method [7]. The GO concentration used here is 4 mg/cm^3. As shown in Fig. 1.2, the fabrication process of rGO_x-coated fabrics follows three simple steps. The textile is first left wetting overnight in GO solution to facilitate complete coverage of the nylon fibers by graphene flakes. Next, the textile is taken out of the GO suspension, placed on a hydrophobic surface, such as polytetrafluoroethylene (Teflon®), and placed in an oven to dry (60 °C) for a few hours until water is fully evaporated. Using a hydrophobic surface prevents the solution from running off the fabric, which could potentially leave exposed regions of nylon. As the water evaporates, GO flakes are deposited onto the nylon fibers, producing an even coating of graphene oxide. To be functional, the GO coating must be reduced to rGO_x through either thermal or chemical processes. Thermal oxidation of GO takes place at high temperatures, typically between 200 and 1000 degrees [1, 5]. Because the fabric is not suited for high temperatures, the GO layers are chemically reduced rather than thermally reduced. This is done by reducing the GO textiles in hydrogen iodide (HI) acid for 1 h. The textile is then taken out of the HI and washed in distilled water. This process creates once-coated textile composite [14] (Figs. 1.3 and 1.4).

Prior to chemical reduction, multiple layers of GO can be applied by simply dipping the dried fabric into GO solution, and drying as seen in Fig. 1.2. The hydrophilic nature of GO causes the solution to be quickly and evenly distributed along the coated nylon fibers, creating uniform layers. This process can be repeated any number of times; however, as the resulting layers become increasingly thicker, they may become brittle and flake off. Each additional coating increases the electrical conductivity of the composite, while also making it more rigid and brittle. The optimization of number of layers was obtained at six coatings.

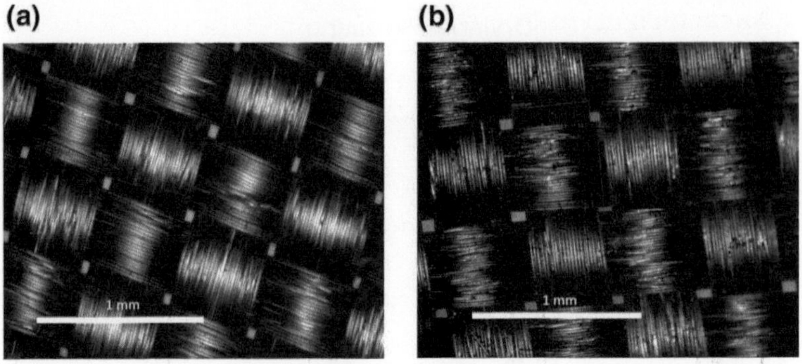

Fig. 1.3 50x magnification optical microscope image of uncoated nylon fabric **a** and six times coated nylon fabric **b**. Hatching pattern comes from weaving of the nylon fabric

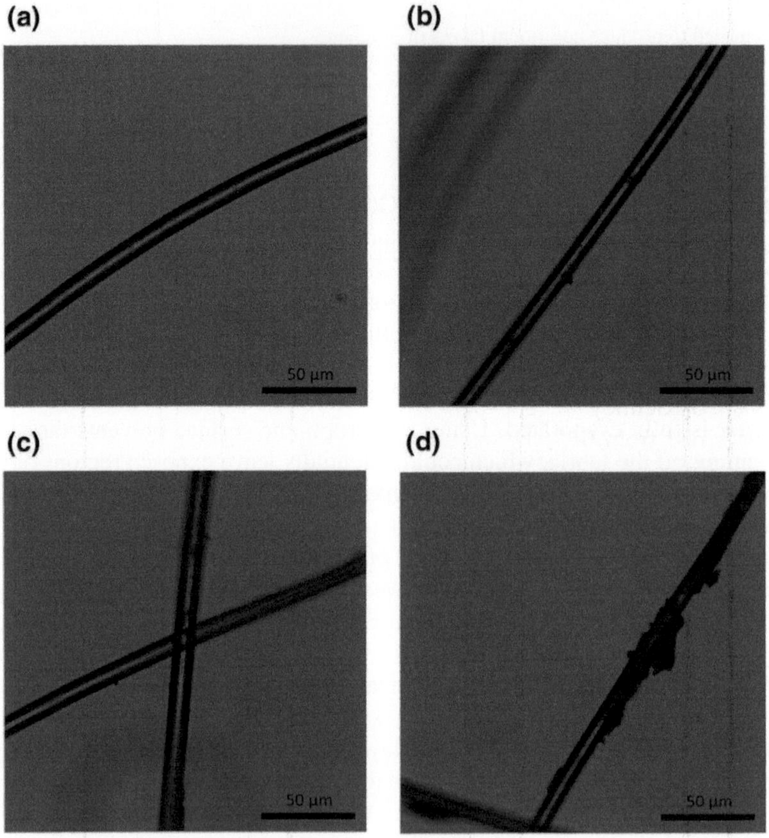

Fig. 1.4 500x magnification optical microscope image of individual nylon fibers coated in **a** one layer of unreduced graphene oxide, **b** one layer of reduced graphene oxide, **c** two layers of unreduced graphene oxide, and **d** two layers of reduced graphene oxide

1.3 Characterization of Coated Fabric

This section discusses the fabrication, design, testing, and implementation of the rGO_x-based sensors for ECG and possibly other wearable, biomedical detectors. Understanding the biomedical data collection based on the novel sensors is challenging for a variety of reasons, including:

1. The need to obtain a comparable signal as collected by the existing ones (otherwise, the medically accepted standards may be hard to adhere to).
2. The need to understand the nature of the signal, as it often may rely on a novel material, which, on one hand, has properties that make it attractive to wearable electronics criteria, but, on the other hand, may rely on insufficiently understood transport mechanism.

While the first point is relevant to most novel sensors, the second one is specifically relevant to the rGO_x sensors. In order to better understand the range of design parameters and the design-relevant physical properties, the I-V curves were recorded at each successive re-coating step. The results are shown in Figs. 1.5 and 1.6. We see that there are at least two distinctive regimes of transport in relation to the range of electric fields (voltages) applied to the sensor. The current undergoes rapid changes around 0.5 V and then changes more steadily for higher voltages, above 1 V range. These differences may be useful in the future exploitation of minimal energy electronic system standards (MEES standards) and are more likely to become the focus of a future research once the best-possible amplitude of the biomedical signal is obtained.

Six samples of rGO_x-coated fabric were produced, varying from one to six coats of rGO_x. The samples were cut into $2\times2\,cm^2$ and analyzed on a Semiprobe probing station and a Kiethley™ 4200-SCS characterization system. A DC voltage sweep was performed from 0–5–0 V, and current across the fabric was measured. From

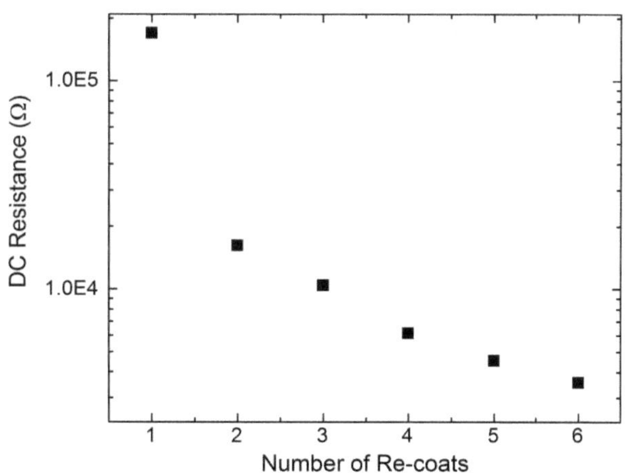

Fig. 1.5 DC resistance of fabric samples with various numbers of rGO_x coatings

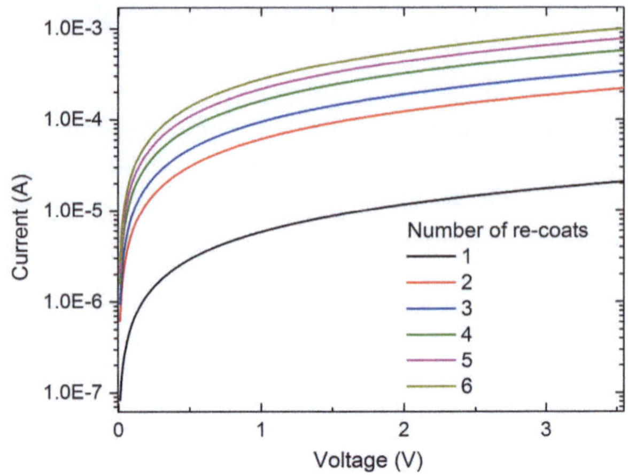

Fig. 1.6 DC current vs voltage across fabric samples with various numbers of GOx coatings

the resulting I-V curve, a resistance measurement was obtained for each sample of fabric.

The DC resistance of each successive re-coating is plotted in Fig. 1.5. These measurements were made with the same electrode and on the same sample. The size of the error bars is comparable to the size of the symbol. There are several features of the behavior of DC resistance worth noting:

1. The value of the resistance drops sharply after the first re-coating (i.e., the second coating).
2. Successive re-coatings also measurably contribute to the resistance decrease, but not as dramatically as the first re-coating.

The behavior described here is the subject of ongoing research at Khalifa University. One can reasonably speculate that:

- The presence of a strongly pronounced substrate pattern, the previously discussed cross-thatched nylon weave, which accounts for limited, domain-like directionality of the carrier transport during the measurement.
- The addition of the layers of rGO_x leads to the increased number of optional carrier pathways, possibly relying on a version of a resistor network model of the carrier transport.

The latter point is understandable from the purely geometric standpoint: When not enough rGO_x is available for the transport, the charge carriers need to "follow" the available path, which is along the weave. We do not consider the preceding discussion as a complete explanation of the unusual behavior of the resistance in a rGO_x film, and the tests of this hypothesis are underway.

rGO_x Fabric vs Ag/AgCl

The rGO_x fabric is utilized as dry electrode ECG sensor placed on the skin [11]. Two 3×3 cm^2 six-coated samples are used as the ECG sensor. Single-sided copper tape

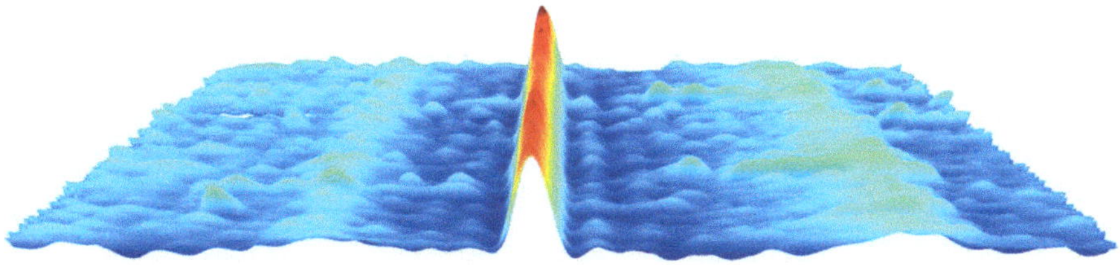

Fig. 1.7 Electrocardiogram waterfall graph of data obtained with six-coated rGO_x electrodes

and a button, similar to commercial ECG electrodes, are used to make an interface with the standard ECG data acquisition leads. The ECG sensors are tape-secured to the right and left sides of the neck, with no additional pressure, of a healthy 22-year-old male subject to capture ECG biopotential. The neck is chosen rather than the arm wrists because of the close distance of the ECG electrodes. This is applicable for wearable devices as minimal sensor device distance is required. The subject was at rest and sitting to minimize motion artifacts. At the same time, 3×3 cm^2 pre-gelled Ag/AgCl commercial electrodes (Skintact®) are used in parallel and placed in close proximity for neck ECG signals to establish a baseline for comparison. The ECG signals were fed into an AD Instruments PowerLab ECG acquisition system to be captured, filtered, and displayed. A 2–30 Hz band-pass filter on both electrode signals is used to filter out 50 Hz powerlines noise or other biosignals such as the electromyogram (EMG). ECG recordings were taken for 207 heartbeats in 3 minutes with minimal motion and displayed with LabChart™ software. Figure 1.7 shows a 3D graph of stacked ECG signals averaged on four beats from the six-coated rGO_x electrode. The characteristic ECG rhythm containing the P wave, QRS complex, and T wave is visible over the whole recording range. This provides an indication on the long-term reliability and repeatability of the sensor in acquiring the characteristic rhythm.

Figure 1.8 is depicting the ECG signal acquired from the two electrodes on the neck. Seen in blue is the Ag/AgCl wet electrode signal overlayed, in red, by the six-coated rGO_x dry electrode. It is clearly seen that the amplitude of the rGO_x electrode is higher than Ag/AgCl electrode. The P wave, QRS complex, and T wave are correctly captured, and each feature is more distinctly captured by the rGO_x fabric. The mean R wave amplitude of Ag/AgCl and rGO_x electrodes were 47 μV and 95 μV, respectively. This marks a factor of 2 improvement in R wave amplitude for the rGO_x fabric over the Ag/AgCl. This is significant from the perspective of designing sensor interface circuitry, especially at such low μ range, as it allows a more detectable signal and to an extent less signal amplification. This improved performance of rGO_x fabric could be attributed to the reduced resistance of the re-coated fabric since the electrode is resistive [3].

Figure 1.9 shows the heart activity in terms of beats in mV against the isoelectric noise. The isoelectric noise is the mean absolute deviation of data points across their

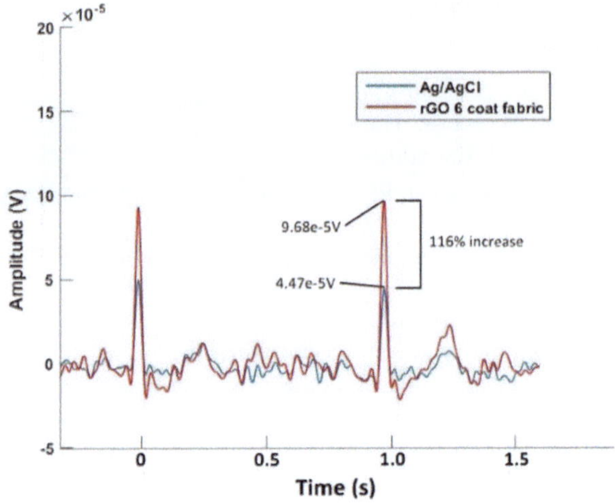

Fig. 1.8 Raw ECG signal overlay of Ag/AgCl wet electrode (blue) and rGO_x six-coated dry electrode (red)

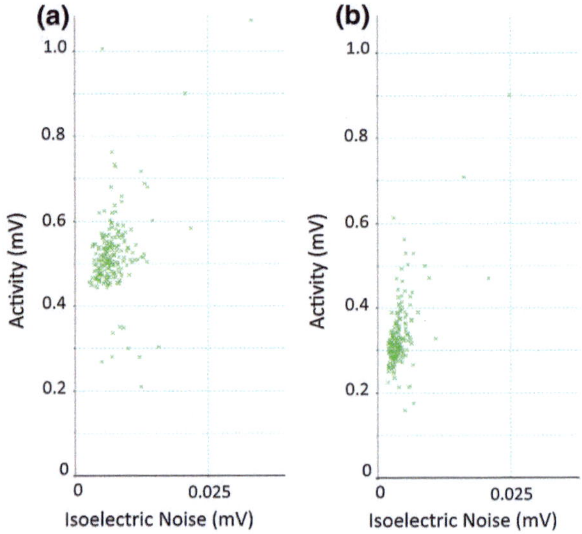

Fig. 1.9 Activity versus isoelectric noise. **a** Ag/AgCl wet electrode and **b** rGO_x six-coated dry electrode

median. It is used here because it indicates contamination from noise and previous beat T wave on the isoelectric line, the zero baseline of the signal referenced in the TP interval. The Ag/AgCl is showing less isoelectric noise with less dispersed data points than rGO_x. However, the effect of isoelectric noise could be attributed to the larger amplitude of the rGO_x signal. The heart activity is detected further by the rGO_x fabric and with very comparable isoelectric noise performance to the Ag/AgCl electrode.

1.4 Conclusions

A conductive fabric composite of nylon and graphene can be produced through a simple process. The conductivity can be controlled by repeatedly re-coating the nylon, applying additional layers of graphene. This composite fabric is well suited to application as an ECG electrode, giving good signal acquisition and relatively low motion artifact. Comparison with the generally accepted Ag/AgCl electrode shows an increase of signal amplitude, while maintaining signal shape. This composite displays many of the desired properties for a long-term, wearable biosignal sensor. In the future, this type of sensor could be incorporated into a wireless continuous monitoring system, with further applications as an Internet of Things device.

Acknowledgements We acknowledge assistance of Dr. Y. A. Samad in early stages of rGO_x growth. This work was supported by the Mubadala and SRC through 2013-HJ-2440 and in part through the Mubadala SRC task 2011-KJ-2190.

References

1. J.Y. Baek, J.H. An, J.M. Choi, K.S. Park, S.H. Lee, Flexible polymeric dry electrodes for the long-term monitoring of ECG. Sens. Actuators A Phys. **143**(2), 423–429 (2008)
2. W. Chen, L. Yan, P.R. Bangal, Preparation of graphene by the rapid and mild thermal reduction of graphene oxide induced by microwaves. Carbon **48**(4), 1146–1152 (2010)
3. Y.M. Chi, T.P. Jung, G. Cauwenberghs, Dry-contact and noncontact biopotential electrodes: methodological review. IEEE Rev. Biomed. Eng. **3**, 106–119 (2010)
4. S. Gabriel, R.W. Lau, C. Gabriel, The dielectric properties of biological tissues: ii. measurements in the frequency range 10 Hz to 20 GHz. Phys. Med. Biol. **41**(11), 2251 (1996)
5. S.H. Huh, Thermal reduction of graphene oxide. INTECH Open Access Publisher (2011)
6. E. Huigen, A. Peper, C.A. Grimbergen, Investigation into the origin of the noise of surface electrodes. Med. Biol. Eng. Comput. **40**(3), 332–338 (2002)
7. X. Li, G. Zhang, X. Bai, X. Sun, X. Wang, E. Wang, H. Dai, Highly conducting graphene sheets and Langmuir—Blodgett films. Nature Nanotechnol. **3**(9), 538–542 (2008)
8. E. Nemati, M.J. Deen, T. Mondal, A wireless wearable ECG sensor for long-term applications. IEEE Commun. Mag. **50**(1), 36–43 (2012)
9. W.H. Organization, Global Status Report on Noncommunicable Diseases (2014), (http://www.who.int/nmh/publications/ncd-status-report-2014/en/). Online, Accessed 13 June 2016
10. M. Rivera-Ruiz, C. Cajavilca, J. Varon, Einthoven's string galvanometer. Tex. Heart Inst. J. **35**(2), 174–78 (2009)
11. Y.A. Samad, Y. Li, S.M. Alhassan, K. Liao, Non-destroyable graphene cladding on a range of textile and other fibers and fiber mats. RSC Adv. **4**(33), 16935–16938 (2014)
12. P.W. Wilson, R.B. D'Agostino, D. Levy, A.M. Belanger, H. Silbershatz, W.B. Kannel, Prediction of coronary heart disease using risk factor categories. Circulation, 97(18), 1837-1847.M. (1998)
13. S. Yao, Y. Zhu, Nanomaterial-enabled dry electrodes for electrophysiological sensing: a Review. JOM, 68(4), 1145–1155 (2016)
14. M.K. Yapici, T. Alkhidir, Y.A. Samad, K. Liao, Graphene-clad textile electrodes for electrocardiogram monitoring. Sens. Actuators B Chem. **221**, 1469–1474 (2015)

Chapter 2
Evaluation of Continuous, Shoe-Integrated Weight Sensors for Heart Failure Patients

Ahmed Elian, Nisrein Sada, Ibrahim (Abe) M. Elfadel, Mohammed Saeed and Ayman Shabra

2.1 Introduction

Congestive heart failure (CHF) is one of the leading causes of death and morbidity in developed countries and is expected to reach endemic levels by the year 2040 [1]. CHF patients are prone to frequent hospitalizations, most commonly due to edema, the buildup of fluids in their bodies. This was the major contributor to the total $37.2 billion annual cost of caring for these patients in 2009 in the USA [1]. A study conducted by Chaudhry et al. [2] has shown that patients diagnosed with CHF experience a detectable increase in body weight at least one week before hospitalization. Regularly monitoring a patient weight change might prove beneficial in a predictive model of the risk of hospitalization. However, due to the limited patient compliance with the daily weight measurement procedures [2], it is inadequate to include user-driven weight measurements in a predictive model. It is important to mention that recent algorithms for predicting CHF decompensation that relied on automatic mul-

A. Elian
Department of Electrical and Computer Engineering, University of Toronto, Toronto, Canada
e-mail: ahmed.elian@mail.utoronto.ca

N. Sada · Ibrahim (Abe). M. Elfadel (✉)
Masdar Institute at Khalifa University, Abu Dhabi, UAE
e-mail: ielfadel@masdar.ac.ae

N. Sada
e-mail: nsada@masdar.ac.ae

M. Saeed
University of Michigan, Ann Arbor, MI, USA
e-mail: msaeed@alum.mit.edu

A. Shabra
MediaTek, Woburn, MA, USA
e-mail: aushabra@yahoo.com

Ibrahim (Abe) M. Elfadel and M. Ismail (eds.), *The IoT Physical Layer*,
https://doi.org/10.1007/978-3-319-93100-5_2

tiple sensor measurements did not include weight [1, 3]. In these works, edema was measured using impedance measurements through electrodes that must be attached to the body. In an effort to alleviate the discomfort caused by these electrodes, [4] explored the use of stretching in socks as an alternative.

In this chapter, we investigate the potential of footwear-mounted pressure sensors as a system that automatically measures weight changes with minimal patient intervention while adhering to standard medical care protocols. To design an in-shoe system that reliably measures patient weight, it is important to understand the relationship between the plantar pressure and the weight of a patient. The literature shows how plantar pressure patterns change with different activities [5, 6], pace [7], shoe type [8], age [9], foot deformities [10], and health conditions affecting gait such as Parkinson's disease and diabetes. This chapter examines the impact of walking and standing on weight estimates of a human subject based on plantar pressure. The data we use are from the PhysioNet database [11] and from our own measurements on a commercial pressure mapping system from Tekscan [12].

For accurate weight estimates, it is not only important to select pressure sensors for patient comfort and ease of shoe integration but also for their electromechanical properties. E-textiles are natural candidates for this application since they can act as substrate enabling the integration of flexible electronic components. E-textiles have been used in various fields such as strain sensors to recognize upper postures [13] and to measure pressure [14, 15]. However, their suitability for weight measurement has not been studied. In addition to the e-textile Velostat, we investigate other widely used piezoresistive sensors in wearable applications [16] such as Interlink force-sensitive resistors (FSR) and Flexiforce sensors from Tekscan.

The chapter begins by outlining some of the important considerations for the design of such a system. The impact of human subject movement on the accuracy of the weight estimate is presented in Sect. 2.3 while Sect. 2.4 presents results for a standing human subject. It finally presents measurement results from the evaluation a variety of flexible sensors. Since the medical literature uses weight in place of mass [2], this chapter will use weight and mass interchangeably.

2.2 Hardware Requirements for Weight Estimation

For the CHF monitoring application, the weight sensor must satisfy a number of key requirements. In addition to providing easy integration within the shoe, the sensors should provide accurate pressure measurements. Moreover, the sensor placement underneath the foot should enable an accurate estimation of the weight. From the data presented in [1], we estimate that the system should allow for the detection of a few hundred grams of weight change. This represents an accuracy of about 0.1%, which fortunately does not need to be satisfied in an absolute measurement of weight but only in a change of weight. The expected range of pressure that a sensor may see is less than 400 kPa.

To achieve the desired accuracy, it is necessary to use an array of sensors that can capture the variation in pressure at different locations under the foot. For a healthy human subject, the pressure distribution is mainly under the heal and the metatarsal regions. For a CHF patient, on the other hand, the pressure is likely to follow more complex patterns due to edema and possibly foot deformities in elderly patients. It is therefore reasonable to uniformly place sensors in the entire area underneath the foot.

If the shoe-mounted sensors measure force or pressure (force per unit area), then the weight can be simply estimated using Newton's second law of motion:

$$m\,g = \sum_{i=1}^{N} F_i = \sum_{i=1}^{N} P_i A_i \tag{2.1}$$

where m is the mass of the body, F_i and P_i are, respectively, the force and pressure measured by sensor i which is assumed to cover an area A_i. N is the total number of sensors, and g is the gravitational acceleration. The measurements of pressure in Eq. 2.1 need to be synchronized in time across both feet. The estimation of weight requires pressure measurements from both feet. Of course, weight estimation based only on single-foot measurements is likely to be more difficult.

2.3 Weight Estimation Under Motion

A relevant consideration is the impact of motion on the weight estimation accuracy. To understand this, we utilized data from three separate studies investigating gaits of Parkinson's disease patients compared to healthy subjects [11]. The data include a total of 301 experiments on 91 Parkinson patients and 72 healthy controls. The studies utilize eight shoe-integrated sensors distributed underneath each foot. The sensors are thin capacitive force sensors from Infotronic, Inc. that provide a range of 0 to 1000 N and a resolution of 2.5 N. An example of force waveforms during walking is illustrated in Fig. 2.1a. The figure shows the force seen by the left foot, right foot, and both feet. The impact of movement results in a regular peak in total force that is more that 50% larger than mg. This makes it difficult to extract an estimate of the patient mass using Eq. 2.1 as shown in Fig. 2.1b, where force is translated into apparent mass at any instant by dividing by g. Figure 2.1b additionally illustrates a close match between mass estimated from force using the following equation and the actual mass, a match that holds true empirically in all waveforms observed:

$$\begin{aligned} Mass &= Max_{time}[(Single\,foot\,average\,force)/g] \\ &= Max_{time}[(Total\,force\,from\,both\,feet)/(2g)] \end{aligned} \tag{2.2}$$

Using this formula as a mass estimation, the percent error in estimated mass for the 301 experiments is shown in the histogram in Fig. 2.2. The movement of the human

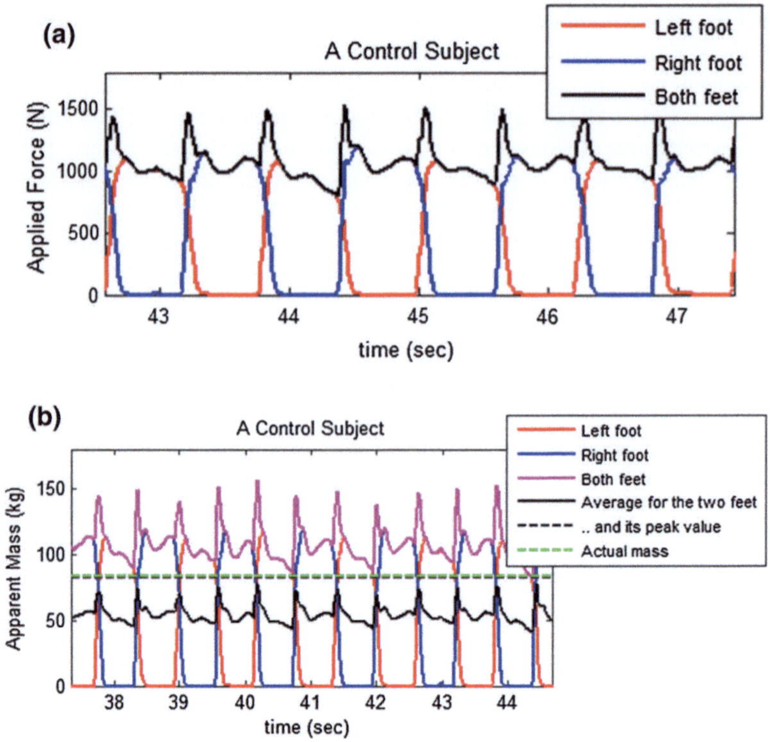

Fig. 2.1 **a** Example waveforms for force underneath feet of a control subject. **b** Apparent mass waveforms of the same subject calculated using m = F/g

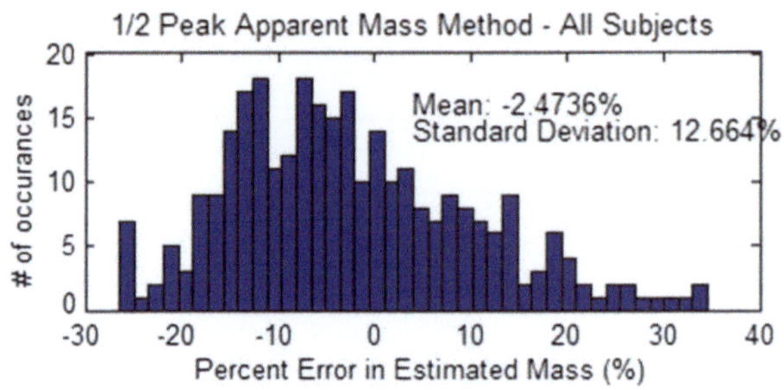

Fig. 2.2 Distribution of experiments in [11] in terms of percent error in mass estimated using Eq. (2.2)

subject results in a significant mass estimation error with a mean error of −2.5% and a standard deviation of 12.7%. It is worth noting that the standard deviation for the healthy subjects and Parkinson patients was 11 and 13.3%, respectively, which does not represent a significant difference. The pace of the human subjects in these datasets ranges from 0.9 to 2.3 steps per second. We did not observe any meaningful improvement in estimation accuracy as the human subject pace decreased.

Force and pressure measurements during walking include contributions of large impact forces that make the weight estimation inaccurate. High peak forces during walking, climbing stairs, and running have been reported by other researchers [6], and it has also been shown that the ground surface material also has an important role [17]. Considering these results, we next explore the force and pressure measurements performed when the human subject is standing.

2.4 Weight Estimation While Standing

To explore the estimation of weight from a 2D map of force or pressure measurements, we utilized the Tekscan Walkway [12], which is a commercial medical pressure mapping system. It is composed of seven mats, each 6 sensors. This

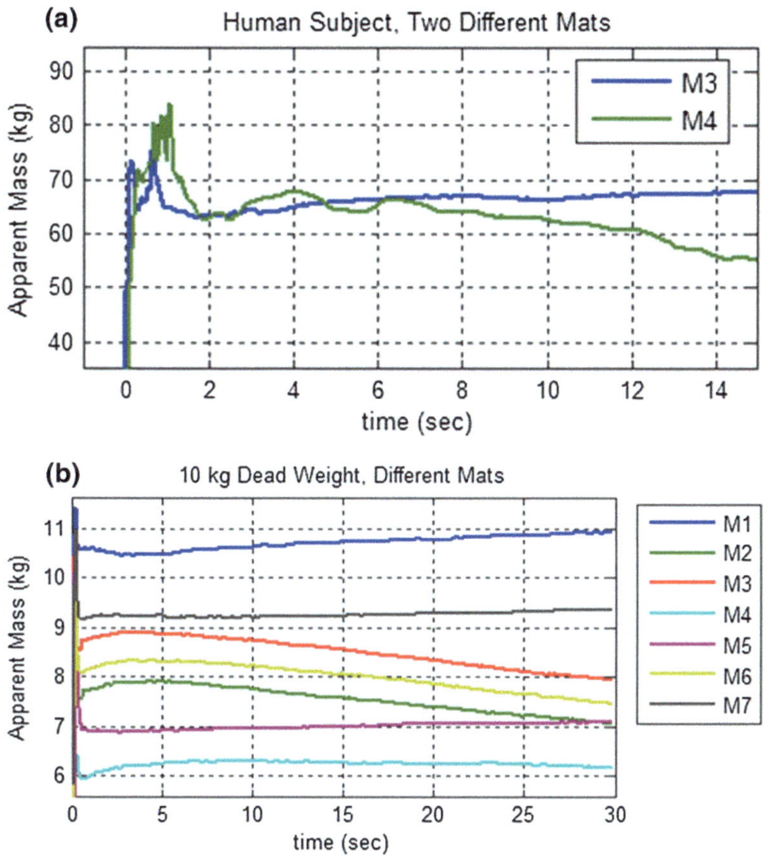

Fig. 2.3 Walkway system weight estimate for: **a** human subject estimate when standing on two different mats from the walkway system. **b** 10 kg dead-weight estimate when placed on mats from 1 to 7

instrument is used in plantar pressure measurements for a wide variety of application, including identification of gait patterns for the medical diagnostics of motor conditions such as Parkinson's.

During our experiments, a human subject stepped on a mat and remained static for 15 s. Figure 2.3a shows the human subject body mass estimation results taken on Mat 3 (M3) and Mat 4 (M4) where Eq. 2.1 was used. An obvious drift in mass reading over time is observed for both mats after the subject steps on the mat. Additionally, the drift characteristic and average mass reading are obviously different between the two mats. In order to determine whether the human subject caused this measurement drift, we used a 10 kg deadweight to calibrate the instrument. Data were collected from each of the seven mats for 30 s after the 10 kg weight was placed on the mat. Figure 2.3b shows the results where the weight estimates still reveal significant variability between mats and significant drift with time. This demonstrates that most of the error and drift in body mass readings are due to the resistive sensors, which is consistent with our characterization of the stand-alone Flexiforce sensors as we will show next.

2.5 Pressure Sensors' Exploration

From the results of the last section, it is apparent that the weight estimates based on a pressure map taken while the human subject is standing are limited by the quality of the sensors. Since it is necessary for any shoe-mounted systems to utilize a very precise pressure sensor, we investigated the suitability of some of the sensors that are widely used in wearable applications.

2.5.1 Rationale for Sensor Search Space

Even though strain gauges are the standard for precision weight measurements, they are not included here for two reasons. First, strain gauges are not intended to measure forces normal to the sensor surface. Second, the typical usage model for strain gauges involves mounting them on a rigid beam that flexes slightly due to the applied force. Both these considerations make strain gauges difficult to apply in a shoe. Our exploration also does not include capacitive pressure sensors such as the ones used in the Infotronic and Pedar systems, mainly due to availability limitations.

The sensors we investigated in this work include FSR400 and FSR402 from Interlink Electronics, Flexiforce from Tekscan and our own textile sensors. FSR sensors have been utilized in shoe-mounted systems for the purpose of gait measurements [16] and are flexible and low cost. The Flexiforce sensor is manufactured by Tekscan the producers of the Walkway system. Finally, textile sensors have found wide utilization for pressure measurements due to their flexibility and comfort. Examples include [14, 15].

These sensors are all piezoresistive sensors, which means that their electrical resistivity varies with applied pressure or strain. These sensors can be fabricated using a diaphragm on a conductive polymer as in FSR sensors or using pressure sensitive ink as in Flexiforce sensors. Examples of a conductive polymer include Velostat or Linqstat.

2.5.2 *Textile Sensor Designs*

We constructed textile sensors using Velostat, that is sandwiched between two conductive layers. Although conductive fabrics such as MedTex180 or MedTex130 can be used for these sensors, sensor flexing and vibration due to the applied force and the movement of the leads introduced artifacts and noise and limited our ability to isolate the properties for the sensing material. Consequently, our measurements were based on a structure that uses ceramic plates coated with copper as the electrodes. Our measurements show that these sensors suffer from poor repeatability, a drift over time, and sensitivity to location of the force on the sensor. In [17], it is reported that the use of neoprene on top of Velostat minimizes drift due to neoprene surface compliance, which mitigates the stress on the piezoresistive material. Additionally, [17] reports that neoprene increases sensitivity relative to sensors using rigid materials. Here we present results of Velostat sensors with and without a neoprene layer.

2.6 Sensor Testing Results

Figure 2.4a shows the estimated decay with time for different Interlink FSR sensors when a force of 50 N is applied. From the figure we can observe the following. First, the sensors continue to exhibit decay in resistance even after a period greater than 1600 s. FSR402 exhibits more decay compared to FSR400 which can be attributed to the smaller pressure range allowed for FSR400 that places the pressure applied near the maximum allowable. Increasing the applied force in both sensors decreases the resulting resistance and decay. A similar behavior is observed with Flexiforce, namely an increase in the applied force decreases the resulting resistance and the continuous decay with time. Flexiforce exhibits a decay of 14% for the same period and same force applied. For the repeatability test, the worst-case relative difference for Flexiforce was 26% compared to FSR sensor which has a minimum relative difference of 8% for FSR400 and 3% for FSR402, and a maximum relative difference of 32% for FSR400 and 70% for FSR402.

The Velostat structure described in Sect. 2.5 achieves a difference of 26% between four trials after introducing a wait window of more than an hour between trials. The decay ranged from 10–13.8% across four trials as shown in Fig. 2.4b. These results are produced when a 2 kg weight is applied on this structure. Moreover, the resistance continued to decrease compared with the initial resistance, implying a long recovery

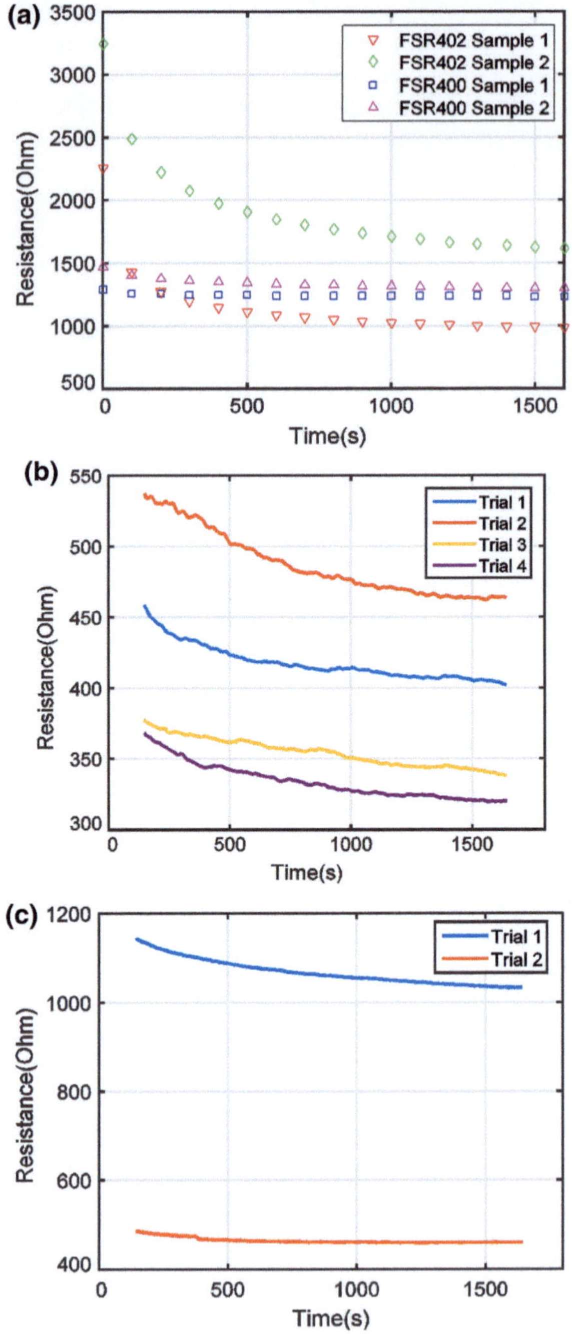

Fig. 2.4 Resistance decay for: **a** two samples of Interlink's FSR402 and FSR400 when a force of 50 N is applied. **b** Velostat sensor when a deadweight of 2 kg is applied, repeated over several trials using the same sensor where the only difference between trials is the time parameter. **c** Neoprene-packaged sensor across two trials when a deadweight of 2 kg is applied, repeating the test is twice using the same sensor

time. This long-term decrease is independent of the waiting time between trials. Comparing the results from this structure with the neoprene structure, the decay achieved is around 5–9% for the same applied deadweight as seen in Fig. 2.4c.

2.7 Conclusions

The purpose of this work is to investigate the potential of shoe-integrated sensors for body weight estimation for CHF patient monitoring. Utilizing PhyisoNet data for plantar pressure, we show that weight estimation accuracy is degraded due to the motion of the human subject. In addition, measurements on a standing human subject show that weight estimates exhibit drift due to the characteristics of the sensors used in the pressure mapping system. Mechanical testing of commercially available pressure sensors shows a continuous long-term resistance decay and inconsistent readings across independent trials. It has also revealed a memory effect for both Interlink FSR and Flexiforce sensors. We have found that the use of neoprene in Velostat sensors decreased drift, but not sufficiently for weight estimation. An alternative, that is worth exploring in future work, is the use of rigid piezoresistive materials such as silicon but packaged appropriately for shoe integration.

The Human Subjects Research Ethics Committee at Masdar Institute approved this work.

Acknowledgements This work is supported by SRC under the ACE4S program, contract HJ2440, with customized funding from Mubadala, Abu Dhabi, UAE. The authors thank Dr. H. Alsafar and several other colleagues at KUSTAR for access to the Tekscan system.

References

1. I.S. Anand et al., Design and performance of a multisensor heart failure monitoring algorithm: results from the multisensor monitoring in congestive heart failure (music) study. J. Card. Fail. **18**(4), (2012)
2. S. Chaudhry et al., Patterns of weight change preceding hospitalization for heart failure. Circulation **116**, (2007)
3. S. Lee et al., Congestive heart failure patient monitoring using wearable bio-impedance sensor technology, in *Proceedings of the Annual International Conference of the IEEE EMBS* (2015)
4. S. Zhang, R. Rajamani, Sensor on instrumentation socks for detection of lower leg edema – an in vitro study, in *Proceedings of the Annual International Conference of the IEEE EMBS* (2015)
5. A. Rozema, J.S. Ulbrecht, S.E. Pammer, P.R. Cavanagh, In-shoe plantar pressures during activities of daily living: implications for therapeutic footwear design. Foot Ankle Int. **17**, 352–359 (1996)
6. S. Rao, S. Carter, Regional plantar pressure during walking, stair ascent and descent. Gait and Posture **36**, 265–270 (2012)
7. I.-J. Ho, Y.-Y. Hou, C.-H. Yang, W.-L. Wu, S.-K. Chen, L.-Y. Guo, Comparison of plantar pressure distribution between different speed and incline during treadmill jogging. J. Sports Sci. Med. **9**, 154 (2010)

8. L.-Y. Guo et al., Effect on plantar pressure distribution with wearing different base size of high-heel shoes during walking and slow running. J. Mech. Med. Biol. **12**, (2012)
9. M. Saito et al., An in-shoe device to measure plantar pressure during daily human activity. Med. Eng. Phys. **33**, 638–645 (2011)
10. K.J. Mickle, B.J. Munro, S.R. Lord, H.B. Menz, J.R. Steele, Gait, balance and plantar pressures in older people with toe deformities. Gait and Posture **34**, 347–351 (2011)
11. Gait in Parkinson's disease, http://physionet.org/pn3/gaitpdb/
12. Strideway system, https://www.tekscan.com/products-solutions/systems/walkway-system
13. C. Mattmann, O. Amft, H. Harms, G. Troster, F. Clemens, Recognizing upper body postures using textile strain sensors, in *Proceedings of the 11th IEEE International Symposium on Wearable Computers* (2007), pp. 29–36
14. W. Xu et al., eCushion: a textile pressure sensor array design and calibration for sitting posture analysis (2013)
15. L. Shu et al., In-shoe plantar pressure measurement and analysis system based on fabric pressure sensing array. IEEE Trans. Inf. Technol. Biomed. **14**, 767–775 (2010)
16. S.J.M. Bamberg et al., Gait analysis using a shoe-integrated wireless sensor system. IEEE Trans. Inf. Technol. Biomed. **12**, 413–423 (2008)
17. J. Meyer. Textile pressure sensor: design, error modeling and evaluation. *ETH* (2008)

Chapter 3
ALD Al-doped ZnO Thin Film as Semiconductor and Piezoelectric Material: Process Synthesis

Ayman Rezk and Irfan Saadat

3.1 Overview

Electronic devices fabricated on flexible substrates have been a research hotspot over the past decade [57]. Multiple motivations and driving forces are behind this interest: first, the viability of producing new devices based on this technology, and consequently, expanding the electronics' market, and second, the immensely lower material and production cost for the flexible devices in comparison to that of the integrated circuit on rigid silicon or glass wafer. The unique 'roll-to-roll' technology, where the devices are formed on a roll of a moving plastic tape/sheets, quite similar to newspaper printing, with electronic devices and components are printed, assembled and fabricated on these sheets.

It is amazing how all the studies found during the literature review, dating only 2–3 years back, had mere concept pictures of futuristic flexible and transparent electronic devices. However, today, these concepts are actually either available as consumer products including curved and fully transparent TVs (LG [39], Samsung [55]) or prototypes such as flexible phones, tablets (Lenovo [37]) and displays (LG [38]) and even a T-shirt that monitors the wearer's breathing rate in real time [22]. This does not only indicate the fast-paced growth of flexible electronics but its leaping and paradigm changing market appeal as well. This can be seen in recent reports such as IHS Markit declaring that revenues from flexible active-matrix organic light-emitting diode (AMOLED) displays were on pace to exceed rigid AMOLED panel in Q3 2017 [28]. Indeed, science fiction has never been closer to reality; we have come a long way since such devices were first imagined by H.G. Wells in his speculative

A. Rezk · I. Saadat (✉)
Masdar Institute at Khalifa University, Abu Dhabi, UAE
e-mail: isaadat@masdar.ac.ae

A. Rezk
e-mail: arizk@masdar.ac.ae

Ibrahim (Abe) M. Elfadel and M. Ismail (eds.), *The IoT Physical Layer*,
https://doi.org/10.1007/978-3-319-93100-5_3

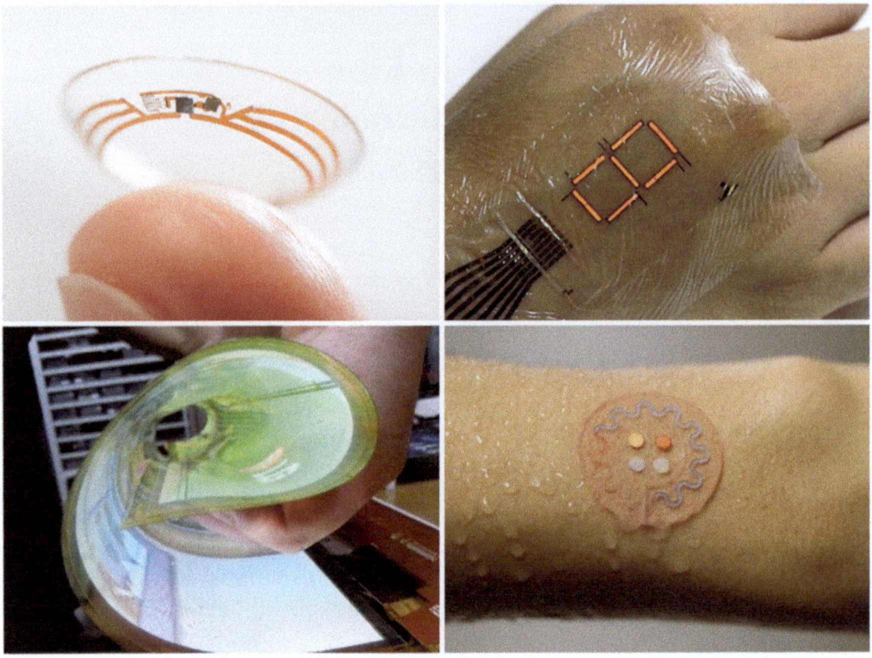

Fig. 3.1 Flexible and transparent electronics. Clockwise from top left: Glucose-sensing contact lens (Google), light-emitting diodes on electronic skin (Someya Group, University of Tokyo), battery-free microfluidic systems for capture, storage and chemical analysis of sweat (John A. Rogers, University of Illinois), 18-inch rollable newspaper-like display (LG)

fiction novel The Shape of Things to Come, published in September 1933. Figure 3.1 highlights some of these flexible and transparent electronics.

The consequential fact that circuits based on conventional semiconductors such as silicon and conductors such as copper can be turned transparent and flexible using different materials, like the transparent semiconducting and conducting oxides (TSOs and TCOs, respectively), enables several innovative and high added-value application fields. The viability of this technology largely depends on the performance, reliability and cost of the flexible transistors. Transistors are the primary components in most modern electronic circuits, commonly used to amplify or to switch electronic analog signals and to process digital data. Besides the high-performance silicon transistors used in microprocessors or amplifiers, designated by metal–oxide–semiconductor field-effect transistors (MOSFETs) and requiring processing temperatures exceeding 1000 °C, other types of transistors are available for large-area electronics, where lower temperatures and costs are required. Perhaps, the most relevant ones are the thin-film transistors (TFTs), which are intimately associated with display and sensor applications. This goes hand in hand with the rising interest in the Internet of Things (IoT) applications where flexible TFTs and sensors here become a major focus of research and development. Flexible devices have become achievable with the recent advent of reliable flexible materials, TSOs and TCOs, which, besides transparency and flexibility, allow for low temperature, low-processing costs and high performance.

3.2 Motivation

The field of flexible electronics has a promising future ahead as illustrated in the previous section, yet, further advances in materials, devices and processes are needed to help realize its fuller potential. Specifically, more progress is needed in developing high-performance electronic materials to be used in uniform devices over a large area to suit applications like displays and sensors, with maximum processing temperatures that are compatible with inexpensive, flexible and plastic substrates ($\leq$250 °C). To pave the way for these technologies to be used in pioneering flexible and transparent devices, the ability to fine-tune their electrical and mechanical properties along with stability under electrical and mechanical stress require further investigation.

Metal–oxide–semiconductors, typically made of combinations of zinc, gallium, indium and tin oxides, form a unique family of materials that has been extensively studied and believed to be behind this boom in flexible and transparent electronics. The key enabler was that their field-effect mobility was shown not to degrade severely in amorphous films compared to crystalline films, which resulted in massive interest and rapid advances in this field. Three main advantages encouraged these significant research efforts to be focused on the fundamental device physics of amorphous metal oxides. First, metal oxides have sufficiently high mobility for fast switching and high current. Second, they can achieve this boost in performance with low, or even room temperature, fabrication processes. Third, metal oxides are transparent. However, it became clear later that to achieve high channel mobilities, the most promising candidates are based on systems including indium, specifically indium gallium zinc oxide ($IGZO$) which dominated the investigation of oxide TFTs. Nonetheless, indium is earth-scarce and therefore has been costly. This caused the oxide TFTs interest to shift to indium-free oxide TFTs instead. Conductive oxides such as indium tin oxide (*ITO*) and oxide semiconductors such as $IGZO$ have been used for decades, but there has been renewed interest in other amorphous metal oxides for applications such as displays and sensors.

ZnO thin films for instance are already widely investigated especially as transparent conductive oxides (TCOs) to function as transparent top electrodes for many of those applications. These films were promoted as an alternative low-cost transparent conductive oxide material instead of indium tin oxide (*ITO*) in photovoltaic cells and displays [9, 10]. In addition, they are being considered as an attractive alternative to $IGZO$ as a semiconductor oxide in high-quality TFTs. One key motivation is abundance, where it is estimated that there is about 1000 times more zinc (132 ppm) than indium (0.1 ppm); this abundance of zinc in turn contributes to the low cost of TFT's major constituents.

ZnO, especially when doped, has a relatively high electromechanical coupling coefficients among semiconductors which allows it to be successfully used as both a piezoelectric and semiconductor thin film devices. The ability to significantly change Al-doped *ZnO* films conductive and piezoelectric properties through small adjustment of the Al doping makes it versatile for designing various high-quality flexible thin film devices. This is in addition to being well known for its non-toxic nature, low extraction cost due to its abundance in the earth crust, its propensity for high field effect mobility (among indium-free metal oxide systems), along with its high piezoresponse and the ability to enhance and tune these properties through doping. In addition, atomic layer deposition (ALD) was chosen to grow these thin films of undoped and doped zinc oxides. Doping allows to precisely control these films characteristics (their composition, conductivity, crystal orientation, etc.), and thus can produce high-quality thin films with the desired and required properties for transparent and flexible electronics. Understanding the interconnected impact of the growth conditions and doping on device performance will be critical to the successful implementation of large area, low cost, flexible substrate with low-temperature metal oxide TFTs and sensors.

Most of the documented and published studies tackle pulsed laser deposition (PLD) or sputtered Al-doped *ZnO* while TSO is under-visited. The vast majority of the previous reports on ALD Al-doped *ZnO* films were grown on rigid glass substrates and have treated TCO, mostly in terms of its growth conditions [32] and doping mechanism [35, 36]. In addition, prior reports on fabricating TFTs using ALD Al-doped *ZnO* with various Al doping concentrations [3] focused on the properties of the TFTs, such as the variation in device stability with different Al concentrations. However, a broader and deeper investigation of the properties of thin films semiconductive and piezoelectric characteristics deposited with various growth conditions and Al concentrations are yet to be reported.

3.3 ALD-grown-Al-doped ZnO Thin Films

This section explores the governing mechanics of the Al-doped *ZnO* growth on flexible substrates using the ALD method and how incorporating Al dopant in the *ZnO* lattice affects the films growth at different deposition temperatures. It begins with discussing our choice of Al-doped *ZnO* as the thin-film material and ALD as the deposition method.

Zinc oxide is a very popular intrinsically n-type semiconductor. Aside from having great potential as a bulk crystal, there is significant interest in it as thin-film material and as a nanostructured material, nanowires and nanoparticles. The corresponding usages range from optoelectronic applications such as laser diodes [27, 52], light-emitting diodes (LEDs) [9, 44] to TFT [5, 6] and TCO [17, 56] and many others. This popularity mainly stems from its non-toxic nature, its abundance in the earth crust, facilitates obtaining it at low cost, and its properties which are highly tunable through doping. Additionally, *ZnO*'s high transparency, even with heavy doping, in the visible region of light which gives it the highest prospects and applicability as a transparent oxide. Accordingly, *ZnO* thin films are being widely investigated especially as transparent conductive oxides (TCOs) to function as transparent top electrodes for many of those applications. The high cost of indium and its lower wet etch rate, compared to *ZnO* or indium gallium zinc oxide (*IGZO*), that can lead to damage in the active layer during the etch process [41], have opened the door for promoting *ZnO* thin films as a low-cost alternative transparent conductive oxide material instead of indium tin oxide (*ITO*) in photovoltaic cells and displays.

The doping of *ZnO* films was initially investigated because most undoped TCOs including *ZnO* have relatively low carrier concentrations and extrinsic doping has been adapted to improve their electrical conductivity by increasing the carrier concentrations. These dopants have to be shallow donors that provide extra ionized electrons to the system. For this purpose, a wide range of different metal atoms have been extensively studied as dopants such as B [18], In [26, 29], Co [62], Zr [20], Ge [7], Hf [2], Sn [60], while the group-III elements Al [10, 12, 19, 46] and Ga [8, 54, 63], are preferred in applications that require good mobility and piezoelectric films. As the extrinsic dopants of these elements substitute into host cation sites thus providing an extra electron, Al is also a very effective n-type dopant and it is preferred to other dopants in some devices as it is less oxygen reactive, and its ionic radius minimizes lattice deformations [42].

3.3.1 Why Al-doped ZnO?

Doped *ZnO*, particularly aluminum-doped *ZnO*, has enabled various new applications aside from TCO due to the versatile properties brought forth by doping, which allows accurate control of the film's electrical, mechanical and optical behavior. However, most, if not all, current Al-doped *ZnO* research targets optimizing the film to achieve minimum resistivity for TCO applications. Our work focuses on other two promising applications of Al-doped ZnO. First, its usage as a semiconductor material for active channel layers in transparent and flexible TFTs. This application takes advantage of Al-doped *ZnO* films specific characteristics such as high transmittance in the visible wavelengths and wide range in electrical conductivity modulated via the growth conditions. Second, It's usage as a thin-film piezoelectric material for micro electromechanical systems, along with the benefits of its low power requirement and available high piezoelectric response to function as a strain sensor. Al-doped *ZnO* has high piezoelectric coupling coefficient, compared to aluminium nitride (*AlN*), and low dielectric constant, compared to lead zirconate titanate (*PZT*), two of the most common piezoelectric material used in strain sensors, both which are achievable at low deposition temperature. Moreover, *ZnO* production is environmental-friendly compared to these materials. Employing Al-doped *ZnO* films as a piezoelectric material in sensors or a semiconductor in active channel layers for transparent and flexible TFTs has been previously proven viable.

High-performance transparent and flexible TFTs require a channel layer with good conductivity and high field-effect mobility. Numerous studies have recommended using oxide semiconductors, such as *IGZO* and *HfInZnO* [45, 49] to achieve these characteristics for current and future devices. To achieve high channel mobilities, the most promising candidates are based on material systems consisting of indium. Many reports currently focus on specifically on *IGZO* for TFTs. However, indium is very costly and does not fit in with the bigger picture of the industry's economic landscape. Therefore, there is a need to further investigate oxide TFTs that do not include indium. *ZnO* has shown high electron Hall mobility and high electron saturation velocity at room temperature. In aggregate, these advantageous electronic properties enables its use in high-performance transistors.

In summary, doped zinc oxide is of interest as a transparent semiconductor oxide (TSO) and piezoelectric material, due to the abundance of its major constituents, its good conductivity, high transparency, high piezoresponse, high transparency and wide bandgap (3.37 eV [27]). The current work focuses on the properties required for TSO applications including resistivity of about $10^{-8}\Omega$.cm, carrier density of $\geq 10^{18}\text{cm}^{-3}$ and high mobility. Aluminum (Al^{3+}) was chosen as the dopant in the current work due to its abundance and because it can act as a donor providing up to one free electron per ion when substituted for Zn^{2+}. The doping process can be controlled using atomic layer deposition (ALD), with the doped films resulting in an enhanced yet well controlled conductivity and mobility.

3.3.2 Why ALD?

To date, the deposition of doped and undoped *ZnO* films has been carried out through several deposition techniques, including reactive magnetron sputtering [30, 30, 46], chemical vapor deposition (CVD) [25, 61], pulsed laser deposition (PLD) [33, 65], spray pyrolysis [4], electron-beam evaporation and sol-gel process [59]. In addition, efforts have been focused to optimize the process parameters, such as doping concentration, deposition temperature and pressure to optimize the required attributes needed for the desired application. However, impurity-doped *ZnO* films usually have some limitations in terms of the dopant distribution and electrical instability. In addition, it is not easy to have a fine control over the dopant concentration and distribution. Therefore, atomic layer deposition (ALD) stands out as a versatile film growth method when compared to other techniques to grow and improve the quality of these films. It uniquely enables the growth of precisely controlled sub-nano, uniform, pinhole-free films and high-quality thin films on a variety of large-area surfaces at relatively low-processing temperatures. This atomic-scale precision is achieved by simply controlling the number of ALD cycles. Using ALD allows the thickness and the uniformity of the *ZnO* films to be independent of temperature changes as long as the deposition is done within the ALD growth window. In addition, adjusting the parameters of the ALD process, such as the precursor exposure time and the exposure ratio between the precursors, the growth rate can be reliably and accurately controlled due to the intrinsically self-limiting surface chemistry of ALD half-reactions mechanism. Moreover, ALD does not require line of sight for deposition which allows conformable coating of high surface area to volume ratio structures and complex geometries with extremely high aspect ratio. Also, with only one reactant present in the chamber at a time prevents any unwanted gas-phase reactions as may occur in chemical vapor deposition (CVD), can lead to particle formation and inferior device performance.

The ALD technique has been successfully used in the past in several attempts to deposit undoped and Al-doped ZnO thin films using diethylzinc (DEZn), trimethylaluminum (TMAl), and water as the precursors [1, 10, 15, 25, 32, 43]. However, as reported earlier, Al-doped ZnO deposited by ALD has not been studied as extensively as the other deposition methods. Previous studies were limited only to the growth conditions and doping mechanism of Al-doped ZnO deposited by ALD.

The microstructure, electrical and mechanical properties of Al-doped ZnO thin films prepared by ALD depend on several factors, including the deposition temperature, the DEZn-H_2O:TMAl-H_2O cycle ratio and their exposure time. As our proposed applications require the material to be prepared in thin-film form, the ALD technique mentioned above with precisely controlled layer sequences is the ideal choice for realizing channel layers with optimized mobility and carrier concentration values; or achieving the required crystalline orientation for maximum piezoresponse. Therefore, an in-depth study of the deposited film growth as a function of various deposition temperature and Al concentration needs to be conducted.

Aside from these well-known advantages of the ALD method and its high reproducibility and reliability, compared to PVD techniques, there are additional advantages that make ALD a unique and empowering method for these applications. ALD can be easily used to dope ZnO and other oxide films by simply changing the sequence of precursor pulses. No target preparation, as in sputtering or PLD, or flow optimization, as in CVD, are required to get precise and uniform films. ALD provides a unique tool to manipulate the crystallographic orientation of Al-doped ZnO films which can be done through varying the deposition temperature and/or pulse and purge times. Given the low temperature of deposition and the ability of ALD to place monolayers of disparate materials contiguously to one another, multiple phases could be accessed which are harder to obtain using other deposition processes. Considering the specific requirements of our devices, ALD enables depositing high-quality films at temperatures below 250 °C which is below the limit imposed by the usage of flexible substrate. The ALD process also allows depositing Al-doped ZnO films that have semiconducting properties with a large variation of the carrier concentration at relatively low temperatures. This variation is unattainable using other techniques. As mention above, ZnO grows as a conducting, polycrystalline film with a rough surface topography resulting from the individual crystallites using ALD method [47, 64]. In contrast, Al_2O_3 grown by ALD is an excellent insulating, amorphous film, deposited with high conformity [16, 48, 51]. The wide disparity in the physical properties of ZnO and Al_2O_3 suggests that an Al-doped ZnO film formed from alternating a various amount of ZnO cycles and a single cycle of Al_2O_3 layers will possess properties that span a wide range of surface roughness and conductivity. In particular, introducing nanolayers of amorphous Al_2O_3 into a polycrystalline ZnO film will provide a means to control the film surface structural and electrical properties.

Therefore, the precise control of doping ALD provides allows to condition the deposited Al-doped ZnO films as a high-quality semiconductor channel for a TFT or a piezoelectric film for a strain sensor with a high gauge factor. The conditioning of the conductivity and Hall mobilities attained using ALD can be modulated with deposition temperatures to obtain semiconducting properties, without post-deposition treatments or annealing of the flexible substrate, avoiding thermal bending. The advantages of using ALD also extend to post-deposition as well. It can provide a superior interface quality between the TFT's gate dielectric and the channel which is critical with ZnO; since Fermi pinning is highly influenced by the interface traps, this can significantly impact the device's reproducibility and reliability. It is worth note that the ALD slow growth rate is not a concern for a TFT with 14/20 nm channel/oxide.

3.3.3 Methods

Zinc Oxide (ZnO)

The deposition of ZnO films using the ALD method is well known and understood. It can be grown using dimethyl-zinc (DMZn, $Zn(CH_3)_2$) or diethyl-zinc (DEZn, $Zn(C_2H_5)_2$) metal organic precursor, and the oxidant can be DI water vapor for thermal ALD (Th-ALD), oxygen (O_2) or ozone (O_3) for plasma-enhanced ALD (PE-ALD). ZnO layers have been reported to grow well with ALD between 100 and 250 °C, although a growth at low temperature from DEZn and H_2O has been performed [58] and the ALD window can technically be extended to room temperature growth [34]. Within the so-called ALD window, the growth rate will be stable and steady but may slightly depend on the used precursors and the deposition pressures and temperatures as well. However, based on the literature, it never exceeded the value of 0.2 nm/cycle [31]. The most commonly used precursor and oxidant for the ALD of ZnO thin films are DEZn and H_2O due to their excellent self-limiting ALD mechanism. Such growth takes place according to the net reaction

$$Zn(C_2H_5)_2 + H_2O \rightarrow \text{ZnO} + 2C_2H_6$$

And the expected elementary reaction sequences are:
During the DEZn pulse

$$Surface - OH * + Zn(C_2H_5)_2 \rightarrow Surface - O - Zn - (C_2H_5) * + (CH_3)_2$$

During the H_2O pulse

$$Surface - O - Zn - (C_2H_5) * + H_2O \rightarrow Surface - O - Zn - OH * + (CH_3)_2$$

where species with the * refer to surface species and Surface indicates a surface site. The growth of a ZnO film is obtained by repeating this sequence up to the desired thickness.

Thermally grown films with H_2O as oxidants were shown to have higher O/Zn ratios than those deposited using oxygen (O_2) or ozone (O_3) plasma. This has been attributed to the residual hydroxyl (OH) groups in the film persisting due to the insufficient purging of the excess H_2O at low-growth temperatures. However, it is believed that this excess oxygen can be decreased as the deposition temperature increases because at higher temperatures less surface hydroxyl groups persist and above 200 ^oC their thermal energy exceeds the energy needed to desorb from the surface. Therefore, for films deposited at higher temperature, the purging of residual H_2O is more effective which allows the purging times to be shortened to 1 s (from 5 s for films deposited at lower temperature). On the contrary, films grown using plasma have their O/Zn ratio slightly increased with increasing temperature due to the stronger oxidation of the DEZn molecules. This higher oxygen concentration contributes to lowering the carrier concentration in the films. Hence, the resistivity of the plasma-grown films is expected to be around 10–1000 times higher than that of water vapor grown films. ALD-grown ZnO films have a very low intrinsic resistivity of $10^{-2}\Omega$.cm, depending on the number of deposition parameters. This low value is believed to be the result of the presence of Zn interstitials, oxygen vacancies and hydrogen contamination [11, 24]. This resistivity can be further reduced by up to two orders of magnitude using Al doping [14, 53].

Al-doped Zinc Oxide

As mentioned earlier, ZnO has been doped before with various materials using various deposition methods, and ALD is no exception. The ALD method makes it quite straightforward to grow doped and multilayer structures from different compounds. There are multiple reports investigating composite thin-film parameters such as the refractive index, dielectric constant, lattice parameter and a number of other physical properties which can be easily controlled by alloying or doping ZnO with a different material [23].

Al_2O_3 films using the ALD method can be grown using aluminum isopropoxide (AIP, Al$[OCH(CH_3)_2]_3$) or trimethyl-aluminum (TMAl, $Al(CH_3)_3$) metal organic precursor and the oxidant can also be DI water vapor, oxygen (O_2) or ozone (O_3). Al_2O_3 films have a much wider ALD window than ZnO with a steady growth rate of about 0.1 nm/cycle over a large range of deposition temperatures around 17 °C. The Al_2O_3 ALD films are deposited by alternating the TMAl with H_2O exposures, and the self-limiting surface reactions that define the cycle for Al_2O_3 ALD are given by

During the TMAl pulse

$$Surface - OH * + Al(CH_3)_3 \rightarrow Surface - O - Al - (CH_3)_2 * + CH_4$$

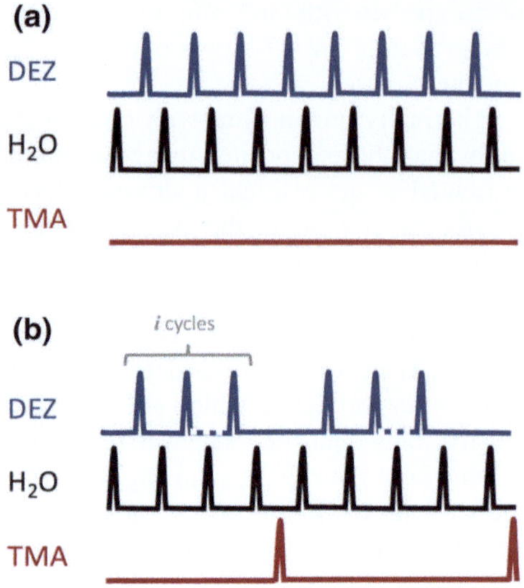

Fig. 3.2 Schematic drawings of the exposure sequence for ALD based **a** undoped *ZnO* and **b** Al-doped *ZnO* with *i* cycle ratio

During the H_2O pulse

$$Surface - O - Al - (CH_3) * + H_2O \rightarrow Surface - O - Al - OH * + CH_4$$

where species with the * refer to surface species and Surface indicates a surface site.

The doping of *ZnO* with Al using the ALD method is generally done by inserting a TMAl precursor pulse(s) among the DEZn and the oxidant exposures with the TMAl pulse followed by an oxidant exposure, which raises the question whether the Al is then built in as Al_2O_3 or the Al alone is placed as a substitutional dopant into the Zn location of the crystal. That is, whether the resulting layer is an alloy of the two oxides or an Al-doped *ZnO* layer. This we will explain in more details in the following chapters.

Figure 3.2 shows ALD-grown Al-doped *ZnO* film created by replacing a percentage of the DEZ exposures by exposures of TMAl instead. Figure 3.2a first illustrates the exposure sequence for pure *ZnO* ALD, where DEZn exposures are conventionally alternated with H_2O exposures. However, an Al-doped *ZnO* film with *i* cycle ratio can be grown by inserting a TMAl exposure after every *i* DEZn pulses as shown in Fig. 3.2b. By evenly distributing the TMAl exposures, the ALD-grown Al-doped *ZnO* film becomes as homogeneous as possible.

A specific atomic percentage (at. %) of Al doping into *ZnO* films can be achieved by adjusting the number of DEZn-H_2O cycles inserted after the TMAl-H_2O cycle. This forms what has become known in ALD as a "supercycle" of Al-doped *ZnO* where each one of these supercycles consists of *i* cycles of DEZn-H_2O followed by a single cycle of TMAl-H_2O. Figure 3.3 shows a schematic diagram of the concept of supercycle for the film layers used to deposit Al-doped *ZnO* films by ALD.

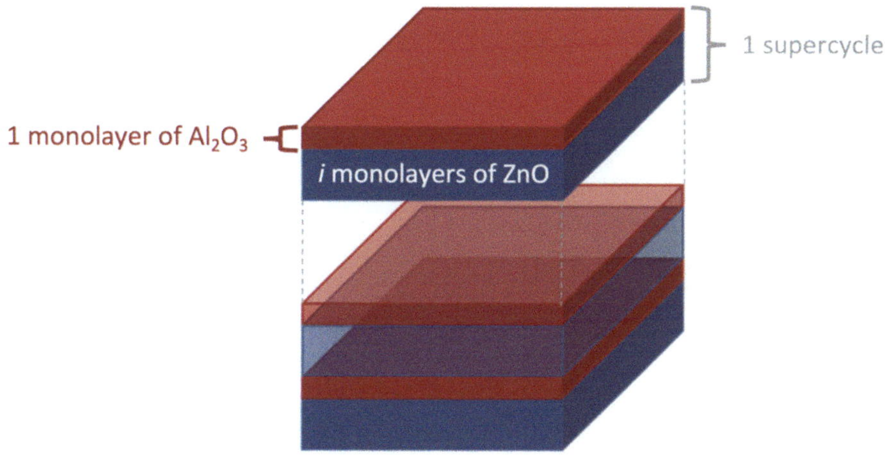

Fig. 3.3 Schematic diagram of the film layers used to make various Al-doped ZnO films by ALD

The growth of Al_2O_3 layers along with ZnO is an interesting issue. As various reports investigating a multi-layer of ZnO and Al_2O_3 have shown it to have a much lower growth rate than the one calculated from pure ZnO and Al_2O_3 growth. This is because both materials have indicated incubation periods, and when the pulse sequences are shifted between them, one must nucleate on the other.

3.3.4 Deposition of Al-doped ZnO Films on Flexible Substrates

The doped and undoped ZnO layers were deposited using a FlexAL ALD reactor from Oxford Instruments Plasma Technology. The doped ZnO films with a nominal Al dopant concentration between 0 and 10 at. % were targeted. A flexible substrate film with a 250 micro-m thickness was used for both ZnO and doped ZnO film growth. The Polyethylene terephthalate (PET) substrate was selected as the substrate due to its excellent flexibility and mechanical tolerance. During the deposition process, the pressure in the chamber was kept around 200 mtorr. A series of depositions with different substrate temperatures in the range of 60–250 °C were performed. The carrier and purging gas was high-purity nitrogen (N_2) with a gas flow rate of 200 sccm. It is used as the carrier to deliver the DEZ and TMA molecules into the chamber, and as purging gas to take the undesired products away from the chamber. All precursor reservoirs used were electronic grade purity, manufactured by Sigma Aldrich. A diethyl-zinc (DEZn, $Zn(C_2H_5)_2$) metal precursor was used to deposit the intrinsic ZnO films. Trimethyl-aluminium (TMA, $Al(CH_3)3$) metal precursor was used as the source of the Al dopant introduced into the ZnO films, in addition to growing Al_2O_3 for the TFT gate dielectric. The oxygen source was high-purity H_2O

Fig. 3.4 The pulsing sequence for ALD ZnO

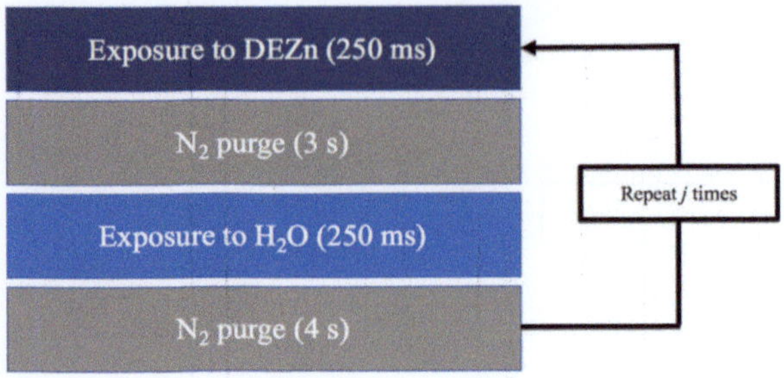

vapor (deionized water) as the oxidant reactant. All precursors were unheated and kept at room temperature. During the ALD process, the DEZn and TMA sources were not intentionally heated, and the precursor delivery lines were kept at substrate temperature (or always 150 °C).

Flow rates of the precursor gases and water were selected to be 150 sccm. The pulse time for all precursor injections was varied between 250 ms and few seconds, while the purging time after each metalorganic precursor pulse and water pulse was kept at 3 and 4 s, respectively. These exposure and purge times were chosen conservatively for each deposition temperature to ensure saturation of the ALD surface reactions while preventing inadvertent mixing of the reactive gases. In the standard operating mode for intrinsic ZnO, the DEZn and H_2O were alternatively introduced into the chamber to grow the ZnO films. The pulsing sequence of one cycle of ALD deposition, where one cycle means the injection of one precursor and H_2O, with N_2 purges after each precursor pulse is designated in Fig. 3.4.

The pulse/purge times indicated inside the brackets for each step. One complete reaction cycle was around 7–10 s. This sequence is repeated j times up to the desired film thickness. The growth per cycle (GPC) for ZnO and Al_2O_3 was established during deposition of pure, individual ZnO and Al_2O_3 ALD films, in each case the process recipe achieving the characteristics (uniformity, precursor dose saturation, etc.) expected for ALD.

The film's doping was realized by alternating pulses of DEZ and H_2O with pulses of TMA and H_2O in a ratio according to the targeted dopant content, in a manner similar to that used for the DEZ-H_2O deposition, with i numbers of ZnO ALD cycles and one Al_2O_3 ALD cycle repeatedly performed for the deposition of thin films. The deposition was designed so that only the dopant layer would be in the middle of the ZnO layers and evenly spaced. The pulse sequence is given in Fig. 3.5.

Note that we always followed the Al_2O_3 cycle by the ZnO cycles. The desired film thickness and doping level can be modulated by changing the DEZn- H_2O sequence i times and the whole cycle j times. The i and j values were controlled after the assessment of the ZnO and Al_2O_3 GPC, and adjusted as per the expected reduction in their growth, compared to pure ZnO, as explained above. We deposited multiple Al-doped ZnO thin films consisting of ZnO interspersed with regularly spaced single layers of Al_2O_3 with an appropriate pulse ratio to achieve ZnO films with a specific Al dopant content between 2 and 10 at. %. The process of adjusting the value of i

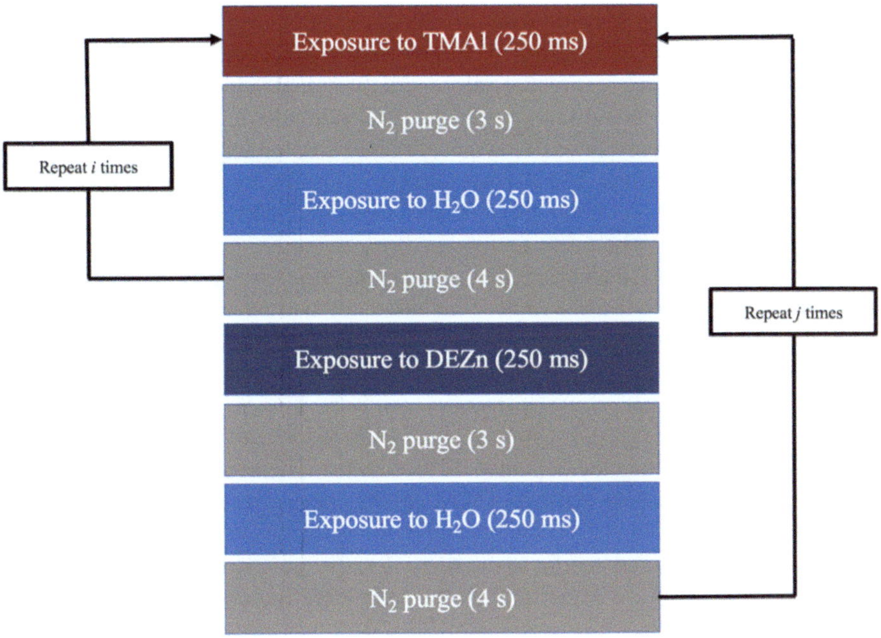

Fig. 3.5 The pulsing sequence for ALD Al-doped ZnO

and j to obtain these Al concentrations and the desired thicknesses are detailed in the following sections.

The pure ZnO and Al_2O_3 films were also deposited on thermal oxide/Si substrates with a different growth temperature ranging from 60 to 350 °C in order to determine the ALD windows and the growth rate of each film at these temperatures. Then, optimized doped and undoped films were deposited on the flexible substrates for the characterization. The number of ALD cycles of ZnO plus Al_2O_3 layers was adjusted for each doping and growth temperature value to ensure that all the deposited films have the same thickness. Plasma- enhanced ALD (PEALD) Al-doped ZnO films were also investigated but were not adopted due to their low conductivity, which made them unsuitable to function as a channel for our TFTs.

3.3.5 Modulating Al Content in ZnO Films

The reported studies on the ALD-grown Al-doped ZnO lacked consensus on the implemented doping techniques and the determination of dopant percentage in the films, which makes it quite difficult to associate what doping level contributes to which effect. Some of these reports also did not include the complete details for the growth process which adds to the confusion. In most cases, these seemingly different percentages are thought to be representations of the same results due to the different doping techniques and calculation methods used in each study, in addition to the differences between these estimated values and the actual values incorporated in the film during the ALD deposition. Not only the determination of the dopant level differ, but the doping technique can be performed either by implementing a fixed number of DEZ-H_2O exposures while TMAl-H_2O number of cycles are manipulated to adjust Al content in the film or instead a single TMAl-H_2O cycle is used while the DEZ-H_2O exposures is changed. The latter option has been chosen to grow all the films used in this research mainly to ensure maximum doping efficiency and that the resulting layers are Al-doped ZnO films rather than alloys of two oxides ZnO and Al_2O_3. It also helps minimizing the fallback in the GPC associated with the introduction of TMAl-H_2O pulse(s) and how it influences the growth of consequence ZnO layers and vice versa as discussed in the previous sections. This is done by reducing the interruption in the ZnO growth which ensures a similar overall growth rate for all the Al-doped ZnO film, when the dopant level is changed. Therefore, by fitting i number of DEZn-H_2O cycles between two TMAl-H_2O cycles, a more homogeneous film can be grown.

It became conventional to refer to the Al at. % by the percentage of the TMAl-H_2O deposition cycles used. Although it does not reflect the actual doping content in the film:

$$Al\ content(at.\%) = \frac{TMAl - H_2O\ cycles}{TMAl - H_2O\ cycles + DEZn - H_2O\ cycles} \tag{3.1}$$

For our case

$$Al\ content(at.\%) = \frac{1}{1+i} \tag{3.2}$$

where i refers to the number of DEZn-H_2O cycles inserted between two TMAl-H_2O cycles or simply as the ALD cycle ratio. However, this percentage does not represent either the ideal nor the actual Al at. % incorporated in the ZnO film as the Zn and Al content within each DEZn-H_2O and TMAl-H_2O deposition cycle are different. The Al at. % can also be determined using the GPC associated with each cycle:

$$Al\ content(at.\%) = \frac{GPC_{TMA}}{GPC_{TMA} + iGPC_{DEZ}} \tag{3.3}$$

The estimated at. % Al given by the above formula was reported to deviate at high doping concentrations as well [13], showing much higher Al content than estimated. This appeared in the measured thickness of the films being much lower than estimated for these higher at. % Al samples. Thus, other factors can influence this ratio. Including the extended nucleation of the layers and the etching of the deposited film by the TMAl pulses that appears to selectively remove Zn atoms as discussed above. ZnO encounters barriers to growth on Al_2O_3, and thus, both the thickness and at. %Al are affected in these samples. We have tried to adjust the Al content (at. %) equation to closely predict the actual Al content based on the findings discussed

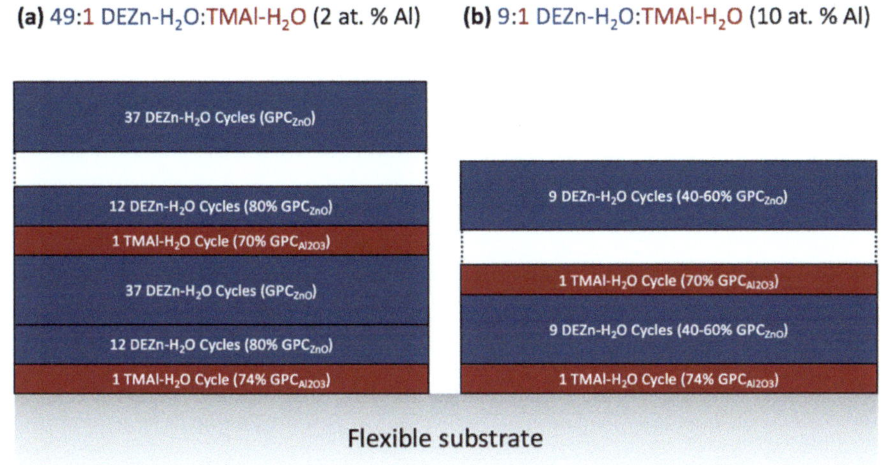

Fig. 3.6 Schematic diagram of adjusted GPC for growing Al-doped *ZnO* using ALD method on flexible substrates at **a** low doping of 2 at. % Al and **b** high doping of 10 at. % Al

in the previous section. These findings are summarized in the schematic shown in Fig. 3.6 for the case of low doping with 2 at. % Al in (a) and high doping of 10 at. % Al in (b).

From our measurements (shown later in Fig. 3.9), the initial TMAl-H_2O cycles will have an average growth around 75% of the stable growth rate of Al_2O_3 on the flexible substrate ($GPC_{Al_2O_3}$). The structure diagram of the adjusted GPC for growing Al-doped *ZnO* using ALD method on flexible substrates at low doping of 2 at. % Al and high doping of 10 at. % Al is shown in Fig. 3.6. The film deposition consists of several supercycles, while one super cycle consists of one monolayer of Al_2O_3 and *i* monolayers of ZnO. For the low doping case in Figure 6(a), *i* is 49 cycles, 12 of them will be deposited at 80% GPC_{ZnO} while the other 37 cycles at the steady-state GPC_{ZnO}. At the high doping in Fig. 3.6b, *i* is only 9 cycles. Therefore, all of them will be deposited at rate greater than 80% GPC_{ZnO}.

Doping of Al into the *ZnO* films to deposit various compositions of Al-doped *ZnO* films ranging from 2 to 10 at. % Al was conducted by alternate deposition of various *i* numbers of DEZn-H_2O cycles and one TMAl-H_2O cycle. This was done by changing the *i* value between 49 and 9. Figure 3.7 shows the estimated Al content for this range of cycle ratios using the three methods discussed above, cycles %, GPCs % and adjusted GPCs % as a function of the cycle ratio *i*.

Figure 3.7 shows that the adjusted GPC method was able to estimate the Al content to be close to the normal GPC percentage at low doping values with higher estimated Al content at higher doping values.

Six versions of the *ZnO* film were grown with various Al content added to the film during the deposition by following *i* *ZnO* cycles by single Al_2O_3 cycle. The at. % Al

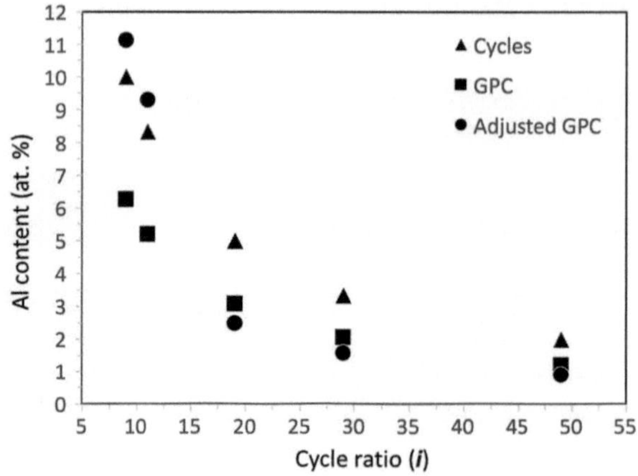

Fig. 3.7 The estimated Al content using the three methods discussed above as a function of the cycle ratio i

Table 3.1 Ratio of Zn to Al cycles used to obtain various Al doping concentration at 150 °C

i	Al content (at. %)			Supercycle (nm)	*j*	Thickness (nm)		
	Estimated	Adjusted	Measured			Estimated	Adjusted	Measured
–	0	0	0	0.19	505	99.99	99.99	101
49	2	1	2.5	9.8	10	98.2	93.9	102
29	3.3	1.7	4	5.9	17	99.6	92.3	98.9
19	5	2.8	6.5	3.9	26	100.9	89.7	96.7
11	8.3	9.3	10.5	2.3	44	101	80.3	89.6
9	10	11.1	14	1.9	53	100.7	80	86.5

was 0, 2, 3, 5, 8 and 10% by setting the value to 49, 29, 19, 11 and 9, respectively, which were grown at deposition temperature of seven different substrate temperatures between 60 and 250 °C. The precursor pulse and purge times were kept the same for all deposited films.

Table 3.1 lists samples used for obtaining data, showing the ratio of Zn to Al cycles used to obtain various Al doping concentration. The supercycles were calculated for each film and repeated *j* number of times to target a resultant thickness of 100 nm for each Al-doped *ZnO* film. The measured total thickness was obtained using Ellipsometry. The estimated at. % Al is calculated by the method shown above.

The thickness values were later adjusted to be as close as possible to the desired value for each device fabricated based on this data.

3.3.6 Film Growth Characterization

The film thickness data was primarily acquired using a JA Woollam spectroscopic ellipsometer, and for secondary measurements, either a Veeco (Dektak) 150 stylus profilometer or a Filmetrics F40-UV reflectometer were used. Steps required for the profilometer measurements were created in the films using lithography to pattern a line on top of the film, and the rest of the film was etched away in 5:95 sulfuric acid : DI-water. These thickness data were used to calculate the deposition rates or the so-called growth per cycle (GPC) values for the *ZnO* and Al_2O_3 films at different deposition temperatures between 60 and 250 °C as well as for the Al-doped films deposited with different doping levels, shown later in (Fig. 3.10) and (Fig. 3.11), respectively. First, all films with various doping level were deposited with a total fixed number of deposition cycles. Then for each doping level, the deposition cycles were readjusted and calculated to maintain a similar thickness for all samples using the data from the previous step. The GPC is calculated by dividing the measured film thickness by the number of ALD cycles used to grow the measured film where the GPC is simply extracted from the slope of the linear fitting of the thicknesses against the number of cycles plot. The GPC to three samples of Al_2O_3, *ZnO* and 3 at. % Al-doped films, deposited at 150 °C are shown in Fig. 3.8.

In Fig. 3.8, the thicknesses of the three films show a linear dependence on the number of deposition cycles. This linear behavior suggests a stable self-limiting growth for Al_2O_3, *ZnO* and Al-doped *ZnO* films which is believed to be successfully obtained using ALD at the low deposition temperature of 15 °C. The black circles, squares and rectangles in Fig. 3.8 show the results of the measurements conducted on the Al_2O_3, *ZnO* and Al-doped *ZnO* films by an ex situ ellipsometer, assuming a refractive index *n* around 1.67, 1.95 and 1.84, respectively. The measurements done by the stylus profilometer are designated by the white circles for Al_2O_3, white squares for *ZnO* and white rectangles for Al-doped *ZnO* in the same figure. The corresponding dotted line shows the fitting to these film thickness measurements from which the average film growth rate can be calculated for each of the three

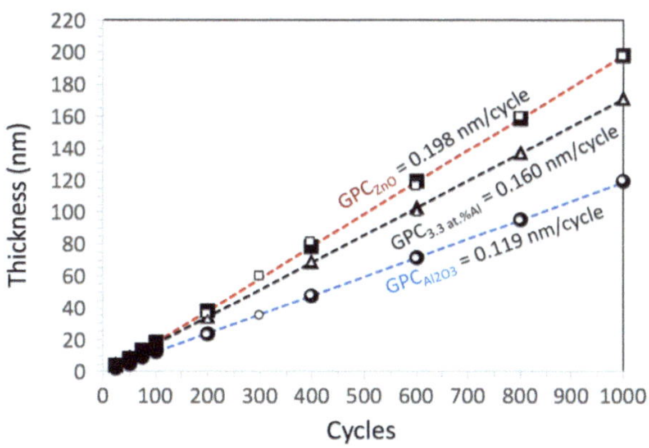

Fig. 3.8 Relationship between the number of cycles and the thickness of *ZnO* (squares), Al_2O_3 (circles) and 3 at. % Al-doped *ZnO* ALD films (triangles) measured using Ellipsometry (Black) and stylus profilometry (White)

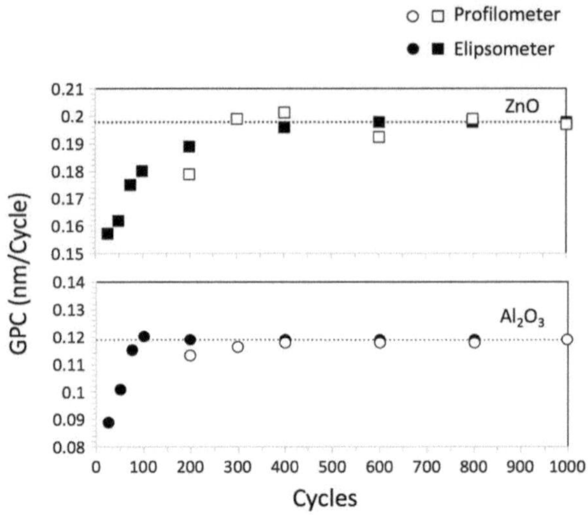

Fig. 3.9 Evaluation of growth rates of pure Al_2O_3 and ZnO deposited by ALD as a function of the number of growth cycles

samples. The dotted blue line slope characterizes the growth per cycle for Al_2O_3 film ($GPC_{Al_2O_3}$) and show it to be 0.119 nm/cycle, the dotted red line is consistent with a ZnO growth rate (GPC_{ZnO}) of 1.98 nm/cycle and the dotted black line with Al-doped ZnO growth rate ($GPC_{3\ at.\%}$) of 0.16 nm/cycle. All thicknesses lower than 10 nm were exclusively acquired by the ellipsometer. The ellipsometry and profilometry measurements are consistent with the value of 0.21 nm/cycle measured recently under similar conditions [66] and a growth rate of 0.11/cycle measured previously for Al_2O_3 ALD at 177 °C.

Figure 3.9 shows that at the first few deposition cycles, the growth rate is not as linear as shown in Fig. 3.8. Figure 3.9 highlights a relatively narrow incubation period for the Al_2O_3 film. Measurements conducted by ellipsometry show that the GPC increases from about 0.089 nm/cycle, around 74% of the steady-state GPC, at 25 deposition cycles to a more steady and stable GPC of 0.119 nm/cycle for more than 100 growth cycles. This nucleation period is believed to be related to the low reactivity of TMAl on the flexible substrate where no hydroxyl groups are initially on its surface [21]. A longer TMAl and H_2O pulse duration can make this incubation period even narrower.

In order to confirm the ALD growth window for Al-doped ZnO on the flexible substrates, the GPC dependence on the deposition temperatures was studied for pure ZnO and Al_2O_3 films, respectively, as depicted by Fig. 3.10.

Growth rates of pure Al_2O_3 and ZnO deposited by ALD as a function of the deposition temperature. Arrowed region indicates an ALD window which exhibits a constant film growth rate. The ellipsometric and stylus profiler growth rate measurements obtained from plots similar to Fig. 3.8. The error bars show the values of three independent measurements conducted on three different films deposited together under the exact same growth conditions. Figure 3.10 shows the growth rates of pure ZnO and Al_2O_3 deposited by ALD at various growth temperatures from 60

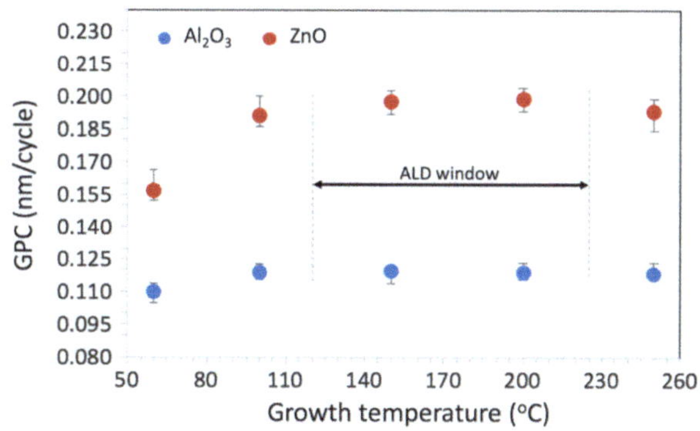

Fig. 3.10 Evaluation of growth rates of pure Al_2O_3 and ZnO deposited by ALD as a function of the growth temperature

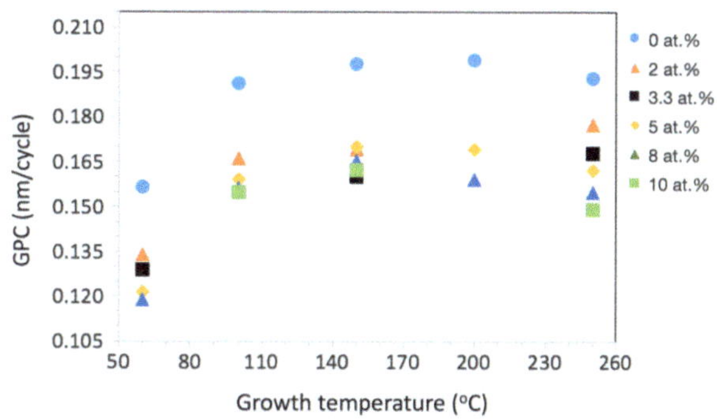

Fig. 3.11 Temperature dependence of the GPC of Al-doped ZnO films with different Al contents

to 25 °C. In both cases, a relatively steady decrease in the GPC is visible in the very low-temperature section. These deposition characteristics have been seen before in the case where both ZnO [40] and Al_2O_3 [50] are deposited by the ALD method using DEZn and TMAl with H_2O as the oxidant. This is believed to be the normal behavior for ALD oxides deposited from a metal precursor and H_2O. The decrease seen in the GPC with decreasing temperatures is likely a sign of less stable growth and indicates that these temperatures are not within the ALD window, and is probably a result of incomplete surface reaction caused by the very low deposition temperature. However, in the higher temperature region, the growth rates remained almost constant, indicating the existence of an ALD process window.

To investigate the dependence of the growth rate of our ZnO films on doping levels, we deposited ZnO at six different Al at. % of 2, 3, 5, 8 and 10 each grown at growth temperature ranging from 60 to 250 °C while keeping constant other growth conditions such precursors' exposition times, purging times and the number of cycles.

Figure 3.11 shows the growth rates of different doping levels at various deposition temperatures. The growth rate of almost all the films have a maximum at around 150–200 °C, which corresponds to the location of the ZnO ALD window described in Fig. 3.10. The Al-doped samples, on the other hand, have a somewhat wider window, as the growth rates are sustained till temperatures close to 250 °C. The growth rate also correlates with the amount of introduced Al, as shown before, where it decreases with

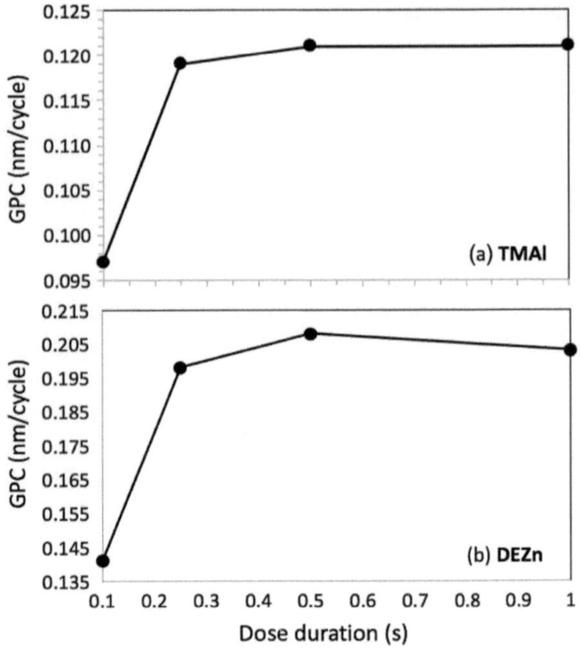

Fig. 3.12 Film growth rate of **a** Al_2O_3 **b** ZnO films as a function of exposure time

the increasing Al content. These general characteristics are similar at each growth temperature. The growth slows down with the increasing number of the Al pulses. This is because again the Al_2O_3 has to nucleate on ZnO surface and vice versa. Also, a more drastic change in GPC is noticed at 250 °C. A likely explanation is that the deposition temperature of 250 °C is too high for perfectly controlled, stable ZnO growth, resulting in fluctuations in the final thickness of the deposited thin films at this growth temperature.

Figure 3.12a shows the GPC for Al_2O_3 deposited on flexible substrates at 150 °C as the TMAl pulse duration (TMA_{pulse}) ranges from 0.1 to 1 s. The films exhibited an almost constant growth rate value of around 0.12 nm/cycle when TMA_{pulse} was longer than 0.25 s. Meanwhile, the film growth rate decreased below 0.1 nm/cycle if TMA_{pulse} was shorter than 0.25 s due to the limited time available for the reaction to occur. Figure 3.12b shows the dependence of the GPC of ZnO film on DEZn exposure time at 150 °C. The growth rate of ZnO increases with dose time and is stabilized at about 0.2 nm/cycle at a dose time longer than 0.25 s, confirming the self-limiting growth.

3.4 Conclusions

This chapter investigates the precise control of doping ALD provides, allows to condition the deposited Al-doped *ZnO* films as a high-quality semiconductor channel for a TFT or a piezoelectric film for a strain sensor with a high gauge factor. The conditioning of the conductivity and Hall mobilities attained using ALD can be modulated with deposition temperatures to obtain semiconducting properties, without post-deposition treatments or annealing of the flexible substrate, avoiding thermal bending. The advantages of using ALD also extend to post-deposition as well. It can provide a superior interface quality between the TFT's gate dielectric and the channel which is critical with *ZnO*; since Fermi pinning is highly influenced by the interface traps, this can significantly impact the device's reproducibility and reliability.

Acknowledgements This work has been supported by the Semiconductor Research Corporation (SRC) under the Abu Dhabi SRC Center of Excellence on Energy-Efficient Electronic Systems (ACE^4S), Contract 2013 HJ2440, with funding from the Mubadala Development Company, Abu Dhabi, UAE.

References

1. C.H. Ahn, H. Kim, H.K. Cho, Deposition of Al doped ZnO layers with various electrical types by atomic layer deposition. Thin Solid Films **519**(2), 747–750 (2010)
2. C.H. Ahn, J.H. Kim, H.K. Cho, Tunable electrical and optical properties in composition controlled Hf:ZnO thin films grown by atomic layer deposition. J. Electrochem. Soc. **159**(4), H384–H387 (2012)
3. C.H. Ahn, B.H. Kong, H. Kim, H.K. Cho, Improved electrical stability in the Al Doped ZnO thin-film-transistors grown by atomic layer deposition. J. Electrochem. Soc. **158**(2), H170 (2011)
4. A.F. Aktaruzzaman, G.L. Sharma, L.K. Malhotra, Electrical, optical and annealing characteristics of ZnO: al films prepared by spray pyrolysis. Thin Solid Films **198**(1–2), 67–74 (1991)
5. H.S. Bae, S. Im, Ultraviolet detecting properties of ZnO-based thin film transistors. Thin Solid Films **469–470**(SPEC. ISS), 75–79 (2004)
6. P.F. Carcia, R.S. McLean, M.H. Reilly, G. Nunes, Transparent ZnO thin-film transistor fabricated by rf magnetron sputtering. Appl. Phys. Lett. **82**(7), 1117–1119 (2003)
7. P.R. Chalker, P.A. Marshall, P.J. King, K. Dawson, S. Romani, P.A. Williams, J. Ridealgh, M.J. Rosseinsky, Atomic layer deposition of germanium-doped zinc oxide films with tuneable ultraviolet emission. J. Mater. Chem. **22**(25), 12824–12829 (2012)
8. P.R. Chalker, P.A. Marshall, S. Romani, J.W. Roberts, S.J.C. Irvine, D.A. Lamb, A.J. Clayton, P. Williams, Atomic layer deposition of Ga-doped ZnO transparent conducting oxide substrates for CdTe-based photovoltaics. J. Vac. Sci. Technol. A **31**(1), 01A120 (2013)
9. C.H. Chen, S.J. Chang, S.P. Chang, M.J. Li, I. Cherng Chen, T.J. Hsueh, A. Di Hsu, C.L. Hsu, Fabrication of a white-light-emitting diode by doping gallium into zno nanowire on a p-gan substrate. J. Phys. Chem. **114**(29), 12422–12426 (2010)
10. N.P. Dasgupta, S. Neubert, W. Lee, O. Trejo, J.R. Lee, F.B. Prinz, Atomic layer deposition of Al-doped ZnO films: effect of grain orientation on conductivity. Chem. Mater. **22**(16), 4769–4775 (2010)
11. E.A. Davis, S.P. Cottrell, P.J. King, J.S. Lord, J.M. Gil, H.V. Alberto, R.C. Vilão, J. Piroto Duarte, N. Ayres de Campos, A. Weidinger, R.L. Lichti, S.J. Irvine, S.F. Cox, S.F. Cox, Exper-

imental confirmation of the predicted shallow donor hydrogen state in zinc oxide. Phys. Rev. Lett. **86**(12), 2601–2604 (2001)
12. T. Dhakal, D. Vanhart, R. Christian, A. Nandur, A. Sharma, C.R. Westgate, Growth morphology and electrical/optical properties of Al-doped ZnO thin films grown by atomic layer deposition. J. Vac. Sci. Technol. A: Vac. Surf. Films **30**(2), 021202 (2012)
13. J.W. Elam, D. Routkevitch, S.M. George, Properties of ZnO/Al_2O_3 alloy films grown using atomic layer deposition techniques. J. Electrochem. Soc. **150**(6), G339 (2003)
14. J.W. Elam, Z.A. Sechrist, S.M. George, ZnO/Al_2O_3 nanolaminates fabricated by atomic layer deposition: growth and surface roughness measurements. Thin Solid Films **414**(1), 43–55 (2002)
15. J.W. Elam, D.A. Baker, A.B.F. Martinson, M.J. Pellin, J.T. Hupp, Atomic layer deposition of indium tin oxide thin films using nonhalogenated precursors. J. Phys. Chem. C **112**(6), 1938–1945 (2008)
16. P. Ericsson, S. Bengtsson, J. Skarp, Properties of A1203-films deposited on silicon by atomic layer epitaxy. Microelectron. Eng. **36**(1–4):91–94 (1997)
17. S. Faÿ, J. Steinhauser, S. Nicolay, C. Ballif, Polycrystalline ZnO: B grown by LPCVD as TCO for thin film silicon solar cells. Thin Solid Films **518**(11), 2961–2966 (2010)
18. S. Faÿ, J. Steinhauser, N. Oliveira, E. Vallat-Sauvain, C. Ballif, Opto-electronic properties of rough LP-CVD ZnO: B for use as TCO in thin-film silicon solar cells. Thin Solid Films **515**(24 SPEC. ISS.), 8558–8561 (2007)
19. Y. Geng, L. Guo, S.S. Xu, Q.-Q. Sun, S.-J. Ding, H.-L. Lu, D.W. Zhang, Influence of Al doping on the properties of ZnO thin films grown by atomic layer deposition. J. Phys. Chem. C **115**(25), 12317–12321 (2011)
20. V. Gokulakrishnan, S. Parthiban, K. Jeganathan, K. Ramamurthi, Investigation on the effect of Zr doping in ZnO thin films by spray pyrolysis. Appl. Surf. Sci. **257**(21), 9068–9072 (2011)
21. L.G. Gosset, J.F. Damlencourt, O. Renault, D. Rouchon, A. Ph Holliger, I. Ermolieff, J.J. Trimaille, F.M. Ganem, M.N. Séméria, Interface and material characterization of thin Al_2O_3 layers deposited by ALD using TMA/H2O. J. Non-Cryst. Solids **303**(1), 17–23 (2002)
22. P. Guay, S. Gorgutsa, S. LaRochelle, Y. Messaddeq, Wearable contactless respiration sensor based on multi-material fibers integrated into textile. Sensors **17**(5), 1050 (2017)
23. P.Q. Jiaping, M. Han, A.M R Senos, Effect of A1 and Mn doping on the electrical conductivity of ZnO. Journal of the European Ceramic Society **21**(10–11), 1883–1886 (2001)
24. D.M. Hofmann, A. Hofstaetter, F. Leiter, H. Zhou, F. Henecker, B.K. Meyer, S.B. Orlinskii, J. Schmidt, P.G. Baranov, Hydrogen: a relevant shallow donor in Zinc Oxide. Phys. Rev. Lett. **88**(4), 045504 (2002)
25. H. Jianhua, R.G. Gordon, Textured aluminum-doped zinc oxide thin films from atmospheric pressure chemical-vapor deposition. J. Appl. Phys. **71**(2), 880–890 (1992)
26. C. Huang, M. Wang, Z. Deng, Y. Cao, Q. Liu, Z. Huang, Y. Liu, W. Guo, Q. Huang, Low content indium-doped zinc oxide films with tunable work function fabricated through magnetron sputtering. Semicond. Sci. Technol. **25**(4), 45008 (2010)
27. J. Huang, S. Chu, J. Kong, L. Zhang, C.M. Schwarz, G. Wang, L. Chernyak, Z. Chen, J. Liu, ZnO p-n homojunction random laser diode based on nitrogen-doped p-type nanowires. Adv. Opt. Mater. **1**(2), 179–185 (2013)
28. Ihs. Revenues from Flexible AMOLED Display on Pace to Exceed Rigid AMOLED Panel in Q3 2017
29. S. Ilican, Y. Caglar, M. Caglar, B. Demirci, Polycrystalline indium-doped ZnO thin films: preparation and characterization. J. Optoelectron. Adv. Mater. **10**(10), 2592–2598 (2008)
30. X. Jiang, F.L. Wong, M.K. Fung, S.T. Lee, Aluminum-doped zinc oxide films as transparent conductive electrode for organic light-emitting devices. Appl. Phys. Lett. **83**(9):1875–1877 (2003)
31. S.K. Kim, C.S. Hwang, S.-H. Ko Park, S.J. Yun, Comparison between ZnO films grown by atomic layer deposition using H_2O or O_3 as oxidant. Thin Solid Films **478**(1—-2), 103–108 (2005)

32. E.H. Kim, D.H. Lee, B.H. Chung, H.S. Kim, Y. Kim, S.J. Noh, Low-temperature growth of ZnO thin films by atomic layer deposition. J. Korean Phys. Soc. **50**(6), 1716 (2007)
33. H. Kim, A. Piqué, J.S. Horwitz, H. Murata, Z.H. Kafafi, C.M. Gilmore, D.B. Chrisey, Effect of aluminum doping on zinc oxide thin films grown by pulsed laser deposition for organic light-emitting devices. Thin Solid Films **377–378**, 798–802 (2000)
34. K. Ching-Shun, H.-Y. Lee, J.-M. Huang, C.-M. Lin, Epitaxial growth of ZnO films at extremely low temperature by atomic layer deposition with interrupted flow. Mater. Chem. Phys. **120**(2–3), 236–239 (2010)
35. D.-J. Lee, H.-M. Kim, J.-Y. Kwon, H. Choi, S.-H. Kim, K.-B. Kim, Structural and electrical properties of atomic layer deposited Al-doped ZnO films. Adv. Funct. Mater. **21**(3):n/a–n/a (2010)
36. D.-J. Lee, J.-Y. Kwon, S.-H. Kim, H.-M. Kim, K.-B. Kim, Effect of Al distribution on carrier generation of atomic layer deposited Al-doped ZnO films. J. Electrochem. Soc. **158**(5), D277–D281 (2011)
37. Lenovo, Lenovo's flexible phone and tablet
38. LG, LG Display Unveils the Latest Cutting-edge Displays at CES (2015)
39. LG, 49WFB | Transparent Special Display (2016)
40. S.J. Lim, S. Kwon, H. Kim, ZnO thin films prepared by atomic layer deposition and rf sputtering as an active layer for thin film transistor. Thin Solid Films **516**(7), 1523–1528 (2008)
41. Y.C. Lin, Y.C. Jian, J.H. Jiang, A study on the wet etching behavior of AZO (ZnO:Al) transparent conducting film. Appl. Surf. Sci. **254**(9), 2671–2677 (2008)
42. Y. Liu, Y. Li, H. Zeng, ZnO-based transparent conductive thin films: doping, performance, and processing. J. Nanomater. **2013**, 1–9 (2013)
43. V. Lujala, J. Skarp, M. Tammenmaa, T. Suntola, Atomic layer epitaxy growth of doped zinc oxide thin films from organometals. Appl. Surf. Sci. **82-83**(C), 34–40 (1994)
44. O. Lupan, T. Pauporté, T. Le Bahers, B. Viana, I. Ciofini, Wavelength-emission tuning of zno nanowire-based light-emitting diodes by Cu doping: experimental and computational insights. Adv. Funct. Mater. **21**(18), 3564–3572 (2011)
45. W.J. Maeng, J. Seok Park, H. Suk Kim, Kwang Hee Lee, Kyung Bae Park, Kyoung Seok Son, Tae Sang Kim, Eok Su Kim, Yong Nam Ham, Myungkwan Ryu, and Sang Yoon Lee. Photo and thermal stability enhancement of amorphous Hf-In-Zn-O thin-film transistors by the modulation of back channel composition. Appl. Phys. Lett. **98**(7), 073503 (2011)
46. T. Minami, H. Nanto, S. Takata, Optical properties of aluminum doped zinc oxide thin films prepared by rf magnetron sputtering. Jpn. J. Appl. Phys. **24**(8), L605–L607 (1985)
47. A.W. Ott, R.P.H. Chang, Atomic layer-controlled growth of transparent conducting ZnO on plastic substrates. Mater. Chem. Phys. **58**(2), 132–138 (1999)
48. A.W. Ott, J.W. Klaus, J.M. Johnson, S.M. George, Al_3O_3 thin film growth on Si(100) using binary reaction sequence chemistry. Thin Solid Films **292**(1–2), 135–144 (1997)
49. W.J. Maeng, H.S. Kim, J.S. Park, Review of recent developments in amorphous oxide semiconductor thin-film transistor devices. Thin Solid Films **520**(6), 1679–1693 (2012)
50. A. Rahtu, T. Alaranta, M. Ritala, In situ quartz crystal microbalance and quadrupole mass spectrometry studies of atomic layer deposition of aluminum oxide from trimethylaluminum and water. Langmuir **17**(21), 6506–6509 (2001)
51. M. Ritala, H. Saloniemi, M. Leskelä, T. Prohaska, G. Friedbacher, M. Grasserbauer, Studies on the morphology of Al_2O_3 thin films grown by atomic layer epitaxy. Thin Solid Films **286**(1–2), 54–58 (1996)
52. Y.R. Ryu, J.A. Lubguban, T.S. Lee, H.W. White, T.S. Jeong, C.J. Youn, B.J. Kim, Excitonic ultraviolet lasing in ZnO-based light emitting devices. Appl. Phys. Lett. **90**(13), 131115 (2007)
53. H. Saarenpää, T. Niemi, A. Tukiainen, H. Lemmetyinen, N. Tkachenko, Aluminum doped zinc oxide films grown by atomic layer deposition for organic photovoltaic devices. Solar Energy Mater. Solar Cells **94**(8) (2010)
54. K. Saito, Y. Hiratsuka, A. Omata, H. Makino, S. Kishimoto, T. Yamamoto, N. Horiuchi, H. Hirayama, Atomic layer deposition and characterization of Ga-doped ZnO thin films. Superlattices Microstruct. **42**(1–6), 172–175 (2007)

55. Samsung, Samsung Expanding Transparent Display Market with a New 46-inch LCD Panel
56. H. Sato, T. Minami, T. Miyata, S. Takata, M. Ishii, Transparent conducting ZnO thin films prepared on low temperature substrates by chemical vapour deposition using $Zn(C_5H_7O_2)2$. Thin Solid Films **246**(1–2), 65–70 (1994)
57. A. Sazonov, M Meitine, D Stryakhilev, A. Nathan, Low-temperature materials and thin-film transistors for electronics on flexible substrates. Semiconductors **40**(8), 959–967 (2006)
58. M. Scharrer, X. Wu, A. Yamilov, H. Cao, R.P.H. Chang, Fabrication of inverted opal ZnO photonic crystals by atomic layer deposition. Appl. Phys. Lett. **86**(15), 1–3 (2005)
59. W. Tang, D.C. Cameron, Aluminum-doped zinc oxide transparent conductors deposited by the sol-gel process. Thin Solid Films **238**(1), 83–87 (1994)
60. Current Applied Physics Effect of dopants on the structural, optical and electrical properties of sol-gel derived ZnO semiconductor thin films. **13**(1), 60–65 (2013)
61. I. Volintiru, M. Creatore, B.J. Kniknie, C.I.M.A. Spee, M.C.M. Van De Sanden, Evolution of the electrical and structural properties during the growth of Al doped ZnO films by remote plasma-enhanced metalorganic chemical vapor deposition. J. Appl. Phys. **102**(4), 043709 (2007)
62. C.C. Wang, B.Y. Man, M. Liu, C.S. Chen, S.Z. Jiang, S.Y. Yang, S.C. Xu, X.G. Gao, B. Hu, The intrinsic room-temperature ferromagnetism in ZnO: Co thin films deposited by PLD. Adv. Condens. Matter Phys. **2012**, 1–5 (2012)
63. D.X. Xia, J.B. Xu, High mobility and low operating voltage ZnGaO and ZnGaLiO transistors with spin-coated Al_2O_3 as gate dielectric. J. Phys. D: Appl. Phys. **43**(44), 442001 (2010)
64. A. Yamada, B. Sang, M. Konagai, Atomic layer deposition of ZnO transparent conducting oxides. Appl. Surf. Sci. **112**, 216–222 (1997)
65. S. Yoshioka, F. Oba, R. Huang, I. Tanaka, T. Mizoguchi, T. Yamamoto, Atomic structures of supersaturated ZnO- Al_2 O_3 solid solutions. J. Appl. Phys. **103**(1), 014309 (2008)
66. E.B. Yousfi, J. Fouache, D. Lincot, Study of atomic layer epitaxy of zinc oxide by in-situ quartz crystal microgravimetry. Appl. Surf. Sci. **153**(4), 223–234 (2000)

Chapter 4
ALD Al-doped ZnO Thin Film as Semiconductor and Piezoelectric Material: Characterization

Ayman Rezk and Irfan Saadat

4.1 Electrical Characterization of Al-doped ZnO Thin Films

Following the synthesis of the ZnO film using atomic layer deposition (ALD), this chapter covers the electrical characterization of the undoped and doped ALD ZnO films where the films' resistivity, carrier concentration, and mobility were measured for various growth conditions and Al doping concentrations. This provides us with a better understanding of the films behavior and relevant properties like conductivity and mobility which are essential for growing high-quality films with the desired properties needed for the targeted applications.

As discussed previously, the rising interest in utilizing Al-doped ZnO films as a transparent conductive oxide (TCO), replacing ITO, or as a transparent semiconductive oxide (TSO) for channel material in thin-film transistors (TFTs), rendered a large body of published investigations of modulating Al-doped ZnO films resistivity. This resulted in a number of high-quality films with various resistivities dependent on the underlying processes [1–3]. However, the investigation of the Al-doped ZnO film as a TSO has not studied as extensively as TCO as discussed previously. Therefore, in this work we have focused on growing a high-quality semiconductor Al-doped ZnO films with high mobility instead of trying to achieve the lowest resistivity.

A. Rezk · I. Saadat (✉)
Masdar Institute at Khalifa University, Abu Dhabi, UAE
e-mail: isaadat@masdar.ac.ae

A. Rezk
e-mail: arizk@masdar.ac.ae

Ibrahim (Abe) M. Elfadel and M. Ismail (eds.), *The IoT Physical Layer*,
https://doi.org/10.1007/978-3-319-93100-5_4

4.1.1 Hall Measurement

The specific resistivity, the carrier concentration, and the mobility were measured by a Hall effect measurement system. These assessments were done in the Van der Pauw mode utilizing a LakeShore 7700A Hall Measurement System. All assessments were executed at room temperature in a dark box.

Pure ZnO- and Al-doped ZnO films with different doping concentrations, thicknesses, pulsing/purging durations, and growth temperature were deposited on the flexible substrate for the measurements. All samples' thicknesses were maintained at 100 nm by varying the number of ALD deposition cycles per temperature as explained in the previous chapter.

The samples were prepped by shaping the flexible substrate into 2.5 cm by 2.5 cm square pieces. A 300 nm SiO_2 layer was first deposited to enhance adhesion of the layers with the flexible substrate. Initially, silver contacts were used to function as electrodes for the Van der Pauw configuration during the assessment but it was found to form a Schottky contact with most of the Al-doped ZnO films. Therefore, a shadow mask was employed to deposit 50/100 nm thick Ti/Al layers using e-beam evaporation. Ti/Al was chosen to further improve adhesion and insure an ohmic contact with the Al-doped ZnO films.

Resistivity of Al-doped ZnO Films

The two-dimensional surface in Fig. 4.1 represents the measured-specific resistivity versus the Al doping level ranging from 0 and 10 atomic precentact (at. %) along with the growth temperature. For films with a resistivity lower than $3 * 10^{-2}$ Ω.cm (blue), higher than 0.7 Ω.cm (orange), and in between (yellow) indicated on the graph.

Figure 4.2 is a projection of the surface shown in Fig. 4.1, showing the dependence of resistivity of the ALD Al-doped ZnO films on the atomic ratio of Al (at. % Al) from 0 (pure ZnO) to 10 at. % for various growth temperature between 60 and 250 °C.

In Fig. 4.2, the measured film resistivity shows a parabolic behavior as a function of Al concentrations at a set growth temperature. At 200 °C, the resistivity initially drops with the addition of Al and reaches a minimum value of 3.64×10^{-3} Ω.cm around 3 at. % (indicated by the dotted line). This value is clearly lower than the 4.23×10^{-2} Ω.cm of the undoped ZnO film (almost 10 times lower) grown at the same temperature; signaling that the incorporated Al atoms function as dopants increase the film conductivity. Further insertion of Al (>3 at. %) caused resistivity to rapidly increase reaching 2.39 Ω.cm at 8 at. % up to 4.81 Ω.cm at 10 at. %. Similar tendencies were mentioned by other reports for Al-doped ZnO films. They pointed out that adding Al up to a certain concentration decreases resistivity, and exceeding that concentration resulted in resistivity increase [4].

This behavior occurs when a TMA pulse is introduced to the ALD film during deposition, where one or more of these scenarios can take place:

- Al substituting Zn sublattice (Al^{+Zn})
- Al capturing the hydroxyl group from the ZnO surface (Al_2O_3)

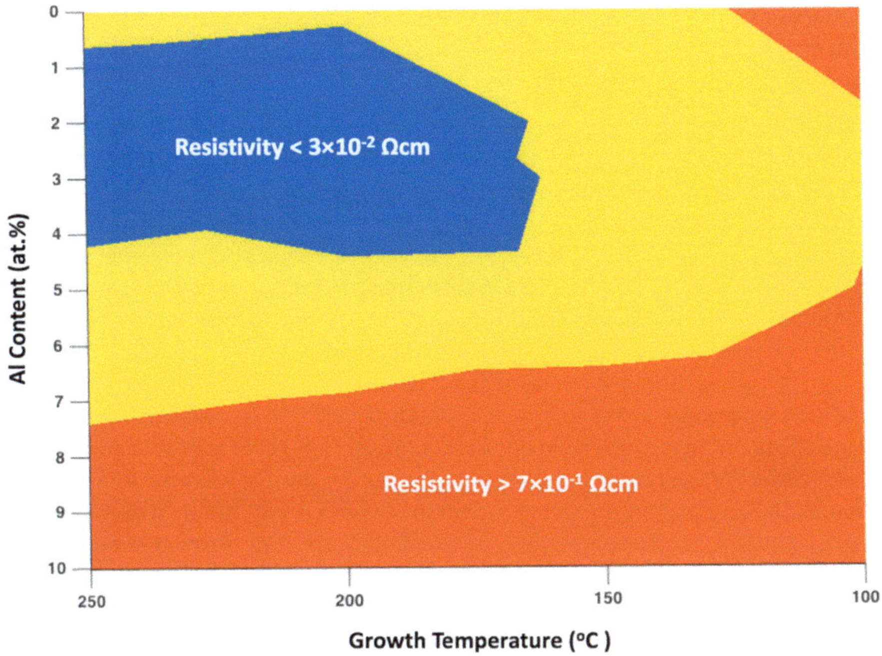

Fig. 4.1 The measured-specific resistivity versus doping levels between 0 and 10 at. % Al along with growth temperatures ranging from 100 to 250 ^{o}C

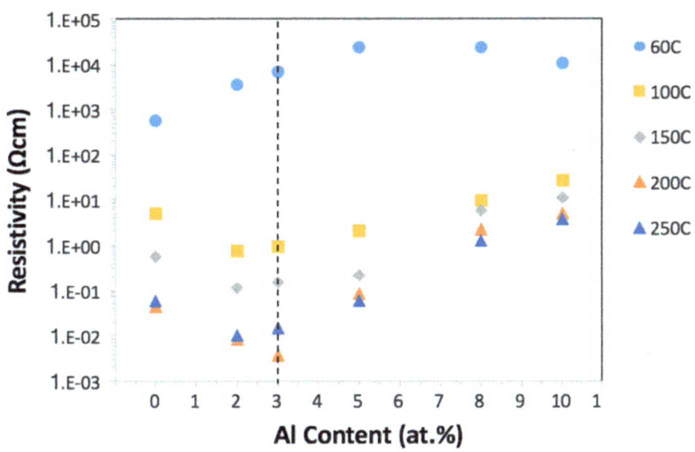

Fig. 4.2 Films resistivity as a function of doping levels between 0 and 10 at. % Al for various growth temperatures

- Generation of oxygen vacancies (V_o^{2+})
- Generation of zinc vacancies (V_{Zn}^{2-})
- Generation of zinc interstitials (Zn_i^{2+})

Zinc vacancies (V_{Zn}^{2-}) are the most probable intrinsic defects that can form during n-type ZnO doping since their formation energy is relatively lower than other defects [5]. Zn vacancies (V_{Zn}^{2-}) where found to occur in modestly doped n-type ZnO and act as a compensating center so V_{Zn}^{2-} as a deep acceptor play an important role in reducing the carrier concentration, alongside with oxygen vacancies which more probable in thermal ALD than the plasma-enhanced ALD. As mention above, some of the extrinsic Al^{3+} ions substitute into host Zn^{2+} sites in the ZnO lattices which generate an extra electron improving the carrier concentration. As more Al^{3+} substitutes into Zn^{2+} sites, the resistivity starts decreasing in the beginning as more and more free carriers are generated. However, when the doping level exceeds a certain value ($\gtrsim$ 3 at. %) and due to the abundance of Al dopants, some of them either start clustering which make it easier for Al_2O_3 (an insulator) to form during the TMAl cycle or be inserted as interstitials instead of replacing a Zn^{2+} ion. Either way the increase in Al_2O_3 clustering and/or interstitials leads to a reduction in the film's carrier concentration along with the rise of film's resistivity as Al doping increases. This is due to the Al in the stoichiometric Al_2O_3 or in an interstitial site aren't able to generate free electrons either by substituting Zn sublattice (Al^{+Zn}) or through the generation of oxygen vacancies (V_o^{2+}) and zinc interstitials (Zn_i^{2+}) by capturing the oxygen from the ZnO instead of the hydroxyl group.

Al-doped ZnO films deposited by sputtering and pulsed laser deposition (PLD) manifested lower resistivities as well when 3 at. % Al was incorporation in the films similar to our study [5]. Though their reported resistivity was about one order lower than the ALD Al-doped ZnO films deposited in our study and other reports using the ALD method. However, the films deposited in this study are only 100 nm thick, while previously mentioned depositions were done on films with larger thicknesses ranging from 300 to 600 nm.

It is also to be noted that looking into the accuracy of the comparison between the reported resistivities, with some groups reporting the lowest resistivity to be at 2–3 at. % while others reported it to be around 5 at. %. As described in detail through Sect. 4.2 in most of these cases, these seemingly different values are thought to be representations of the same results due to the difference in calculation method or the doping technique. Figure 4.3 is a projection of the surface in Fig. 4.1, and it shows the dependence of the ALD Al-doped ZnO films resistivity on the growth temperature from 60 to 250 °C for various atomic ratio of Al (at. % Al) between 0 (pure ZnO) and 10 at. %

The resistivity of ALD ZnO- and Al-doped ZnO films have a typical semiconductor behavior, with decreasing values as deposition temperature increases (Fig. 4.3) in agreement with the literature data [2]. At any given level of doping, the temperature dependence of the resistivity follows a convex curve. The resistivity rapidly decreases between 60 and 150 °C. Above 150 °C, the decrease in resistivity values starts to slow down and heads back toward a slow increase after 200–250 °C, with

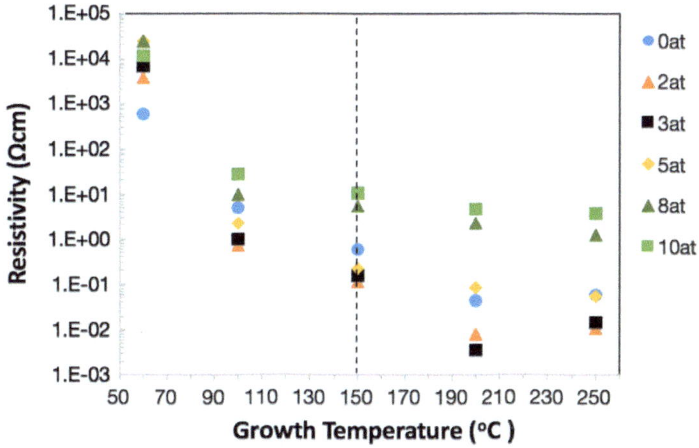

Fig. 4.3 Films resistivity as a function of growth temperatures between 60 and 250 °C for various doping levels

most of these curves having their minimum around 200–250 °C (it undergoes a more aggressive increase after 250 °C when measured on the Si substrates—not shown in these measurements due to limitations imposed by the flexible substrate). The ZnO film has its minimum resistivity at 200 °C with a value of 4.23×10^{-2} Ω.cm without intentional doping which is much lower than the ZnO film grown at 60 °C with a value of 575.4 Ω.cm. The very high resistivity of pure ZnO is due to its low carrier concentration. This low value of the carrier concentration can be explained by compensation effects, i.e., by native formation defect.

The slowdown in resistivity reduction supports that the doping effect increases with the temperature as proposed in [6] up to this region, beyond this, the doping effect slows down and resistivity starts increasing again. Indicating that the growth temperature also impacts the doping efficiency (i.e., the ratio of electrically active Al versus interstitial Al and Al_2O_3); reaching maximum efficiency at around 150 °C (indicated by the dotted line).

The tendency of the resistivity exhibited above is vividly a combinatorial result of the changes in the Hall mobility and in the carrier concentration. Therefore, they are presented in the following subsections.

Carrier Concentration of Al-doped ZnO Films

The carrier concentration of the film's subject of this work was also measured in conjunction to the resistivity measurements to understand the physical background of the changes observed in the resistivity. The two-dimensional surface in Fig. 4.4 represents the measured carrier concentration versus the Al doping level and the growth temperature. This graph shows films with a carrier concentration lower than 10^{18} cm^{-3} (blue), higher than 10^{19} cm^{-3} (orange), and in between (yellow). The border between the indicated orange and yellow regions represents the transition from TCO to TSO, respectively.

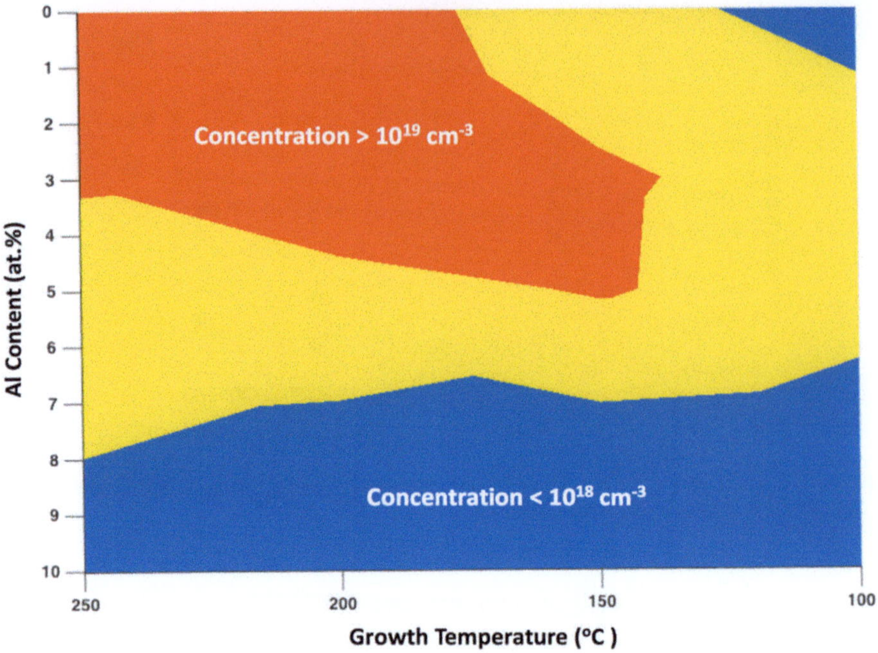

Fig. 4.4 The measured carrier concentration versus doping levels between 0 and 10 at. % Al along with growth temperatures ranging from 100 to 250 °C

Figure 4.5 is a projection of the surface in Fig. 4.4, and it shows the dependence of carrier concentration of the ALD Al-doped ZnO films on the atomic ratio of Al (at. % Al) from 0 (pure ZnO) to 10 at. % for various growth temperature between 60 and 250 °C.

Carrier concentration has been known to strongly correlate with the deposition temperature. However, Al doping has also shown a great effect on modulating the carrier concentration allowing to control the film conductivity to either create a highly conductive oxide or a semiconductive oxide depending on the desired application. Figure 4.5 shows that for almost all growth temperature, the carrier concentration grows with the increasing Al content, passing a maximum in the same region where the resistivity has its minimum, then it falls again.

For films deposited at a growth temperature of 200 °C and as Al atoms is incorporated into the ZnO film, the carrier concentration initially increases rapidly about 6 times from 1.14×10^{19} dcm^{-3} for undoped ZnO films to a maximum concentration of 6.68×10^{19} cm^{-3} at dopant concentration of 3 at. % then it starts to decrease again as more Al dopant is added to the film until it reaches 4.82×10^{17} cm^{-3} at 10 at. % which is in agreement with reported Al-doped ZnO films deposited by ALD and even other deposition methods [7, 8].

This trend of rapid increase in carrier concentration when the Al dopant is incorporated into the films is seen amplified at higher growth temperature where the increase

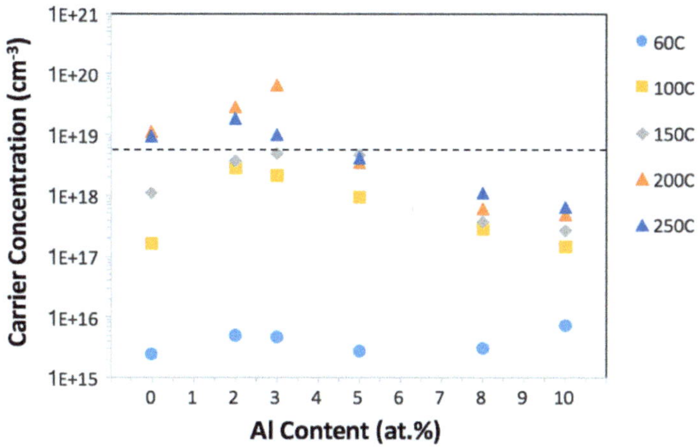

Fig. 4.5 Films carrier concentration as a function of doping levels between 0 and 10 at. % Al for various growth temperatures. The shift from TSO to TCO indicated by the dotted line

in carrier concentration as Al is introduced is more abrupt than the samples deposited at relatively lower temperatures. Which indicate a higher doping efficiency as only a small change in the Al concentration is needed to generate much higher carrier concentration than samples deposited at low temperatures.

All the carrier concentration values obtained by the Hall measurement system were negative; indicating that all the synthesized thin films, both Al-doped ZnO and pure ZnO, have n-type conductivity. ALD ZnO films are naturally n-type oxide semiconductors due to highly efficient natural electron donors generated by O vacancies and Zn interstitials [9].

Mobility

In addition to carrier concentration, the Hall mobility of the films was also measured in conjunction to the resistivity measurements to better understand how the Hall mobility of the Al-doped ZnO films behave under various Al doping concentrations and growth conditions. This is critical for achieving high-quality TFTs where mobility is an essential parameter for high-performance devices and also allows further understanding of the changes observed in the film electrical properties in the previous section. The two-dimensional surface in Fig. 4.6 represents the Hall mobility versus the Al doping level and the growth temperature. In this graph films with a mobility lower than $10\,\text{cm}^2/Vs$ are the blue zone, higher than $20\,\text{cm}^2/Vs$ are the orange zone and in between are yellow zone. Along with the inset showing the changes in the films domain size under the same growth conditions and doping levels.

The figure shows that a correlation between the electron mobility and mean grain size where the decrease in films mobility is associated with a decrease of the mean grain size of the film at these high Al concentrations and low growth temperature. The grain boundary functions as trap sites and discourages electron migration. Decreased

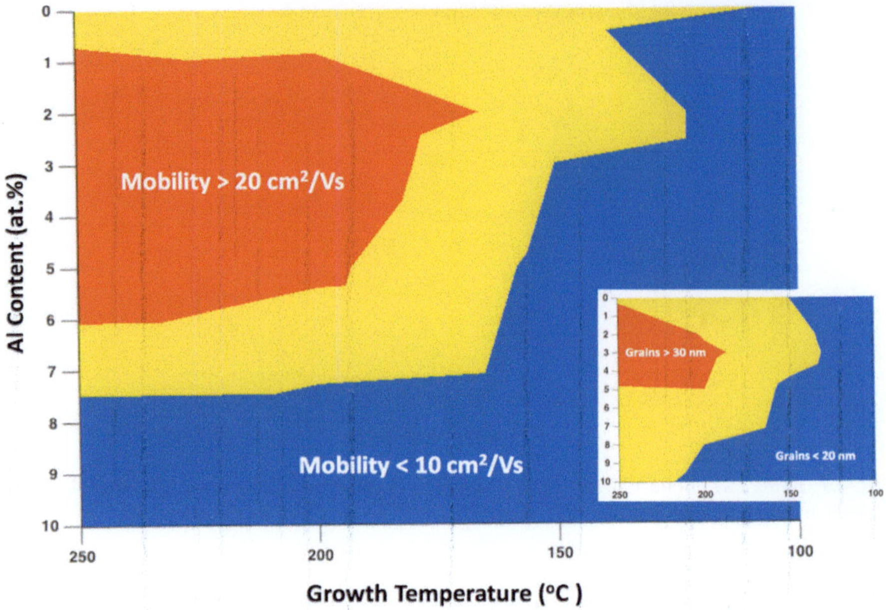

Fig. 4.6 The measured Hall mobility versus doping levels between 0 and 10 at. % Al along with growth temperatures ranging from 100 to 250 °C. The inset shows the grain sizes as a function of the same growth conditions and doping levels (Inset axis are the same as the main plot)

grain size of the thin film breeds increased amount of grain boundaries which reflects a worse mobility due to carrier scattering.

Figure 4.7 is a projection of the surface in Fig. 4.6, and it shows the dependence of mobility of the ALD Al-doped ZnO films on the atomic ratio of Al (at. % Al) from 0 (pure ZnO) to 10 at. % for various growth temperature between 60 and 250 °C.

Figure 4.7 shows that at 250 °C as Al atoms is incorporated into the ZnO matrix, the Hall mobility increases from 11.38 cm^2/Vs for undoped ZnO to 43.11 cm^2/Vs for 3 at. % of Al then start decreasing from 26.63 cm^2/Vs to become as low as 2.75 cm^2/Vs when Al doping increased from 5 to 10 at. %.

This behavior persists for all the films grown at temperatures higher than 150 °C. The initial increase in Hall mobility when a small amount of Al doping is introduced in the film (<2–3 at. %) is primarily caused by enhanced crystal structure at these high growth temperatures allowing the increase seen in the mobility values. Nevertheless, in the case of a low deposition temperature (< 150 °C), mobility undergo a continuous decrease as Al doping concentrations increases. This dramatic decrease in mobility values correlates with the increase in carrier concentration and the poor crystallinity of films grown at low temperatures where the dopants tend to interrupt the carrier transport, thus causing the continuous decrease in mobility as the Al doping increases. In addition to the nucleation or segregation is believed to start at a lower

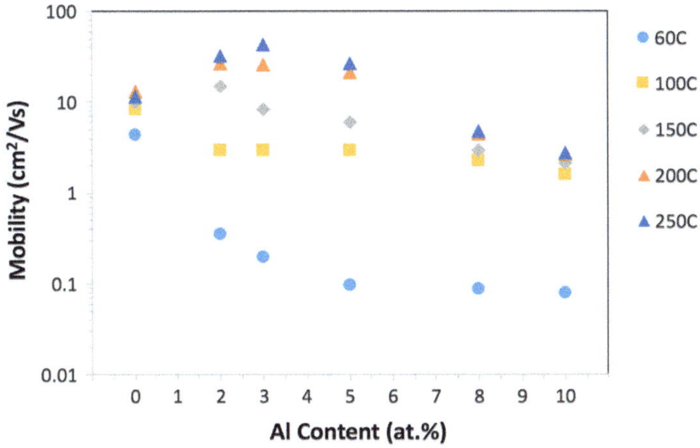

Fig. 4.7 Films Hall mobility as a function of doping levels between 0 and 10 at. % Al for various growth temperatures

Al concentration due to the very low temperatures which lacks the needed energy to effectively rearrange and diffuse to generate carriers. In contrary to the high growth temperatures where those effects occur at a higher Al doping concentration giving the mobility a chance to increase before decreasing again.

For low doping concentrations and growth temperature, the decrease in mobility is mostly associated with the increasing carrier concentration at the same temperatures which promote more scattering. This trend is contrary to the films deposited with the same Al concentration but at higher growth temperature, where the mobility increases with the Al doping until certain ratio (below 3%) then go back to decreasing with the rest of the films deposited at lower temperatures.

The decrease of the electron mobility when the Al dopants in the film exceed a certain level (3 at. %) can be also explained by a decrease in the Zn content during the TMAl pulses as shown in Sect. 4.2 which is associated with an increase in the formation of zinc vacancies. Along with the excess Al dopants that reduce mobility mainly due to the scattering from these ionized impurities [10].

Figure 4.8 is a projection of the surface in Fig. 4.6, and it shows the dependence of mobility of the ALD Al-doped ZnO films on the growth temperature from 60 to 250 °C for various atomic ratio of Al (at. % Al) between 0 (pure ZnO) and 10 at. %.

For intrinsic ZnO films, Fig. 4.8 shows that the deposition temperature doesn't have a significant impact on the mobilities as it does in the Al-doped ones which is almost saturated at all growth temperatures with little increase for undoped ZnO samples. In films grown without any intentional doping, the mobility increasing from 4.43 to 11.38 cm^2/Vs as the deposition temperature increases from 60 °C to 250 °C. Also, these films' mobilities were higher than the Al-doped at low temperatures ($<$150 °C). This is due to the Al dopants which act as ionized impurities, increasing the scattering probability and reducing the overall mobility of the doped films at low

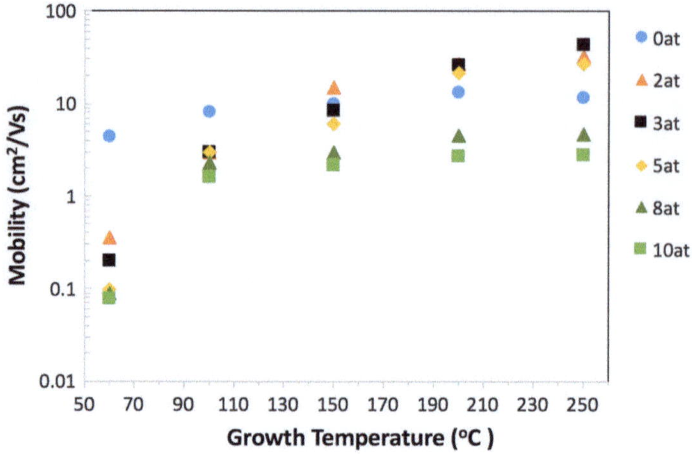

Fig. 4.8 Films carrier concentration as a function of growth temperatures between 60 and 250 °C for various doping levels

deposition temperatures. It can be concluded that the Al dopants interrupt the carrier transport in the case of a low deposition temperature as the Al dopants atoms lack the needed energy to effectively rearrange and diffuse to generate carriers at a very low deposition temperatures.

Nonetheless, at any given level of doping, the doped films mobility rapidly increases for films grown at temperatures lower than 150 °C, then the mobility increase slows down at growth temperature higher than 150 °C especially at very high doping concentrations. The high mobility values seen at high deposition temperature for Al-doped ZnO films is again primarily caused by the donor-like behavior of the Al dopants. For instance, the 3 at. % films mobility increases from 0.24 cm^2/Vs at deposition temperature of 60 °C up to 43.11 cm^2/Vs for 250 °C. The rapid and continuous increase in the mobility values as Al doping increases where a small increase in deposition temperature causes a large increase in the mobility values opposite to the case for pure ZnO films. This enables Al doping to be used to fine-tune the mobility values using doping and the deposition temperatures as process variables. At low temperature, small addition of Al result in a larger increase in carrier concentrations than seen at high temperature. The more carrier generated the lower the mobility values due to scattering.

4.1.2 ALD-grown Versus PVD-grown Al-doped ZnO

The reported resistivity values of Al-doped ZnO films grown by ALD have been in the order of 10^{-3} Ω.cm, [11] which is considered higher than Al-doped ZnO films grown with other deposition techniques. The lowest reported resistivity for Al-doped

ZnO films was deposited by PLD and had resistivities as low as 8.5×10^{-5} Ω.cm with extremely high carrier concentrations of 1.5×10^{21} cm^{-3}. This isn't essential for this work as higher mobility is more important for the TFT channels and highly conductive channel can make it harder to turn off the TFT as higher turn off voltages is required leading to higher power consumption. In this section, we establish the understanding of the reason behind the higher resistivity of the ALD grown Al-doped ZnO films compare to other methods specifically PVD techniques.

Our ALD Al-doped ZnO film with the lowest resistivity grown at 200 °C with an Al doping of 3 at. % had a Hall mobility of 26 cm$^2/Vs$ and a carrier concentration of 6.68×10^{19} cm^{-3}. This Hall mobility was similar to Al-doped ZnO films deposited via sputtering which are around 15^{-25} cm$^2/Vs$ [8]. Meanwhile, the carrier concentration was much lower compared to PVD drown films which can achieve a carrier concentration around 6×10^{20}–1.5×10^{21} cm^{-3} when a similar amount of Al being incorporated [12, 13]. This low carrier concentration for ALD-grown Al-doped ZnO films has been reported before. Nonetheless, these discrepancies in the characteristics of Al-doped ZnO films could be explained by the deposition mechanism of ALD. By large, the ALD method produces Al-doped ZnO thin films with much lower defect density (e.g., oxygen/Zn vacancy and Al/Zn interstitial atom) than that of PVD methods. Furthermore, the films were induced by self-saturated surface reaction without any deposition damages, whereas PVD can cause physical damage and low reactivity compared to ALD. In addition to ALD method allowing stable Al-O / Zn-O bonding and formation of hydroxyl groups at a low deposition temperature which decreases oxygen vacancy formation; lowering the carrier concentration compare to PVD. These characteristics of Al-doped ZnO are related to the deposition mechanism of ALD. Previously, the decrease of ALD ZnO carrier concentrations was explained by the decrease of oxygen vacancies due to the formation of hydroxyl groups at low deposition temperature [14].

4.2 Mechanical Characterization of Al-doped ZnO Thin Films

The structural and electrical properties of the engineered ZnO thin films in this work were shown in previous sections prior to integrating them into functional TFT to minimize their piezoresponse or into sensor devices to maximize it. In the previous sections, we looked into how the various deposition conditions and doping levels could impact the performance of TFTs and strain sensors with Al-doped ZnO layers as the active material. This section addresses our work on the piezoelectric implications and associated features of the Al-doped ZnO semiconductor thin films through optimizing their piezoresistance and piezopotential effects. It details the examinations conducted to study and determine these piezoelectric effects in the ALD-grown films along with explaining the quantitative relationships between the piezoelectric activity and growth conditions including the Al doping of the ZnO

films. Accordingly, a definite understanding of how all these aspects individually and interactively shape the piezoelectric domains is key in our process and device optimization. Therefore, this investigation was conducted by fabricating a proof-of-concept test structure which is then characterized as a piezoelectric strain sensor where these parameters were extracted and assessed.

4.2.1 *Semiconductor Oxides as Piezoelectric Materials*

Due to the increased interest in scalable integrated microfabrication processes, along with thin-film processing and the unprecedented precise engineering control over material properties, the technological compatibility sets piezoelectric semiconductor thin films as the perfect candidate for realizing piezoelectric applications not seen before. This has driven a rising interest seen today in exploring the possibilities of wurtzite structured ZnO, GaN, InN, and ZnS to be used for new applications because of their piezoelectric features along with being semiconductors [15–17]. These materials provided the conceivable possibility of making both the sensing and logic parts from the same material as multiple reports has shown; thus, the unrealized prospects of implementing them in sensors, flexible and transparent electronics, optoelectronics, smart MEMS, and human–machine interfacing were brought to light [18–20]. Table 4.1 shows a comparison of three of the most common inorganic piezoelectric ceramics which with large piezoelectric coefficients are used for micro-electromechanical applications and some important properties of these materials. Lead zirconate titanate, $Pb(Zr_xTi_{1-x})O_3$, (PZT) is the most well-known piezoelectric material for having a number of traditional contributions in electromechanical sensors, actuators, and energy-harvesting devices [21, 22]. This is due to its excellent piezoelectricity of extremely high material coupling and piezocoefficients, k_{33} and d_{33} as shown in Table 4.1. Nonetheless, it has a very high dielec-

Table 4.1 A comparison of three the most common piezoelectric ceramics which are used in micro-electromechanical applications and some important properties of these materials

	ZnO	AlN	PZT
Density, $p(kg.m^{-3})$	5606	3260	7500
Young's modulus, Y (GPa)	201	308	60
Piezocoefficient, $d_{33}(pm.V^{-1})$	12.4	5.5	500
Coupling Coefficient, k_{33}	0.48	0.31	0.69
Dielectric Constant, ε_{r33}	10.9	10.5	2400
Ratio, k^2/ε	21.14×10^{-3}	9.15×10^{-3}	19.84×10^{-3}

tric constant ε_r and consequently is electrically insulating. Therefore, it cannot be optimally used in manufacturing electronic devices [23]. The deposition process of PZT also has its environmental issues because of the lead contamination associated with it. On the other side, non-ferroelectric piezoelectric materials such as ZnO and aluminum nitride (AlN) are also well-known piezoelectric materials with great piezocoefficients and high k_2/ε ratios as shown in Table 4.1. In addition to being non-ferroelectric materials, which does not require polling or post-deposition annealing, these films have excellent bonding to plastic substrates and low deposition temperature. These advantages make them attractive for transparent and flexible devices along with their high mechanical quality factor (Q_m) [24]. Their production is environment-friendlier in comparison to PZT and has well-known standard deposition technique which is highly compatible with semiconductors and CMOS processing due to their low dielectric constant and low power requirement. Because of their semiconducting nature and the presence of piezoelectricity, semiconductor oxides became attractive for integrated electromechanical-based devices and in recent years there has been increased interest in studying semiconductor oxides, specifically ZnO, in hopes of exploiting their physical properties in the fabrication of sensing devices and energy-harvesting systems. There is already a large number of micro-electromechanical devices which utilize the piezoelectric effect in thin films of the polar semiconducting material ZnO. They are widely used to fabricate ultrahigh frequency surface and bulk acoustic wave (SAW) devices that are employed in radar and telecommunications for filtering [25], flexible thin-film transducers for pressure and strain sensing [26], actuators [27], measuring displacement, tactile, force and torque [28, 29], energy-harvesting architectures [30], along with accelerometers and microphones as well [31, 32]. These applications also investigated employing doped ZnO films because intrinsic ZnO conductivity and transparency could be modulated by adding Group III metal dopants. Especially Al, as shown in the previous sections, where the doping of ZnO with aluminum not only allowed the modulation of the film electrical properties but also enhanced its thermal stability and provided a direct knob to control and predict the film texture to compensate for the low growth temperature which results in polycrystalline films. The piezoresponse of these films can be diminished or enhanced by adjusting the Al doping levels in the films. Especially, there is also a proposition that perfect c-axis texture in the film are required to attain adequate piezoelectric activity.

4.2.2 1D and 2D Piezoelectric Structures

At the moment, majority of the demonstrated ZnO piezoelectric-based pressure and strain sensors rely on using one-dimensional (1D) Wurtzite nanostructures, mainly ZnO nanowires (NWs), due to the preparation of ZnO to be in diverse micro- and nanostructures which include micro- and nanowires (NWs), and nanorods (NRs) became less complicated nowdays. Regulating the charge carrier transport properties in ZnO-based nanomaterials was reported to be viable on both fixed and flexible substrates based on one-dimensional (1D) ZnO nanostructures [33]. ZnO nanobelts have been reported to have high piezoelectric coefficient d_{33} values; ranging from 14.3 to 26.7 $pm V^{-1}$. These successful outcomes encouraged the application of ZnO nanostructures as nanosensors and nanogenerators. These 1D semiconductor nanostructures are promising seeds for future electronics and multiple techniques for their synthesis have been reported; nevertheless, various technical challenges and limitations persist which obstruct their anticipated role in practical implementations through immediate applications and commercialization. They require difficult fabrication technology and complicated package process; synthesizing these 1D nanostructures doesn't guarantee sufficient uniformity due to inconsistencies of dimensions, morphology and doping levels. Another major concern is its negative impact on device performance, as the mentioned inconsistencies reflect a lack of reliability; with performance varying from one device to another. Above all, there is the unanswered reproducibility question. Additionally, the available techniques for positioning, aligning, and integrating as grown 1D nanostructures are either cumbersome or incompatible with state-of-the-art microfabrication processes and integrated circuit technology. While 1D nanowire structures also allow for very high sensitivity, 2D thin-film devices provide robust and stable devices due to the redundancy of multiple crystallites in the film. Typically, 1D nanostructures are oriented laterally on a flexible substrate, with the bending applied perpendicular to the direction of the wire. Uniquely, the use of 2D thin-film structures, with their c-axis perpendicular to the substrate, improves device resilience and achieves a gauge factor higher than that previously reported for a single nanowire [34]. While lateral nanowires technically experience a bending load, it is often reasonable to assume that the nanowires are experiencing pure compression or tension, due to the geometry and mechanical properties of the device and substrate [35]. This results in a potential distribution along the c-direction of the nanowire, as opposed to across the diameter of the crystalline grains in thin films as in the case of lateral deflection.

In the particular case of piezoresponse-based applications, aside from feasibly integrating 1D nanostructures with microfabrication of device architectures and peripheral circuits, it is integral to determine and has adequate engineering control of the polar c-axis orientation in as-synthesized nanomaterials for further construction of integrated systems where perfect c-axis texture is essential to obtain adequate piezoelectric activity. However, both determination and associated engineering control are not quite attainable with 1D nanostructures. The piezoresponse effect is expected to be a ubiquitous one, and so it is absolutely advantageous to study and employ it through thin-film-based material systems, in order to conceivably evade the 1D nanostructures shortcomings, fully appreciate cutting-edge microfabrication technologies, and further explore stretching its boundaries. Accordingly, the established thin film technology is considered a more efficient alternative for the synthesis of piezoelectric-based devices, being fully compatible with the semiconductor technology and with the microfabrication processes for now.

4.2.3 Piezoelectric Effects in Wurtzite ZnO Thin Films

It is established that intrinsic ZnO films have the highest piezoelectric tensor among the tetrahedrally bonded semiconductors. As discussed above, these specific characteristics highlight ZnO as a vital material for a lot of practical applications that require a large electromechanical coupling. Consequently, it has been the subject of a large body of experiments, which yielded a mature understanding of its mechanical and electrical properties but still did not fully explain this unusual strong piezoelectric behavior. Especially, the piezoelectric tensor of similar II–VI compounds that share the same wurtzite structure as the ZnO compound (e.g., ZnS, CdS, and CdSe) is not more than one-half of its value. Interestingly, BeO has the same structure and the same anion; yet, it has one of the smallest piezoelectric tensors among the II–VI and III–V semiconductors, which is an order of magnitude smaller than that of ZnO. The three factors which could cause the variable properties of semiconductor thin films under mechanical bending; starting with the piezoresistive effect [36] which is due to changes in the band structure of the semiconductor [37]; in addition to trapped electrons at the interface of the semiconductor. Or the change of grain size where the mobility generally increases with grain size as the grain boundaries are associated with scattering and trapping sites [38].

These piezoelectric properties of the ZnO films are a result of the strong bond polarity [39]. When strained, an electrical polarization is generated; similarly, a strain is created in the thin film under the effect of electric field. The positively Zn^{2+} and negatively O^{2-} charged ions are arranged in a crystalline form to cause the change in polarization with applied stress. Deflecting a ZnO thin film causes a relative displacement of Zn^{2+} cations with respect to O^{2-} anions in the wurtzite structure; rendering a strong piezoelectric potential field inside the channel area. A simplified 2D schematic is shown in Fig. 4.9 that can explain such behavior. In Fig. 4.9a, a square structure with positive and negative ions' arrangement is hypothetically placed under compressive stress. For this structure, the equivalent center of charge of the positive and negative ions is still at the same point and there is no polarization (P = 0) change due to the force applied. On the other hand, for a 2D hexagon structure like in ZnO with the ion arrangements shown in Fig. 4.9b, the applied stress induces a change in the center of charge of the positive and negative ions, which means a change in the polarization ($P > 0$) that causes an effective electrical field. In the non-polar structure, the material exhibits non-ferroelectric properties (as shown in the upper part of Fig. 4.9b). When a strain is applied to the material, the symmetry of the crystal is disrupted, creating a separation of the unbalanced charge centers, which in turn generates the permanent dipole; thus exhibiting a ferroelectric behavior (as shown in lower part of Fig. 4.9b). By applying an electric field, the orientation of the dipole moment can be switched from positive (pointing upwards for instance) to negative (pointing downwards).

In order to achieve the strongest piezoelectric response, ZnO films need uniform polarity orientation of grains in addition to controlled optimization of the deposition conditions to ensure high-quality crystallinity. Additionally, the crystallites in

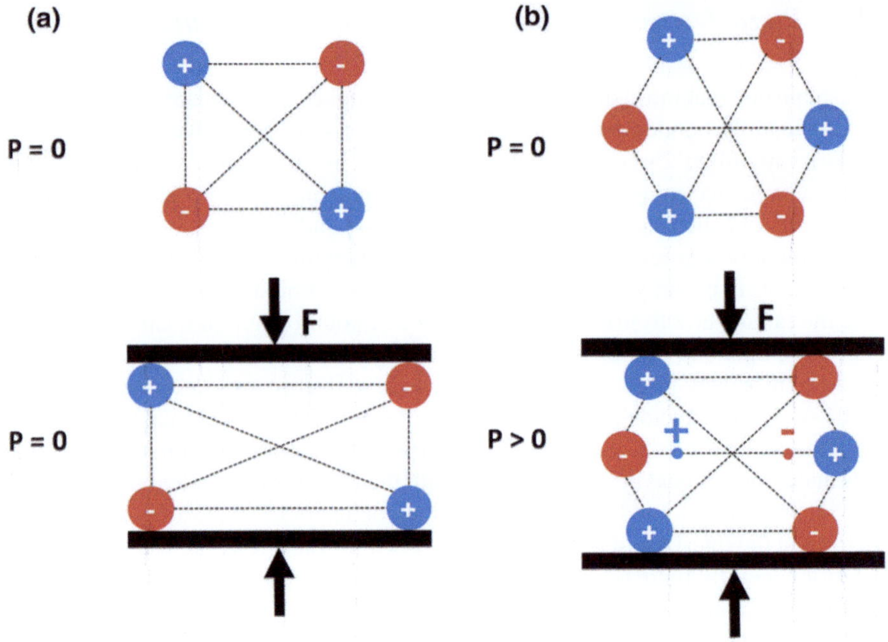

Fig. 4.9 Examples of 2D crystal structures: (**a**) non-piezoelectric square, (**b**) piezoelectric hexagon

ZnO films grow preferentially with the c-axis perpendicular to the substrate surface. Hence, the characterization of the local piezoelectric effect and polarity configuration of preferred c-axis-oriented ZnO films is critical for both the design and performance of any ZnO piezoelectric device. The crystalline quality of ZnO films, particularly the orientation of the crystallites in the film with respect to the substrate surface, is determined by multiple factors. Among the most crucial ones are the substrate nature (chemical composition, crystalline orientation, and surface roughness), deposition specifications (pressure, growth temperature, purging, and deposition cycles) and also doping levels which was discussed in detail in Sect. 4.3.

As shown in Fig. 4.9b, the lack of central symmetry in the wurtzite structure causes ZnO to display strong piezoelectricity and crystal polarity. The oppositely charged ions generate positively charged Zn^{2+} (0002) and negatively charged O^{2-} (000-2) polar faces, causing a normal dipole moment and spontaneous polarization whose orientation always points from the positive Zn face to the negative O face as presented in Fig. 4.10a. This makes the piezoelectric coefficient along the c-axis orientation the largest one for ZnO (only the magnitude, but not the direction), as the piezoresponse vector changes within most individual grains (Fig. 4.11). For perfectly c-axis-oriented ZnO films, the local spontaneous polarization orientation is either from the top surface to the bottom electrode or vice versa, as per the terminating atom at the surface. The polarity of the film will then determine the polarity of the corresponding electric field. So, applying a tensile strain will end up with negative

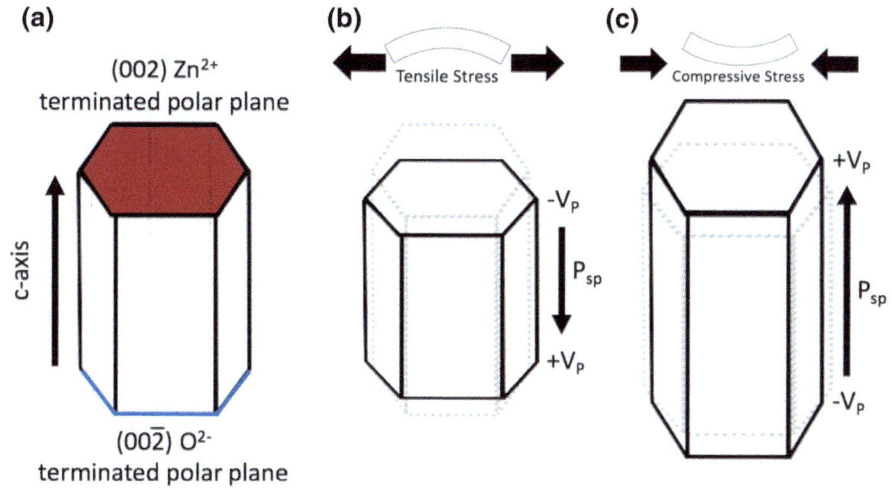

Fig. 4.10 **a** Zn^{2+} (0002) and O_{2-} (000-2) polar faces responsible for the spontaneous polarization in ZnO, **b** inducing a piezopotential along the c-axis under tensile strain or **c** compressive strain

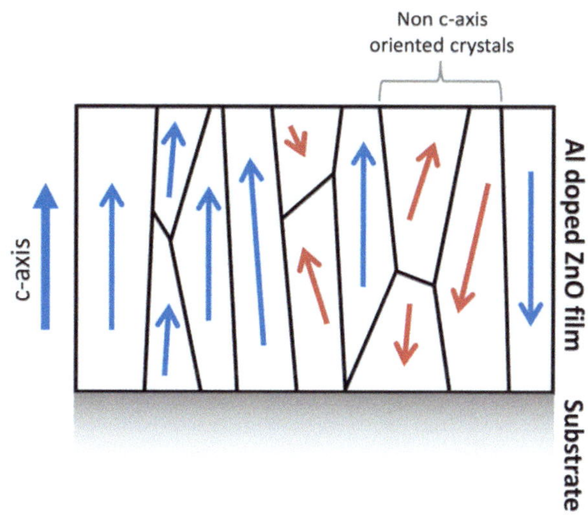

Fig. 4.11 The alignment of the columnar grains responsible for the macroscopically piezoelectricity of polycrystalline films

charges induced on the top surface of the thin film as the case shown in Fig. 4.10b and by default a positive charge under compressive strain as in Fig. 4.10c.

Figure 4.11 schematically depicts the alignment of the c-axes within these columnar grains which gives rise to macroscopically observed piezoelectricity of the polycrystalline undoped and doped ZnO thin film.

The polarization orientation can be then averaged either in z direction along c-axis or -z direction. The polarity of the films can be controlled using the ALD deposition cycles to determine the terminating atom at the surface. This polarity can be determined from the phase scan of the piezoresponse force microscopy (PFM) or from the polarity of the voltage generated across the thin films under strain as shown later in this chapter.

4.2.4 Stress and Strain in Al-doped ZnO Thin Film

To understand the piezoresponse in the Al-doped ZnO films, strain has to be applied to the film and the stress generated has to be calculated. The mechanical behavior of the flexible substrate is not influenced by the grown thin films due to their much smaller thickness. According to the Saint-Venant bending theory which can be utilized for small deflections as a simple estimation of the strain in the devices where the shape of the flexible substrate and the film can be approximated as a beam [40]. Calculating the ε_z component of the strain is sufficient and all that is needed is to assess how the thin film length changes as the substrate is deflected under an external bending force. As the PET substrate has a much smaller Young's modulus than that of all the deposited thin film layer, with Young's modulus Y around 201 GPa for the thin film and 3–6 GPa for the flexible substrate. Additionally, the included metal electrodes have a much smaller area and thickness compared to those of the substrate; hence, the thin-film layers and metal electrodes, which are fabricated on the outer surface of the flexible substrate, do not cause any major change to the underneath substrate's mechanical properties. Accordingly, the strain in the film can be assumed to be uniform and either purely tensile or compressive as per the bending direction of the substrate which is depicted in Fig. 4.12. In view of this immensely small thickness of the film, compared to the thickness of the substrate, an approximate calculation of the axial strain ε along the length of the film is possible.

$$\varepsilon = \frac{t_s + t_f}{2R_c} \tag{4.1}$$

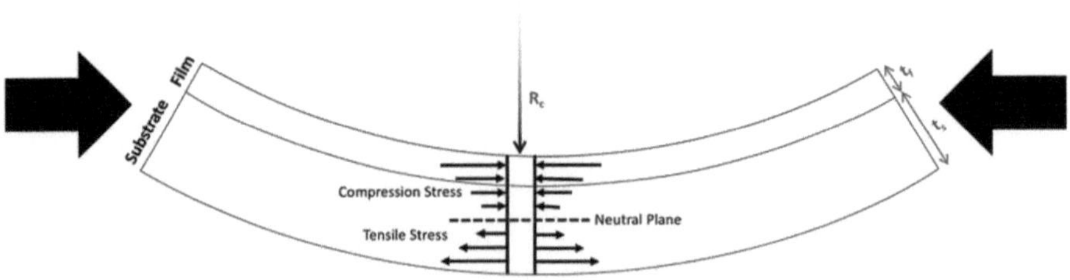

Fig. 4.12 The strain in the film can be assumed to be uniform and purely tensile or compressive as per the bending direction of the substrate (compressive strain in this case) due to immensely small thickness of the film compared to the thickness of the substrate. The arrows indicate the direction and the magnitude of the applied strain

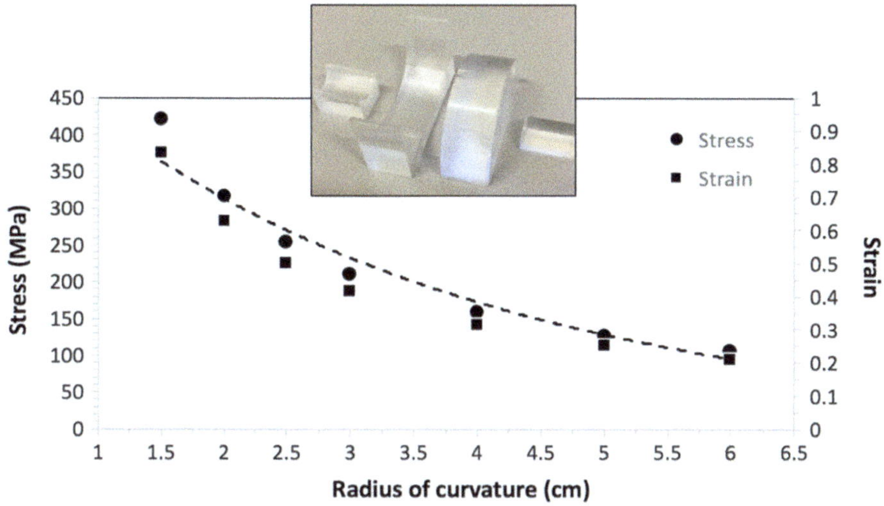

Fig. 4.13 The strain and stress induced in the thin films when bend with curvature radius ranging from 1.5 to 6 cm. Inset shows the custom made stages used to apply these strains on the thin films during the characterization process

Where t_s is the flexible substrate thickness, t_f is thin-film thickness, and R_c is the Radius of Curvature of the deformed film which can have a negative or positive sign depending on whether the film is under compressive or tensile strain, respectively.

This relationship shows that the strain ε has a linear relationship with the deformation bending radius R_c where the strain in the film is effectively uniform to an excellent approximation, when the thickness of the substrate is way larger than the thickness of the film ($t_s >> t_f$). It is to be noted that our estimation of the strain values is based on previous studies' reports of the analytical solutions for elastic bending, with tensile strain defined as positive and compressive strain defined as negative.

The stress values used here are for longitudinal stress generated parallel to the substrate plane. Thus, utilizing substrate properties like Poisson's ratio and Young's Modulus are viable to estimate the transverse stress along the overall effective polar direction of the thin film according to the following:

$$\sigma_c = \frac{Y_s t_s^2}{6R_c(1 - v_s)t_f} \tag{4.2}$$

Where Y_s is the substrate Young's Modulus, t_f is the film thickness, ts is the substrate thickness, R_c is the radius of curvature, and X_s is the Poisson's ratio of the substrate material which is around 0.33 for the flexible substrate used in this study). These stress values generated via employing customized curved stages (Fig. 4.13 inset), that have varying curvature radii and directions, were used to bend the fabricated devices, inducing compressive or tensile strains on each device.

Given that the deformation of the film is much smaller than its length, these 3D mechanical stages are able to introduce uniformly distributed strain and deformation on the devices. As Fig. 4.13 shows changing the curvature radius of these stages between 1.5–6 cm induced a strain in the film from 0.83 to 0.21 %, respectively.

4.3 Conclusions

This chapter covered the electrical and mechanical characterization needed for any thin-film dielectrics to be used for piezoelectric applications. It highlighted the electrical and mechanical responses and demonstrated the use of response surface methodology to model the relationships between different observations with the underlying process parameters. It explained the impact of various process variables, like Al doping and growth temperature, on piezoresponses and the underlying material crystal structures.

Acknowledgements This work has been supported by the Semiconductor Research Corporation (SRC) under the Abu Dhabi SRC Center of Excellence on Energy-Efficient Electronic Systems (ACE^4S), Contract 2013 HJ2440, with funding from the Mubadala Development Company, Abu Dhabi, UAE.

References

1. J.S. Na, G. Scarel, G.N. Parsons, In situ analysis of dopant incorporation, activation, and film growth during thin film ZnO ZnO: Al Atomic layer deposition. J. Phys. Chem. C **114**(1), 383–388 (2010)
2. Renier klenk, Martina Ch Lux-Steiner, Chalcopyrite based solar cells, in *Thin Film Solar Cells Fabrication, Characterization and Applications* (Wiley, UK, 2006), pp. 237–275
3. Su Cheol Gong, Ji Geun Jang, Ho Jung Chang, Jin-Seong Park, The characteristics of organic light emitting diodes with Al doped zinc oxide grown by atomic layer deposition as a transparent conductive anode. Synthetic Metals **161**(9–10), 823–827 (2011)
4. A.K. Pradhan, R.M. Mundle, K. Santiago, J.R. Skuza, B. Xiao, K.D. Song, M. Bahoura, R. Cheaito, P.E. Hopkins, Extreme tunability in aluminum doped zinc oxide plasmonic materials for near-infrared applications. Sci. Rep. **4**(1), 6415 (2014)
5. T. Minami, H. Nanto, S. Takata, Optical properties of aluminum doped zinc oxide thin films prepared by rf magnetron sputtering. Jpn. J. Appl. Phys. **24**(8), L605–L607 (1985)
6. S.D. Kirby, R.B. van Dover, Improved conductivity of ZnO through codoping with In and Al. Thin Solid Films **517**(6), 1958–1960 (2009)
7. C. Agashe, O. Kluth, J. Hüpkes, U. Zastrow, B. Rech, M. Wuttig, Efforts to improve carrier mobility in radio frequency sputtered aluminum doped zinc oxide films. J. Appl. Phys. **95**(4), 1911–1917 (2004)
8. Kun Ho Kim, Ki Cheol Park, Dae Young Ma, Structural, electrical and optical properties of aluminum doped zinc oxide films prepared by radio frequency magnetron sputtering. J. Appl. Phys. **81**(12), 7764 (1997)
9. G. Luka, P. Stakhira, V. Cherpak, D. Volynyuk, Z. Hotra, M. Godlewski, E. Guziewicz, B. Witkowski, W. Paszkowicz, A. Kostruba, The properties of tris (8-hydroxyquinoline) aluminum

organic light emitting diode with undoped zinc oxide anode layer. J. Appl. Phys. **108**(6), 064518 (2010)
10. Z.C. Jin, I. Hamberg, C.G. Granqvist, Optical properties of sputter-deposited ZnO: Al thin films. J. Appl. Phys. **64**(10), 5117–5131 (1988)
11. Sang-Hee Ko Park, Jeong-Ik Lee, Chi-Sun Hwang, Hye Yong Chu, Characteristics of organic light emitting diodes with Al-doped ZnO anode deposited by atomic layer deposition. Jpn. J. Appl. Phys. **44**(7), L242–L245 (2005)
12. H. Agura, A. Suzuki, T. Matsushita, T. Aoki, M. Okuda, Low resistivity transparent conducting Al-doped ZnO films prepared by pulsed laser deposition. Thin Solid Films **445**(2), 263–267 (2003)
13. H. Tanaka, K. Ihara, T. Miyata, H. Sato, T. Minami, Low resistivity polycrystalline ZnO : Al thin films prepared by pulsed laser deposition. J. Vac. Sci. Technol. A **22**(4), 1757–1762 (2004)
14. J.W. Elam, Z.A. Sechrist, S.M. George, ZnO/$Al2_{O3}$ nanolaminates fabricated by atomic layer deposition: Growth and surface roughness measurements. Thin Solid Films **414**(1), 43–55 (2002)
15. Horacio D Espinosa, Rodrigo A Bernal, Majid Minary-Jolandan, A review of mechanical and electromechanical properties of piezoelectric nanowires. Adv. Mater. **24**(34), 4656–4675 (2012)
16. Z.L. Wang, X.Y. Kong, Y. Ding, P. Gao, W.L. Hughes, R. Yang, Y. Zhang, Semiconducting and piezoelectric oxide nanostructures induced by polar surfaces. Adv. Funct. Mater. **14**(10), 943–956 (2004)
17. Z.L. Wang, J. Song, Piezoelectric nanogenerators based on zinc oxide nanowire arrays. Science **312**, 242–246 (2006)
18. Q. Yang, W. Wang, S. Xu, Z.L. Wang, Enhancing light emission of ZnO microwire-based diodes by piezo-phototronic effect. Nano Lett. **11**(9), 4012–4017 (2011)
19. H. Youfan, Y. Zhang, L. Lin, Y. Ding, G. Zhu, Z.L. Wang, Piezo-phototronic effect on electroluminescence properties of p-type GaN thin films. Nano Lett. **12**(7), 3851–3856 (2012)
20. Z.L. Wang, W. Wu, Nanotechnology-enabled energy harvesting for self-powered micro-/nanosystems. Angewandte Chemie - Int. Edition **51**(47), 11700–11721 (2012)
21. N. Izyumskaya, Y.-I. Alivov, S.-J. Cho, H. Morkoc, H. Lee, Y.-S. Kang, Processing, structure, properties, and applications of PZT thin films. Critical Rev. Solid State Mater. Sci. **32**(September), 111–202 (2007)
22. Hong Chen, Chen Jia, Wenhan Hao, Chun Zhang, Zhihua Wang, Chunsheng Liu, Power harvesting with PZT ceramics and circuits design. Analog Integr. Circuits Signal Process. **62**(2), 263–268 (2010)
23. Kiyotaka Wasa, Isaku Kanno, Hidetoshi Kotera, Fundamentals of thin film piezoelectric materials and processing design for a better energy harvesting MEMS. Power MEMS **61**, 61–66 (2009)
24. C.T. Pan, Z.H. Liu, Y.C. Chen, C.F. Liu, Design and fabrication of flexible piezo-microgenerator by depositing ZnO thin films on PET substrates. Sens. Actuators, A: Phys. **159**(1), 96–104 (2010)
25. A. Kuoni, R.L. Holzherr, M. Boillat, N.F. De Rooij, Polyimide membrane with ZnO piezoelectric thin film pressure transducers as a differential pressure liquid flow sensor. J. Micromech. Microeng. **13**(4), S103–S107 (2003)
26. J. Molarius, J. Kaitila, T. Pensala, M. Ylilammi, Piezoelectric ZnO films by r.f. sputtering. J. Mater. Sci. Mater. Electron. **14**(5–7), 431–435 (2003)
27. Takayuki Shibata, Kazuya Unno, Eiji Makino, Yoshiho Ito, Shiro Shimada, Characterization of sputtered ZnO thin film as sensor and actuator for diamond AFM probe. Sens. Actuators, A: Phys. **102**(1–2), 106–113 (2002)
28. H.I. Kuo, J. Guo, W.H. Ko, High performance piezoresistive micro strain sensors, in *Proceedings of the 2nd IEEE International Conference on Nano/Micro Engineered and Molecular Systems* (IEEE NEMS, 2007), pp. 1052–1055
29. H. Gullapalli, V.S.M. Vemuru, Ashavani Kumar, A. Botello-Mendez, R. Vajtai, M. Terrones, S. Nagarajaiah, P.M. Ajayan, Flexible piezoelectric zno-paper nanocomposite strain sensor. Small **6**(15), 1641–1646 (2010)

30. D. Choi, K.Y. Lee, K.H. Lee, Eok Su Kim, T. Sang Kim, S.Y. Lee, S.-W. Kim, J.-Y. Choi, J.M. Kim, Piezoelectric touch-sensitive flexible hybrid energy harvesting nanoarchitectures. Nanotechnology **21**(40), 405503 (2010)
31. Eun Sok Kim, R.S. Muller, IC-processed piezoelectric microphone. IEEE Electron Device Lett. **8**(10), 467–468 (1987)
32. C.J. Van Mullem, F.R. Blom, J.H.J. Fluitman, M. Elwenspoek, Piezoelectrically driven silicon beam force sensor. Sens. Actuators: A. Phys. **26**(1–3), 379–383 (1991)
33. V.F. Rivera, F. Auras, P. Motto, S. Stassi, G. Canavese, E. Celasco, T. Bein, B. Onida, V. Cauda, Length-dependent charge generation from vertical arrays of high-aspect-ratio ZnO nanowires. Chem. - A Eur. J. **19**(43), 14665–14674 (2013)
34. W. Zhang, Ren Zhu, Vu Nguyen, R. Yang, Highly sensitive and flexible strain sensors based on vertical zinc oxide nanowire arrays. Sens. Actuators, A: Phys. **205**, 164–169 (2014)
35. J.M. Wu, C.Y. Chen, Y. Zhang, K.H. Chen, Y. Yang, Y. Hu, H. He, Z.L. Wang, Ultrahigh sensitive piezotronic strain sensors based on a ZnSnO 3 nanowire/microwire. ACS Nano **6**(5), 4369–4374 (2012)
36. Jae Min Kim, Taewook Nam, S.J. Lim, Y.G. Seol, N.E. Lee, Doyoung Kim, Hyungjun Kim, Atomic layer deposition ZnO: N flexible thin film transistors and the effects of bending on device properties. Appl. Phys. Lett. **98**(14), 142113 (2011)
37. Niko Munzenrieder, Kunigunde H. Cherenack, Gerhard Troster, The effects of mechanical bending and illumination on the performance of flexible IGZO TFTs. IEEE Trans. Electron Devices **58**(7), 2041–2048 (2011)
38. Yugang Sun, John A. Rogers, Inorganic semiconductors for flexible electronics. Adv. Mater. **19**(15), 1897–1916 (2007)
39. Claus F Klingshirn, Andreas Waag, Alex Hoffmann, Jean Geurts, *Zinc Oxide: From Fundamental Properties Towards Novel Applications* (Springer, Berlin, 2010)
40. R.W. Soutas-Little, *Elasticity* (Dover Publications, 1999)

Chapter 5
ALD Al-doped ZnO Thin Film as Semiconductor and Piezoelectric Material: Transistors and Sensors

Ayman Rezk and Irfan Saadat

5.1 ALD-grown Al-doped ZnO Thin-Film Transistors

Based on the investigated structural and electrical properties of ALD-grown Al-doped ZnO thin films of the previous two chapters [1, 2], the Al-doped ZnO films can be used as an active channel layer in flexible and transparent TFT devices. They have high mobility and high transparency with low growth temperature while maintaining high performance, which makes them a great candidate for flexible and transparent applications.

We have fabricated and characterized Al-doped ZnO TFTs by employing the ALD method to prepare the Al-doped ZnO active channels and Al_2O_3 gate insulators. In the first iterations, the TFT failed to be operate as a normal device due to the low conductivity of the ZnO channel especially when grown using the Plasma-Enhanced ALD (PEALD) process. However, after more investigations as shown in the previous chapter, we were able to fabricate and demonstrate a high-performance bottom-gate type Al-doped ZnO TFT. Furthermore, the results demonstrate that these TFTs show a superior endurance and reliability in comparison with standard ZnO TFTs. The amount of Al incorporated into the ZnO was varied to investigate its effects on the device characteristics, including bias-stress stabilities. The objective of this work is to design an optimal bottom-gate structure for the Al-doped ZnO TFTs. There were several challenges associated with ZnO TFT that had to tackled and worked around. These were ranging from optimizing the mobility in non-crystalline or polycrystalline ZnO channel while maintaining a good on–off current ratio and the associated transconductance, along with dealing with the high electron concentrations encountered in ZnO channel layers. Due to this high concentration, the

---

A. Rezk · I. Saadat (✉)
Masdar Institute at Khalifa University, Abu Dhabi, UAE
e-mail: isaadat@masdar.ac.ae

A. Rezk
e-mail: arizk@masdar.ac.ae

Ibrahim (Abe) M. Elfadel and M. Ismail (eds.), *The IoT Physical Layer*,
https://doi.org/10.1007/978-3-319-93100-5_5

initial ZnO channels did not show any pinch-off and depletion which resulted in very high turn-off voltages. The other challenges tackled were related with any TFT such as choosing the right gate oxide and metal contacts. Additionally, the use of flexible substrate introduced its own challenges such as low-temperature fabrication requirements, film adhesion, and long-term reliability and bendability.

5.2 TFT Fabrication

Two separate sets of TFTs, Group 1 and Group 2, were fabricated along with their dimensional variations. The main difference between the two sets is the channel width. Both sets had a channel length of 20, 30, and 40 µm. However, the first set, Group 1, channel had a width of 5, 10 and 15 µm while the second set, Group 2, had a width of 30, 45, and 60 µm. Figure 5.1a shows a schematic 3D cross-sectional diagram of the general structure of the fabricated Al-doped ZnO bottom-gate TFT on the flexible substrate. Figure 5.1b shows the details of the process flow used to fabricate the TFT shown in (a) with thickness, layers and deposition methods indicated on the graph.

An optical microscopic image of a single device structure on the Silicon substrate with L/W of 40/15 nm channel from Group 1 is shown in Fig. 5.2a and on the flexible PET plastic substrate with L/W of 40/60 nm channel from Group 2 in Fig. 5.2b.

Initially, The TFT characterization and process flow development were done on a silicon wafer with 30 nm thermal SiO_2 used as the substrate to rule out any variability related to the flexible substrate.

5.2.1 Adhesion Layer and Gate Contact

To begin with, the 250 µm thick flexible substrates were mounted onto glass carriers for ease of handling and mechanical rigidity during device fabrication. The first layer was a relatively thick 300 nm which can be either a SiO_2 or Si_3N_4 layer which can be deposited using ALD or PECVED (FlexAL ALD system and Plasmalab 100 PECVD System both from Oxford Instruments Plasma Technology). This layer was grown directly on the substrate to enhance the adhesion of subsequent layers and to minimize the fallback of the substrate surface roughness in the case of a flexible substrate. Subsequently, 50/100 nm thick titanium/aluminum contact was deposited by Temescal FC/BJD2000 electron beam evaporation bell jar deposition system as a gate metal electrode and patterned by liftoff using negative photoresist along with AlignerKarl Suss MA/BA6 aligner. The choice of the contact metals is to assure an ohmic contact with the films.

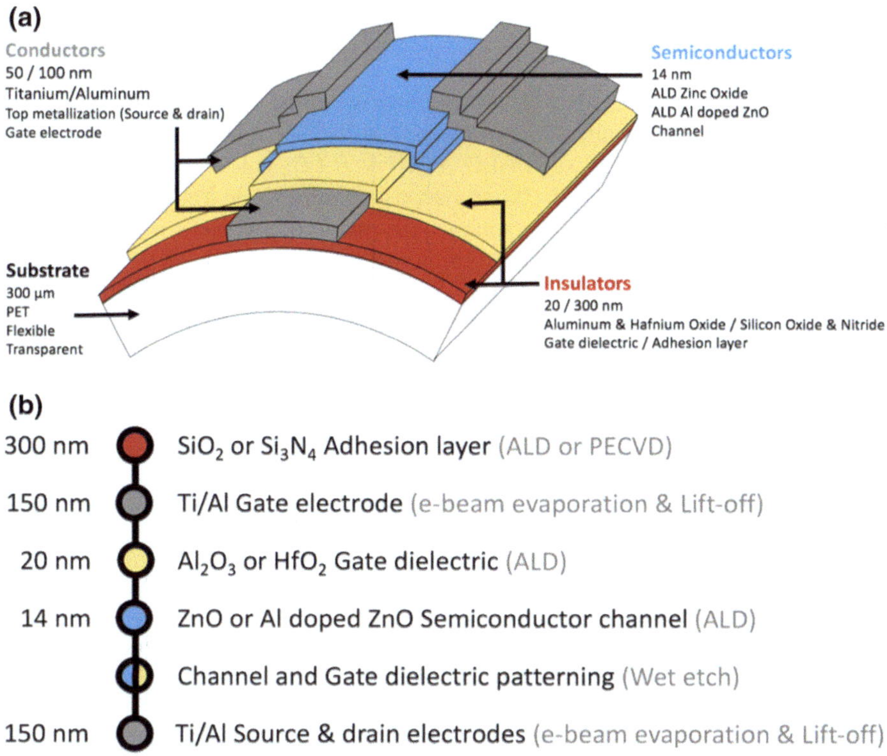

Fig. 5.1 **a** 3D schematic cross-sectional diagram of the fabricated TFT structure and **b** the process flow used to fabricate the bottom-gate type TFT shown in (**a**)

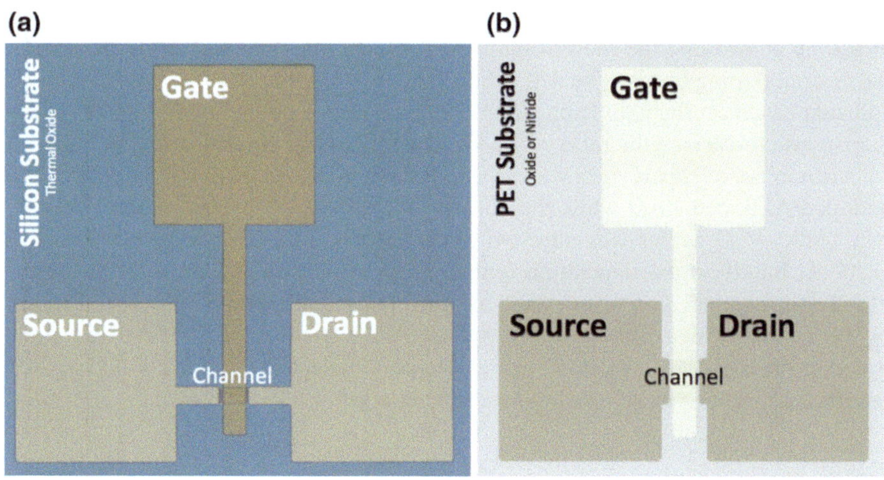

Fig. 5.2 An optical microscopic image of a single device structure on (**a**) silicon substrate and on (**b**) flexible PET plastic substrate

5.2.2 Gate Dielectric

Both ALD Al_2O_3 and HfO_2 films were investigated as the gate dielectric using TEMAHF and TMAl precursors, respectively, with H_2O vapor as an oxidant. Following the gate contact, a 20 nm of ALD Al_2O_3 or HfO_2 layer as the gate dielectric with a growth temperature of 150 °C was deposited. One cycle for deposition of the Al_2O_3 monolayer was designed as follows: the pulsing times were 0.3 s for TMAl, 3 s for N_2 purge, 0.2 s for H_2O, and 4 s for purge. One cycle of the deposited HfO_2 monolayer had the following pulsing sequence: a pulsing time of 0.1 s for TEMAHF, 3 s for N_2 purge, 0.2 s for H_2O, and 4 s for purge. The deposition rate was approximately 0.119 nm/cycle ALD Al_2O_3 films and 0.11 nm/cycle for the ALD HfO_2 films based on the data from Chap. 4.

5.2.3 Channel

As presented in the previous chapters, the most critical component of the TFT, the ALD-grown Al-doped ZnO semiconductor thin films were studied in advance to investigate the physical, electrical, and mechanical properties of the films. We prepared doped ZnO films with Al concentrations of 2, 3, 5, 8, and 10 at. % using DEZn and TMAl precursors with H_2O vapor as an oxidant. The deposition temperatures were varied over 60, 100, 150, 200, and 250 °C. One cycle for deposition of the ZnO monolayer was designed as follows: the pulsing times were 0.25 s for DEZ, 3 s for N_2 purge, 0.25 s for H_2O, and 4 s for purge. For the process of Al incorporation into the ZnO lattice, the pulsing times were 0.25 s for the TMA, 3 s for N_2 purge, 0.25 s for H_2O, and 4 s for purge. The composition of the Al-doped ZnO films was determined by controlling the ratio of number of cycles for ZnO and Al_2O_3 to achieve the desired doping levels. The deposition rate of approximately 0.198 nm/cycle was adjusted based on the data from Chap. 4 to take into account any marked variation in response to varying the ratio of Al dopants to ZnO compositions.

The number of supercycles was adjusted to end up with a 14 nm of the active layer undoped/Al-doped ZnO while the number of DEZn-H_2O cycles inserted between two TMAl-H_2O cycles was adjusted to vary the doping levels between 1 and 10 at. % Al based on the deposition temperatures which ranged from 60 to 250 °C. Finally, the channel area was defined by conventional lithography process, followed by wet etching with a diluted H_2SO_4 solution (H_2SO_4:H_2O $= 5 : 95$). The channel was deposited in the same ALD run as the gate dielectric to assure a high-quality interface and minimize defects and traps generation.

5.2.4 Source and Drain Contact

For these contacts, a film of 50/100 nm thick titanium/aluminum was deposited by e-beam evaporation and patterned by liftoff, following the same process used for the gate contact. After processing, the flexible substrate was easily demounted from the glass substrate and prepped for characterization.

5.2.5 Flexible Substrate

There are several potential flexible substrates that could be used for thin film devices including stainless steel foil, polymeric films, woven fabric, and ultra-thin glass. Besides being low cost and flexible, the ideal substrate should be lightweight, rugged, electrically insulating, transparent to the visible spectrum, impermeable to moisture, chemically and thermally stable and have a smooth surface. Though there is no single ideal flexible substrate that has all these advantages, a low overall processing/fabrication temperature opens up the options and provides more choices. In general, the maximum operation temperatures of the majority of flexible plastics are low, and the glass transition temperatures are in the range from 80 to 250 °C. This constraint will preclude many standard technologies that are traditionally used in the fabrication of Si-based devices, including thermal oxidation, diffusion, or epitaxy. Correspondingly, the choice of semiconductor materials for the flexible devices is very limited.

For this work, a commercially available flexible polyethylene terephthalate substrate (PET) plastics film with a 250 μm thickness is used. The PET substrate was selected as the substrate due to its excellent flexibility and mechanical tolerance. A photograph of the flexible Al-doped ZnO TFTs on the flexible substrate described here is shown in Fig. 5.3, and it shows that the Al-doped ZnO TFTs are flexible, transparent, and bendable.

The use of flexible substrates raised several challenges that had to be assessed and taken into account during the fabrication process. One important challenge is the integrity of thin films during both the fabrication and under the stress of flexing. It is important that the thin films have good adhesion to the substrate on which they are deposited. Both van der Waal forces (physisorption) and chemical interactions (chemisorption) contribute to this adhesion. In addition, adhesion depends on substrate morphology, nucleation processes, and inter-diffusion at the film–substrate interface. It is also important to note that virtually all thin films have internal stress (tensile or compressive) that develops during growth, which can lead to film failure even without external stress especially in case of poor adhesion. Stress can develop during thermal cycles, especially if there is a large mismatch in thermal expansion coefficients between the films and substrate.

Another challenge arises when trying to deposit a high-quality semiconductor and gate dielectric, namely the low-temperature cap imposed by the flexible substrate.

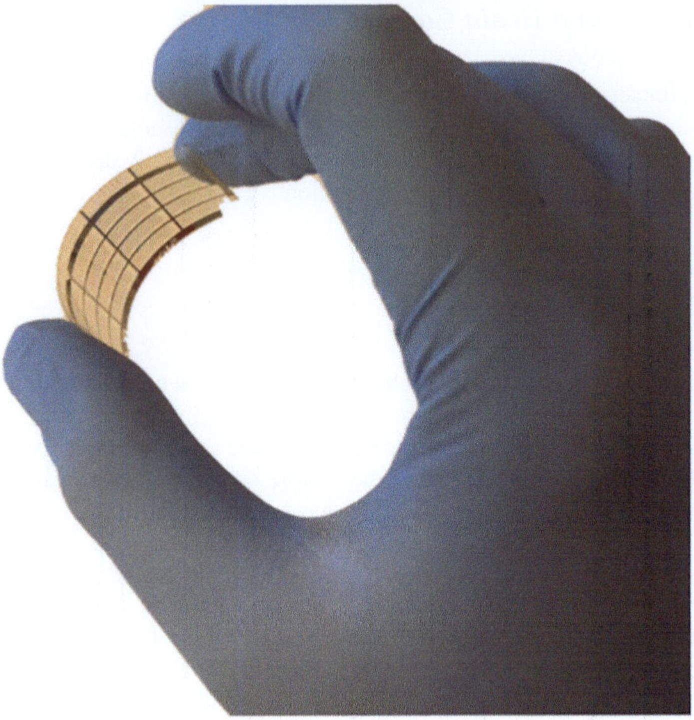

Fig. 5.3 A photograph of the Al-doped ZnO TFTs on a flexible substrate

In this case, ALD provided an ideal process candidate to solve these problems. Its thin film deposition is not only uniform but also conformal. This conformal growth is a necessity especially for multilayer devices on polymeric substrates. The flexible polyimide substrates used here had a melting point temperature around 274 °C and relatively small coefficient of thermal expansion (CTE) of 18×10^{-6} m/m/°C. The substrate's surface was characterized by an AFM, and the RMS surface roughness was found to be around 15 nm. There are many surface features appearing as white dots on the AFM image, representing significant local defects or damages in the substrate.

To grow a uniform and low defect density dielectric that is able to cover the rough surface of the flexible substrate, a typical approach is to deposit a planarization layer. This can be a thick dielectric ($\gtrsim$300 nm) to make the surface smoother for semiconductor and dielectric material growth. This is critical in view of the substrate surface roughness of 15 nm which is of the same order of magnitude as the gate dielectric (20 nm) and the channel semiconductor films (14 nm).

Although the two challenges discussed above may be the most important for the fabrication of a high-quality TFTs, other challenges can arise with flexible substrates such as alignment and thermal challenges. Alignment is challenging on a transparent flexible substrate, especially that ZnO is transparent itself. Thermal challenges impose current limitations in order to avoid damaging the flexible substrate. This is due to the flexible substrate's limited heat dissipation in comparison to Si substrate.

5.3 TFT Characterization

The electrical characteristics of the fabricated TFTs described above were conducted using a Signatone 1160 Manual Prober along with a Keithley 4200-SCS semiconductor parameter analyzer and Agilent B1505A curve tracer in a dark box at room temperature. This included output and transfer curves along with capacitance measurements. The plot of the source–drain current (I_{DS}) versus gate voltage (V_{GS}) transfer characteristics of a TFT from grown at 100 °C and has a ZnO channel without any intentional doping is shown in Fig. 5.4a.

Transfer characteristics were measured while sweeping V_{GS} from −15 to 15 V at a V_{DS} of 8 V with moderate integration time. Channel width and length of the measured TFTs were 60 and 40 μm, respectively. For these TFTs, a positive gate voltage required to induce the conducting channel with the channel conductivity increasing with the increase of the positive gate bias. A pinch-off and current saturation are clearly observed indicating a typical FET behavior. The subthreshold voltage (SS), which is the voltage required to increase the drain current by a decade, is given by:

$$SS = \frac{dV_{GS}}{dlog(I_{DS})} \tag{5.1}$$

which was 262 mV/decade for the TFT under analysis. The SS is given by the slope of the transfer curve (blue) in Fig. 5.4a. The field effect mobility μ_{sat} and the threshold voltage V_T were calculated from the slope of the square root of I_{DS} versus V_{GS} plot (red) shown in Fig. 5.4a by applying the following equations in the saturation region:

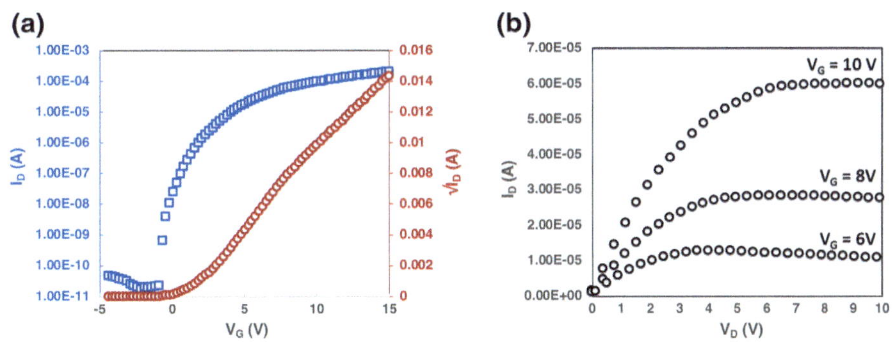

Fig. 5.4 Transfer and (**b**) output characteristics for a TFT grown at 100 °C with no intentional doping

$$I_{DS} = \left(C_{ox}\mu_{sat}\frac{W}{2L}\right)(V_{GS} - V_T)^2 \quad for \ V_{DS} > V_{GS} - V_T \tag{5.2}$$

where C_{ox} is the capacitance per unit area of the gate insulator which is around 350 nF/cm^2. W is the width of the channel of 60 μm, and L is channel length of 40 μm. The obtained μ_{sat} is 26.91 cm^2/Vs, and the V_T is 1.06 V. The magnitude obtained for the threshold voltage V_T is always directly proportional to the gate dielectric thickness. Therefore, these values could be easily adjusted by optimizing the gate dielectric thickness. The off current was shown to be very low of the order of 15 pA, making the I_{ON}/I_{OFF} current ratio about 10^7. Figure 5.4b shows the I_{DS} current curves as a function of the drain voltage V_{DS} for several values of gate voltage V_{GS} with clear saturation of the output characteristics as seen in the plot. Figure 5.4b also shows that the TFT has a good ohmic contact between the channel interface and the Al/Ti electrodes as no current crowding can be found at low V_{DS}. Besides the high value obtained for the saturation current, the device exhibits hard saturation indicating that the entire thickness of the ZnO channel layer is depleted. This kind of behavior is very desirable for most sensor circuit applications because TFT will exhibit large output impedance. High resistivity is important and useful for display and sensors than to other electronic applications.

5.3.1 Al Doping

TFTs fabricated using these films are expected to have their electronic behavior modulated by controlling Al concentrations. Figure 5.5 shows the transfer characteristics curves of the devices from Group 2 fabricated with Al-doped ZnO active channel layers incorporating 2, 3, 5, 8, and 10 at. % Al with growth temperature of 100 °C. Undoped ZnO transfer characteristics is shown again here as a reference. Channel width and length of the measured TFTs were 60 and 40 μm, respectively.

Figure 5.6 shows the parameters of the TFTs with W = 60 μm, L = 40 μm as extracted from log (I_{DS}) versus V_{GS} curves shown in Fig. 5.5. The TFTs have a field effect mobility ranging from 10 cm^2/Vs to 30 cm^2/Vs, a threshold voltage between −1 and 4 V, an average subthreshold slope of about 300 mV/decade and an average current on/off ratio of 10^7.

The above results demonstrate that semiconducting Al-doped ZnO thin films can be obtained by modulating Al concentration. The TFTs with 2 at. % Al doping ZnO exhibited the highest field effect mobility at a saturation region with μ_{sat} of 29.45 cm^2/Vs which is up from 26.91 cm^2/Vs at 0 at. % Al. The high value of the mobility μ_{sat} can be attributed to the high quality of the Al-doped ZnO layer as well as the good channel insulator interface. This in turn is mainly due to improved crystallinity and low oxygen vacancies and Zn interstitials with Al dopants working as donors. However, as the Al concentration increased, the μ_{sat} decreases to 16.48 and 8.92 cm^2/Vs for the 5 and 10 at. % Al devices, respectively.

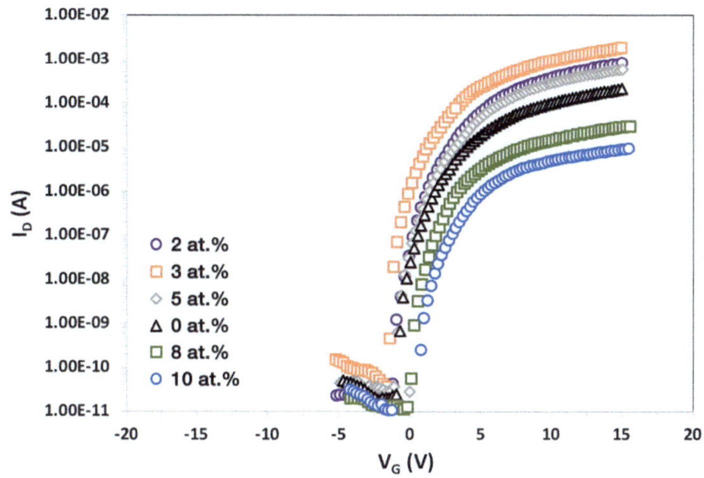

Fig. 5.5 Transfer characteristics (I_{DS} versus V_{GS}) of six TFTs from Group 2 grown at 100 °C with 0, 2, 3, 5, 8, and 10 at. % at Al doping

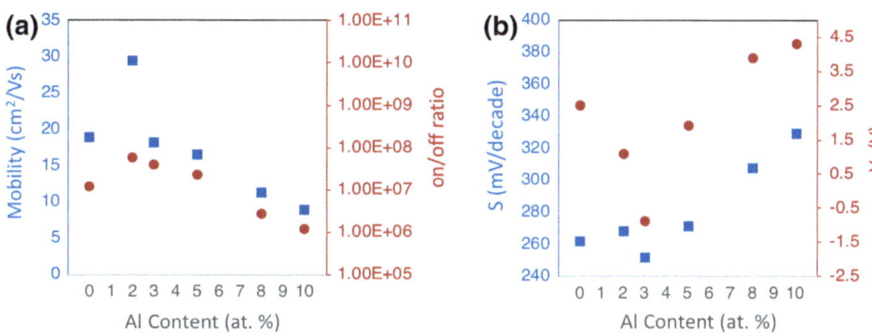

Fig. 5.6 **a** Subthreshold voltage swing SS (Blue) and threshold voltage (Red) and **b** Saturation mobility μ_{sat} (Blue) and on/off ratio (Red), extracted from the plots in Fig. 5.5

All the shown TFTs (except at 3 at. % Al) operate as n-channel enhancement mode devices with a positive gate voltage required to induce the conducting channel with the channel conductivity increasing with increasing the positive gate bias. Enhancement mode is preferable to depletion mode behavior and is very desirable for most sensor and display circuit applications because a gate voltage has to be applied to turn the transistor off which makes the circuits design easier and minimizes power dissipation in the system as a voltage is not required to switch off the transistor.

The turn-on voltage V_{ON} defined as the voltage at which the I_{DS} starts to switch from the off state to the on state was approximately -1.5 V for devices with 3 at. % Al. This was due to a higher degree of conductivity for that particular Al-doped ZnO channel layer. Conversely, V_{ON} values were observed to increase with higher Al concentrations as film conductivity decreases due to the reduction in carrier concentration as seen before in this chapter. In addition, a positive shift of V_T was observed also as the Al concentration increased from 3 to 10 at. % Al.

The ratio of on to off current increases with Al doping from 1.2×10^7 with no doping up to a maximum of 6×10^7 at 2 at. % Al. In contrast, the on/off ratio above the 8 at. % Al was severely degraded because of the reduction in on current. The low values of subthreshold swing (SS) which were estimated to vary from 252 mV/decade at 3 at. % Al to 329 mV/decade at 10 at. %. This low subthreshold swing exhibited by all devices suggests that the interfaces between the Al-doped ZnO active and Al_2O_3 gate insulator layers were well formed with relatively adequate properties for all doping concentrations.

Adding a small amount of Al doping (2–3 at.%) to the ZnO TFTs shown enhanced threshold voltage, mobility and on/off ratio than previously reported Sputtered ALD-grown ZnO TFTs on flexible and transparent substrates [3–5].

5.3.2 Growth Temperature

Previous chapters have also shown a strong correlation between growth temperature and electrical properties of the Al-doped ZnO films. Fabricating TFTs using these Al-doped ZnO films as active channel layers has shown similar trends. Figure 5.7 shows the transfer characteristics curves of the devices fabricated at various deposition temperatures ranging over 60, 100, 150, 200, and 250 °C for an active channel layer deposited with 2 at. % Al. Channel width and length of the measured TFTs were 60 and 40 μm, respectively.

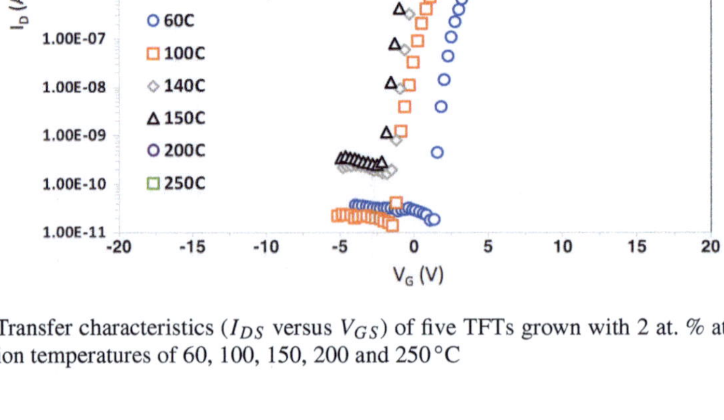

Fig. 5.7 Transfer characteristics (I_{DS} versus V_{GS}) of five TFTs grown with 2 at. % at Al doping at deposition temperatures of 60, 100, 150, 200 and 250 °C

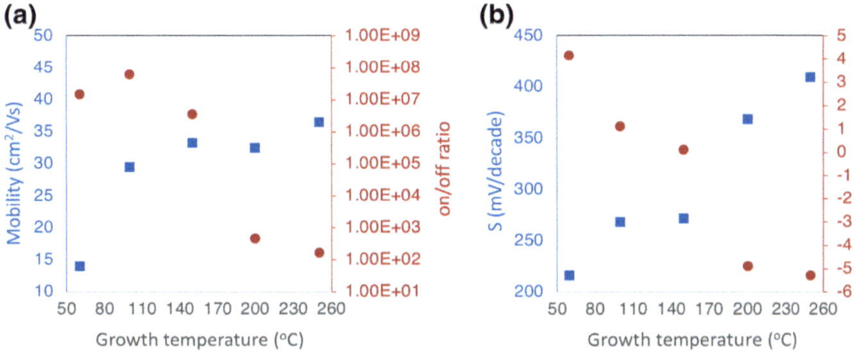

Fig. 5.8 **a** Subthreshold voltage swing SS (Blue) and threshold voltage (Red) and **b** Saturation mobility μ_{sat} (Blue) and on/off ratio (Red), extracted from the plots in Fig. 5.7

Figure 5.8 shows the TFT parameters extracted from log (I_{DS}) versus V_{GS} curves shown in Fig. 5.7. The mobility of these TFTs increases with higher deposition temperature as expected. The mobility of the TFTs deposited at 100 °C is more than twice higher than that of the Al-doped ZnO TFT deposited at 60 °C. This change in mobility corresponds precisely to the decrease in the mobility of the Al-doped ZnO thin film with decreasing deposition temperature. Also, an increase of V_T value with decreasing deposition temperature is observed. This positive shift can be explained by the large decrease in the carrier concentration with decreasing deposition temperature. The V_T value is susceptible to the total number of carriers in the active layer because a more negative gate bias is required to turn off the TFT by depleting the carriers from the channel in the case of a high carrier concentration. On the other hand, in the case of a low carrier concentration, a more positive gate voltage is necessary to achieve significant electron accumulation and conducting path formation in the front channel. Thus, the positive shift of V_T in the low temperature deposited Al-doped ZnO active layer TFT can be attributed to the large decrease in carrier concentration along with the large increase in carrier concentration at high temperatures which deteriorates the TFTs' on/off current ratio.

Increasing the deposition temperature tends to cause a negative threshold voltage shift and a stretched subthreshold slope. This is due to defects generation in ZnO thin film such as oxygen vacancy and zinc interstitial in addition to activation more Al dopant in the ZnO lattice, which causes increased conductivity in ZnO.

5.3.3 *Gate Dielectric*

In addition to the active channel layer semiconductor, dielectric materials play an important role in determining TFT performance. Due to the large bandgap and electron affinity (around 4.5 eV) of Al-doped ZnO, an insulator with a large bandgap

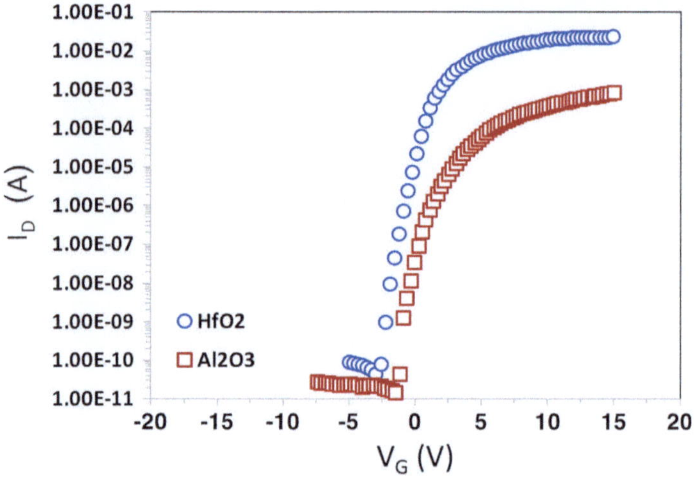

Fig. 5.9 Transfer characteristics (I_{DS} versus V_{GS}) of two TFTs grown with 2 at. % at Al doping at deposition temperatures of 100 °C. One (Red squares) has Al_2O_3 as the gate dielectric and the other (Blue circles) HfO_2

(>6 eV) is needed to provide barriers to both the valence and conduction bands. These requirements largely rule out most organic dielectrics and a large number of inorganic dielectrics. Therefore, Al_2O_3 was chosen in this work due its large bandgap (8.7 eV) and low-temperature deposition although it was a difficult material to dry etch and even wet etch. Figure 5.9 depicts the transfer characteristics (I_{DS} vs. V_{GS}) of two TFTs grown with 2 at. % at Al doping at a deposition temperatures of 100 °C. One (Red squares) has Al_2O_3 as the gate dielectric and the other (Blue circles) HfO_2.

TFTs with HfO_2 as a gate dielectric show lower threshold voltages, higher mobility and as expected better on/off current ratios. However, the gate leakage current was much higher than TFTs with Al_2O_3. The reason for this is not entirely clear but can be assumed to be related to a larger interface state density at the dielectric/semiconductor (HfO_2/ZnO) interface or enhanced charge injection into the Al_2O_3 layer. This difference in performance between TFTs fabricated using HfO_2 and Al_2O_3 is probably linked to a relatively poorer interface with the HfO_2 than with Al_2O_3 and the very thin active layers in these devices (around 14 nm) could make it worse. Additionally, deposited dielectrics charge trapping, particularly at low temperature (<200 °C), is a well-known phenomenon that has significant impact. Another aspect is that devices with HfO_2 and Al_2O_3 were not passivated. Therefore, any charges on the channel back interface can dramatically affect the band bending in the accumulation region.

Figure 5.10 shows the gate capacitance as the gate voltage is swept from −4 V to 8 V for the two TFTs shown in Fig. 5.9. The measurement was conducted using Signatone 1160 Manuel Prober along with Keithley 4200-SCS semiconductor parameter analyzer with a 30 mV small signal and at 500 kHz.

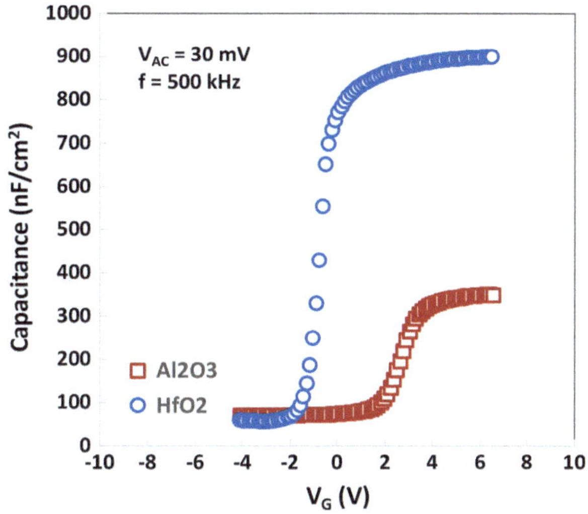

Fig. 5.10 Gate capacitance as a function of gate voltage for the two TFTs shown in Fig. 5.9 with (Red squares) Al_2O_3 or (Blue circles) HfO_2 as the gate dielectric

5.4 Conclusions

This chapter covered the thin-film transistors process development methodology and demonstrated ALD ZnO TFT grown on flexible substrate, optimized using Al doping. A relationship between the Al doping and deposition parameters was established. Adding a small amount of Al doping (2–3 at.%) to the ZnO TFTs shown enhanced threshold voltage, mobility and on/off ratio than previously reported Sputtered ALD-grown ZnO TFTs on flexible and transparent substrates.

Acknowledgements This work has been supported by the Semiconductor Research Corporation (SRC) under the Abu Dhabi SRC Center of Excellence on Energy-Efficient Electronic Systems (ACE^4S), Contract 2013 HJ2440, with funding from the Mubadala Development Company, Abu Dhabi, UAE.

References

1. A. Rizk, I. Saadat, ALD Al-doped ZnO thin film as semiconductor and piezoelectric material: process synthesis, in *The IoT Physical Layer*, ed. by M. Elfadel, M. Ismail (Springer, Berlin, 2018), pp. 23–46
2. A. Rizk, I. Saadat, ALD Al-doped ZnO thin film as semiconductor and piezoelectric material: characterization, in *The IoT Physical Layer*, ed. by M. Elfadel, M. Ismail (Springer, Berlin, 2018), pp. 47–68
3. E. Kim, J. Bak, J. Choi, S. Yoon, Effect of Al concentration on Al-doped ZnO channels fabricated by atomic-layer deposition for top-gate oxide thin-film transistor applications. J. Vac. Sci. Technol. B **32**(2014), 41202 (2014)

4. E. Kim, J. Bak, J. Choi, S. Yoon, Thickness and composition effects of Al-doped ZnO channels prepared by atomic layer deposition on the device behaviors of oxide thin-film transistors. **26502**(2012), 31210
5. K. Jang, H. Park, S. Jung, N. Van Duy, Y. Kim, J. Cho, H. Choi, T. Kwon, W. Lee, D. Gong, S. Park, J. Yi, D. Kim, H. Kim, Optical and electrical properties of 2wt.% Al_2O_3-doped ZnO films and characteristics of Al-doped ZnO thin-film transistors with ultra-thin gate insulators. Thin Solid Films **518**(10), 2808–2811 (2010)

Chapter 6
Lab-on-Chip Silicon Photonic Sensor

Mahmoud S. Rasras and Osama Al Mrayat

6.1 Introduction

Optical sensors are a low-cost solution to detect the presence and concentration of a target analyte. They can achieve a high metering accuracy in a compact chip size. In the literature, various structures have been published to improve the sensing precision and device compactness. Examples include planar waveguide structures [1], micro-ring resonators [2], Mach–Zehnder interferometers (MZIs) [3], photonic sensors [4], surface plasmon resonance (SPR) [5], and disk resonators [6–10]. In particular, micro-ring resonators (MRRs) provide the advantages of compactness, high-sensitivity, and the ability to develop multiplexed sensing platforms. In essence, the shift in resonance frequency corresponds to a change in RI of the region probed by the resonant mode. As a result, the subsequent sensing signal can identify target analyte by comparing it with a reference and determine its concentration [11, 12]. However, the conventional MRR sensing structures have a restrictive free spectral range (FSR) due to the limited minimum bending radius allowed in the corresponding photonic platform. Larger FSR requires smaller effective ring length, which degrades the sensitivity of the sensor. In other words, the magnitude of the spectral shift is small due to a change in refractive index, which makes the detection of small concentrations a challenging process. To overcome such limitation, a cascaded ring resonator in a Vernier configuration has recently been proposed. It improves the sensor performance in terms of high tracking sensitivity and low limit of detection (LOD) [13–16]. In the context of refractive index (RI) sensors, LOD refers to the smallest refractive index change the sensor can detect. To build a portable

M. S. Rasras (✉)
Engineering Division, New York University Abu Dhabi, Abu Dhabi, UAE
e-mail: mrasras@nyu.edu

O. Al Mrayat
Masdar Institute at Khalifa University, Abu Dhabi, UAE
e-mail: oalmrayat@masdar.ac.ae

Ibrahim (Abe) M. Elfadel and M. Ismail (eds.), *The IoT Physical Layer*,
https://doi.org/10.1007/978-3-319-93100-5_6

and cost-effective photonic sensor, an on-chip optical spectrometer analyzer (OSA) is also needed to measure the wavelength response. Small analyte concentrations demand higher-resolution spectrometer, which in turn increase the size of the sensor. The trade-off between device size and sensitivity is the main design challenge. In this chapter, compact, high-sensitive, integrated photonic sensor is presented. The motivation is to quantify the existence of gas molecules like carbon monoxide. The sensor is constructed from two ring resonators coupled to an on-chip low-resolution spectrum analyzer. The latter design is based on a flat-top wide-channel planner echelle grating (PEG) de-multiplexer. The novelty is the use of photonic integrated components with a relatively low spectral resolution and high fabrication tolerance to achieve a high detection sensitivity. Such approach reduces the production cost of the device. The proposed sensor detects small incremental changes in the effective index of the sensor ring in the presence of a target analyte. This change is then transmitted to a photodetectors (PDs) integrated at each output of the de-multiplexer. Only one PD (or channel) will be illuminated at a time, where the corresponding de-multiplexer channel number is proportional to the magnitude shift in the effective index contrast. As a result, the proposed sensor exhibits a binary-like response, where its sensitivity is proportional to the de-multiplexer channel spacing. The limit of detection is determined by the de-multiplexer and the free spectral range (FSR) of the Vernier rings. A very preliminary version of the results presented in this chapter first appeared in [17].

6.2 Operating Principles

The layout of the proposed photonic sensor chip is illustrated in Fig. 6.1a. It consists of two micro-ring resonators (MRRs) connected in a Vernier configuration and coupled to a wide-channel de-multiplexer. The optical power at the output of each de-multiplexer is monitored by on-chip integrated photodetectors, similar to the one of [18]. The first ring acts as a filter and is covered with silica cladding. The free spectral range (FSR) is designed such that it equals the spacing between the de-multiplexer channels. In addition, its resonance frequency is aligned with the center of the de-multiplexer transmission band as seen in Fig. 6.1b.

As for the second ring, it acts a sensor with its cladding being fully etched away. This enables the circulating optical mode to repeatedly interact with the targeted analyte. The sensor ring has a FSR slightly smaller than the filter ring as shown in Fig. 6.2a. Referring to Fig. 6.1b, its resonance frequency at channel zero ($ch = 0$) is perfectly aligned with that of the corresponding filter ring at a desired operational wavelength, λ_o ($=1.53\,\mu$m). Using a broadband light source, such as an amplified spontaneous emission (ASE), the sensor (vernier rings and de-multiplexer) will pass the mth de-multiplexer channel that has a wavelength, matching the vernier resonance frequency. Subsequently, a photocurrent will be detected at that particular channel.

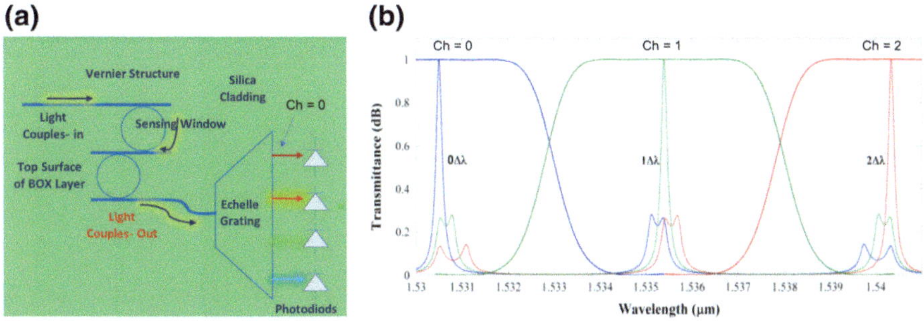

Fig. 6.1 **a** Schematic of cascaded micro-ring resonators based on the Vernier effect. **b** Vernier transmission peak aligned with the center of the de-multiplexer channel

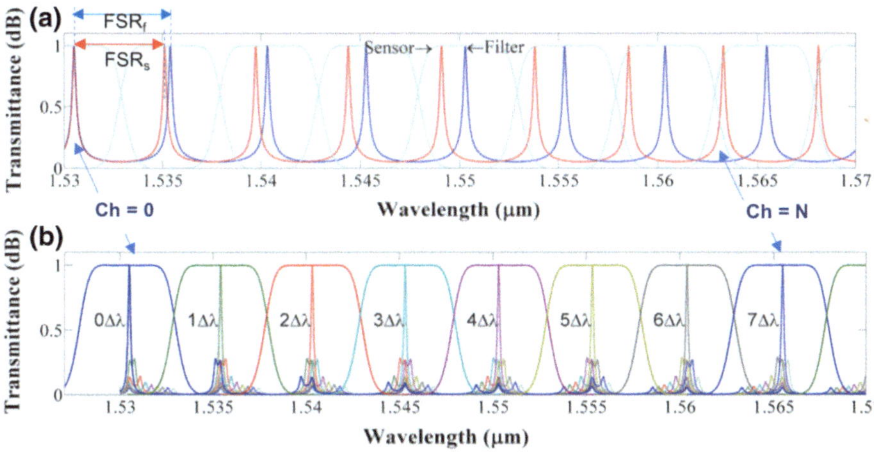

Fig. 6.2 **a** Spectrum of filter MRR, sensor MRR. **b** Vernier spectrum shown for different refractive index changes

6.2.1 Detection Sensitivity

In the presence of a targeted analyte, the output channel index m will change. Note that an increase in the integer m means the detection of higher concentration. This is due to a change of the effective index, Δn_{eff}, which causes a spectral shift in the resonance spectrum of the sensor ring. The corresponding wavelength shift ($\Delta\lambda_{Analyte}$) of the sensor ring is defined as:

$$\Delta\lambda_{Analyte} = m\Delta\lambda + \delta\lambda, \quad where \quad m = 0, 1, 2, \ldots, N-1 \tag{6.1}$$

The variable N is the number of the de-multiplexer output channels, and $\Delta\lambda$ is the required sensor ring spectral shift needed to move the Vernier resonance wavelength by one channel. In addition, $\delta\lambda$ is a fractional spectral shift such that $(0 < \delta\lambda < \Delta\lambda)$.

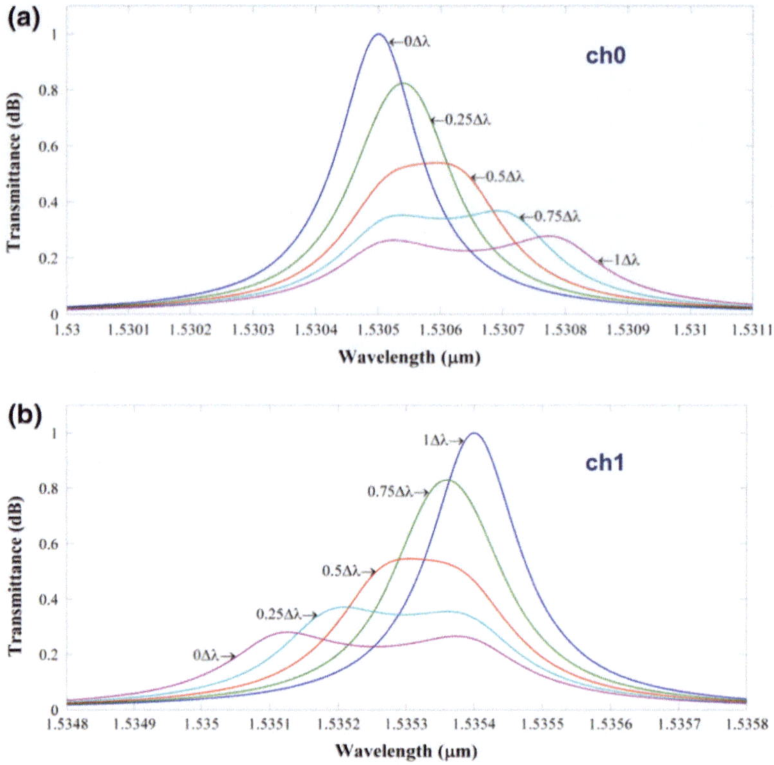

Fig. 6.3 Spectral shift of two adjacent channels induced by various gradual changes of effective index. **a** Spectral response of channel zero. **b** Spectral response of channel one

Referring to (Eq. 6.1), if an analyte is present and the maximum transmittance moves to the m-th channel, the total spectral shift is $m\Delta\lambda$. This one-to-one correspondence between the de-multiplexer channel number and the shift index provides a simple detection mechanism. The corresponding effective index is:

$$\Delta n_{eff} = n_{eff}\frac{\Delta\lambda_{Analyte}}{\lambda o} = n_{eff}\frac{m\Delta\lambda}{\lambda o} + n_{eff}\frac{\delta\lambda}{\lambda o} = \Delta n' + \delta n \qquad (6.2)$$

Note that n_{eff} is the effective refractive index of the circulating fundamental mode in the sensor ring.

Enhancing the Detection Sensitivity

The fractional spectral shift, $\delta\lambda$, of (Eq. 6.1) is likely to be nonzero in actual operation. The resulting photocurrent will then be split between two neighboring channels, causing a reduction in the peak magnitude response of the impacted de-multiplexer. Consequently, the actual concentration of the targeted analyte cannot be accurately detected. For example, assume an analyte concentration generating a wavelength shift in between $ch = 0$ and $ch = 1$, the corresponding photocurrents generated for various intermediate analyte concentrations are illustrated in Fig. 6.3a, b.

We propose to implement a thermo-optic phase shifter at the sensor ring to estimate $\delta\lambda$ of the de-multiplexer channel with the highest peak magnitude. Here, $\delta\lambda$ is equivalent to the resultant error of the resonant frequency determined by the thermo-optic calibrator. The operating principle of the proposed sensor is summarized in Fig. 6.4. For instance, suppose that the concentration of the analyte causes a fractional shift of $\delta\lambda = 0.75\Delta\lambda$ with $m = 0$ in the example shown in Fig. 6.3. By using a thermo-optic phase shifter, the sensor ring resonance wavelength can be shifted by $0.25\Delta\lambda$, aligning the Vernier rings perfectly at channel $m = 1$. The wavelength shift of the sensor ring due to the analyte in (Eq. 6.1) can be reformulated as:

$$\Delta\lambda_{Analyte} = m\Delta\lambda - \delta\lambda_{Thermal} \quad (6.3)$$

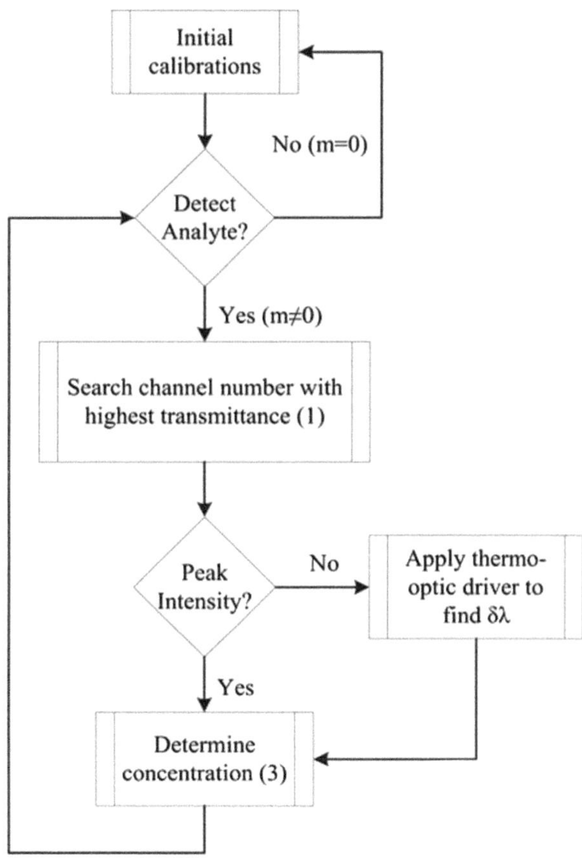

Fig. 6.4 Flowchart of the designed sensor operation

Its corresponding effective index change is computed by:

$$\Delta n_{eff} = n_{eff}\frac{m\Delta\lambda}{\lambda_o} - n_{eff}\frac{\delta\lambda}{\lambda_o} = n_{eff}\frac{m\Delta\lambda}{\lambda_o} - \frac{FSR_s}{\lambda_o}n_{eff}\frac{\Delta p}{p_{2\pi}} \tag{6.4}$$

$$\Delta n_{eff} = \Delta n + \delta n \tag{6.5}$$

where FSR_s is the FSR of the sensor ring, and Δp is the applied phase shifter power. The term $p_{2\pi}$ represents the power required to shift the resonance peak by one FSR. The second term in the right-hand side of (Eq. 6.4) and (Eq. 6.5) refers to the effective index change needed to shift the Vernier resonance peak to the next closest de-multiplexer channel.

In silicon, the thermo-optic coefficient is approximately about 1.8×10^{-4}, and the shift of the resonance frequency with the temperature is 12.4 GHz/°C [19]. Hence, the power of the thermo-optic phase shifter is in the range of mW. Furthermore, the range of Δp is expected to be $< \frac{\Delta\lambda}{FSR_s} p_{2\pi}$ ($\sim$max 0.1 mW). A control accuracy of the resonance frequency to within 0.125 GHz can be achieved using a high-precision thermo-optic driver [20], which corresponds to $\sim 2pm$ wavelength shift. For example, assume a silicon ring with $FSR = 5$ nm, n_{eff} is about 2.4 (transverse electric mode), $\lambda_0 = 1.55\,\mu$m, and $\Delta p/p_{2\pi}$ is swept in coarse steps of $0.05\,\text{mW}/2\pi$. The respective index change will be $\delta n = 80 \times 10^{-6}$ RIU, which defines the resolution of the designed sensor. Note that the search for the maximum transmission can be performed heuristically without using measurements from the spectrum. Of course, based on the application requirements, a more precise thermo-optic calibrator can be implemented to meet the desired resolution. However, the channel spacing needs to be factored to prevent channel cross-talk interference.

Limit of Detection

The limit of detection (LOD) is dependent on the channel spacing of the ring resonator. The LOD of the sensor can be defined as the smallest change in the effective refractive index (Δn_{eff}) that results in the switching the Vernier resonance peak by one channel spacing. Referring to (Eq. 6.5), LOD is:

$$\begin{aligned} LOD = \Delta n_{eff} &= n_{eff}((FSR_F - FSR_S)/\lambda_o) - n_{eff}\delta\lambda/\lambda_o \\ &= n_{eff}\Delta FSR/\lambda_o - n_{eff}\Delta p/p_{2\pi} \end{aligned} \tag{6.6}$$

The term FSR_F is the FSR of the filter ring. Therefore, LOD is proportional to the FSR difference (ΔFSR) of the Vernier cascaded-coupled rings. To detect small analyte concentrations, it is desirable to design both resonators with a small ΔFSR. However, this will increase the cross-talk interferences between channels as illustrated in Fig. 6.3. To achieve a reasonable cross talk of less than 25% (6 dB extinction), we chose a ΔFSR to be at least twice the maximum of the full width at half maximum ($FWHM$) of the sensor or filter rings. Note the $FWHM$ is calculated by:

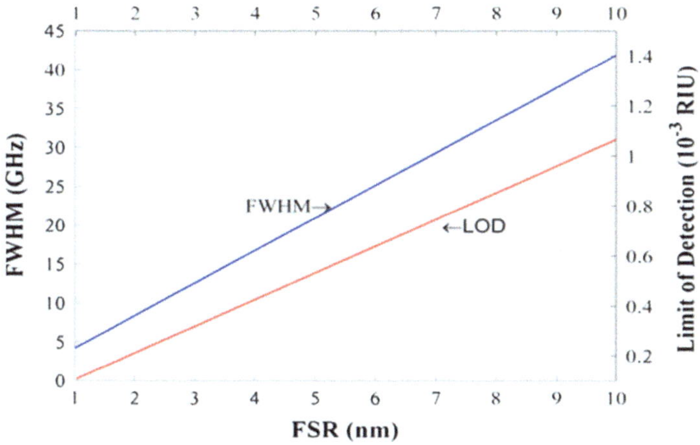

Fig. 6.5 Calculated device $FWHM$ and LOD for different choices of the ring $FSRs$

$$FWHM = \frac{(1 - \gamma\sqrt{\kappa_1\kappa_2})\lambda_0^2}{\pi n_g L_r \sqrt{\gamma\sqrt{\kappa_1\kappa_2}}} \tag{6.7}$$

where $\gamma = 10^{\frac{-\alpha L}{20}}$ and α is the waveguide attenuation coefficient. κ_1 and κ_2 are the filter and sensor ring coupling rations, respectively.

According to simulations shown in Fig. 6.5, the FWHM and LOD improve for rings with smaller $FSRs$. Although increasing the number of channels in the de-multiplexer can achieve a small FSR, it increases the size of the device. Therefore, there is trade-off between the detection limit and number of channels. Another important consideration is the round-trip loss (RTL) of the resonator unit. The RTL reduces the ring's quality factor (Q) and hence increases its $FWHM$. Referring to Fig. 6.6a and Table 6.1, a larger RTL leads to a greater degradation in LOD. Note the LOD is expressed in terms of the ring's normalized transmittance (γ). Under a typical scenario, an RTL of 0.1 dB ($\gamma = 0.98$) can increase the LOD by a factor of 1.05 and a factor of 1.4 for RTL = 0.72 dB. These results are obtained using ring's coupling coefficients of $\kappa_1 = \kappa_2 = 0.1$. Note that these coefficients need to be carefully selected when determining the parameters of the resonators. Meanwhile, Fig. 6.6b summarizes the impact of the coupling strength, $\kappa_r = \kappa_1 = \kappa_2$, on the Vernier peak transmission loss calculated at $\gamma = 0.98$. For small values of κ_r (i.e., small $FWHM$), the RTL of the rings can increase the insertion loss of the device by up to 1.5 dB. In this case, the value of the LOD remains small when the coupling range is [0.06, 0.1]. This demonstrates the ability of the designed sensor to achieve high sensitivity while using a low-resolution spectrum analyzer.

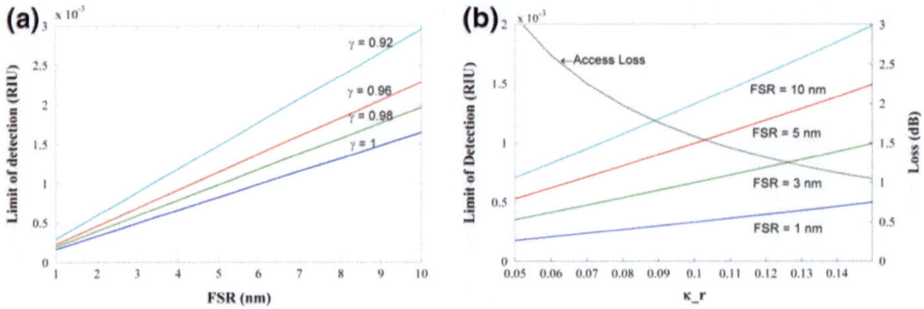

Fig. 6.6 **a** Impact of the RTL on the LOD shown for different ring's FSRs calculated for ring coupling coefficient of $\kappa_1 = \kappa_2 = 0.1$. **b** Impact of the ring's coupling ratio on the LOD shown for different ring's FSRs

Table 6.1 Summary of LOD subjected to various RTLs at different FSRs

FSR (nm)	γ	$LOD(10^{-6}$ RIU)
1	1.0	160
	0.92	300
3	1.0	490
	0.92	890
5	1.0	820
	0.92	1480
10	1	1650
	0.92	2960

6.2.2 *Spectrum Analyzer Integration*

Diffraction gratings are used in a wide variety of applications, including spectrometers [21, 22], external cavities for multi-wavelength LASER sources [23–25], and de-multiplexers for wavelength division multiplexing (WDM) devices in fiber optical telecommunications [26]. In the context of realizing on-chip spectrometer, there are two choices based on integrated diffraction gratings: an arrayed waveguide grating (AWG) and planar echelle grating (PEG). The output of the diffraction-based (e.g., PEG, AWG) spectrometer consists of an array of waveguides with each one of them designed to pick up a diffracted light beam at a specific wavelength and guide it to a separate photodetector. Each photodetector converts the light-beam intensity at a particular wavelength into its commensurable electrical current signal, the enabling interface of the optical sensor's outputs with an ad hoc electronic readout integrated circuit (ROIC) for processing various spectral shifts. Both PEG and AWG devices are used to perform wavelength (de)multiplexing, and their principle of operation is based on multipath interference of incident light. Furthermore, the PEG is considered a reflective grating device; the light is diffracted and defocused, while it propagates in its free propagation region (FPR), and the phase delays of diffracted light modes

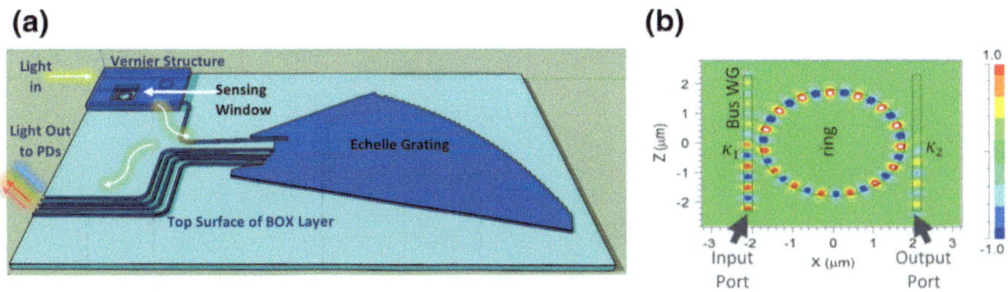

Fig. 6.7 Overview of the proposed Vernier rings integrated with a PEG de-multiplexer. **b** Simulated 3D full-wave response of the ring

at different wavelengths are contributed by the grating facets. On the other hand, AWG is considered as a transmissive device where the incoming light is diffracted in the input star coupler, propagates through an array of equally spaced waveguides contributing to different phase delays for different wavelengths, and defocused at spatially distributed apertures via an output star coupler [27]. Thus, it can be deduced that PEG is a more compact device than AWG. This is attributed to the fact that in PEG devices, a single slab region (a.k.a., FPR) is used for both diffraction and defocusing, reducing the physical size of the PEG by approximately less than 50% with respect to the AWG; this is typically the case when the PEG has a large FSR.

A planar echelle grating (PEG) is an essential building block for realizing a low-cost on-chip optical spectrum analyzer (OSA) in integrated optical sensors. It is capable of measuring spectral shifts caused by effective index changes induced by variations in gas concentration. To this extent, the drop/output channel of the optical sensor is integrated into the PEG input waveguide as shown in Fig. 6.7a. A simulated 3D full-wave response of the resonator is shown in Fig. 6.7b.

Determining the Planar Echelle Grating Design Parameters

A PEG consists of an input waveguide, an array of output waveguides, a slab region, and a set of etched echelle grating facets. Unlike classical blazed gratings, the centers of the grating facets are aligned along a curvature instead of a flatline which gives rise to the name 'Echelle.' Because of the curved feature of the gratings, the PEG is not only capable of spatially dispersing the different wavelength components of the incident light, but it is also able to defocus various diffracted modes with different wavelengths at specific points along the PEG's focal field. The number of output apertures are distributed as shown in Fig. 6.8. The equally spaced output waveguides are spatially distributed along the focal plane, and each one of them is terminated with a germanium (Ge) photodetector for optical–electrical conversion to interface the optical sensor with a readout circuit [19].

The centers of the grating facets are positioned along a curved line using one of several techniques, including recursive approach [28], one- and two-stigmatic point(s) configurations [29], and the Rowland configuration [30–32]. This Rowland configuration has been widely used in the literature for its simplicity and reasonable PEG's performance.

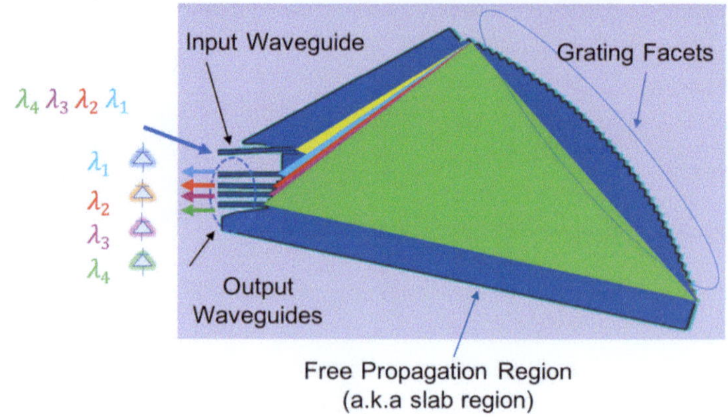

Fig. 6.8 Schematic layout of echelle grating

According to the Rowland configuration, the input and output waveguides are terminated tangentially along the edge of the Rowland circle with radius '*R*' as seen in Fig. 6.9. The input waveguide is positioned at point '*I*', and the array of the output waveguides is equally spaced with respect to their central point 'O', also known as the blazing point. On the other hand, the grating facets' centers are positioned along

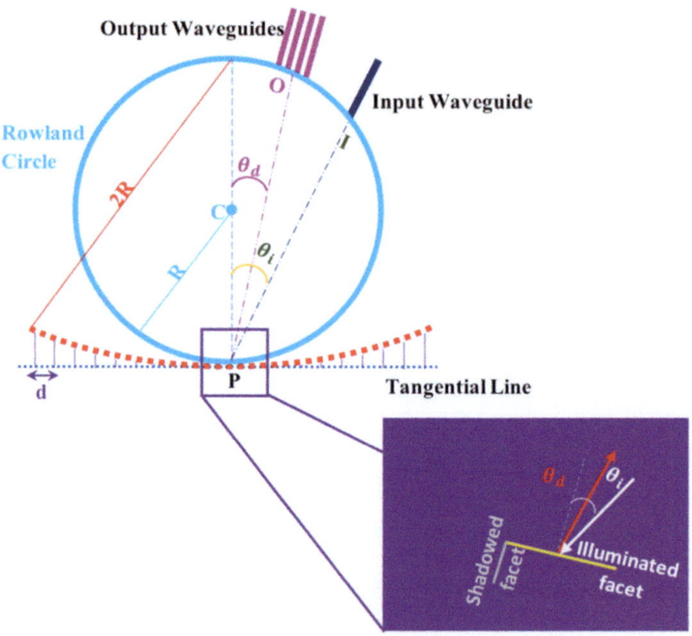

Fig. 6.9 Layout of PEG based on the Rowland configuration

the grating circle, which has a radius twice that of the Rowland circle. The grating and Rowland circles are tangential to each other at the PEG's pole 'P', which represents the central point of the echelle grating. The projections of individual grating facet centers to the tangent line at 'P' are equally spaced by the grating period 'd'.

The incoming light beam emanating at wavelength (λ_o) from input waveguide at point 'I' diffracts, while it propagates in the free propagation region (FPR) toward echelle gratings. Upon reflection off the grating facets, diffracted modes interfere constructively with each other, while being focused at one of different points equally spaced along the Rowland circle and centered at 'O'. The condition for constructive interference is governed by the grating equation,

$$d(sin\theta_i + sin\theta_d) = m\frac{\lambda_o}{n_{eff}} \tag{6.8}$$

where θ_i and θ_d are the angles between grating facet normal at 'P' on the one hand, and the incident and diffracted light beams, respectively, on the other; d is the grating period, λ_o is the free-space central wavelength, m is an integer representing diffraction order, and n_{eff} is the effective refractive index of the slab mode corresponding to λ_o.

Grating Design Parameters

Two key parameters that determine the PEG performance are dispersion (linear and angular) and the free spectral range (FSR). The angular dispersion $((d\theta_d)/d\lambda)$ is derived from grating (Eq. 6.8) by differentiating the diffraction angle θ_d with respect to the incident light wavelength [32]

$$\frac{d\theta_d}{d\lambda} = \frac{1}{cos\theta_d}\frac{m}{d}\frac{n_g}{n_{eff}^2} \tag{6.9}$$

where n_g is the group index of the slab mode and is defined as

$$n_g = n_{eff}(\lambda_0) - \lambda\frac{\partial n_{eff}}{\partial\lambda}|_{(\lambda=\lambda_o)} \tag{6.10}$$

However, the angular dispersion is rarely used in the context of the PEG design; instead, the linear dispersion LD is more useful. The linear dispersion describes the spatial shift (in μm) of the reflected image along the PEG's focal plane in response to the incident light spectral shift (in nm). It relates to the output waveguide pitch (ΔS), which is the spatial separation between the centers of adjacent output waveguides. It is needed to resolve a specific channel spacing ($\Delta\lambda$), i.e.,

$$\begin{aligned} LD &= \frac{Spatial\ Separation\ between\ Centers\ of\ Adjacent\ Output\ Apertures}{PEG\ Channel\ Spacing} \\ &= \frac{\Delta S}{\Delta\lambda} \end{aligned} \tag{6.11}$$

The linear dispersion can be derived from the angular dispersion by multiplying the 2nd expression term given in (Eq. 6.9) by the Rowland circle's diameter

$$LD = 2R\frac{d\theta_d}{d\lambda} = 2R\frac{1}{cos\theta_d}\frac{m}{d}\frac{n_g}{n_{eff}^2} \tag{6.12}$$

Substituting the grating equation 6.8 into 6.12 yields an analytical expression for the *Linear Dispersion*,

$$LD = \frac{\Delta S}{\Delta\lambda} = \frac{2R}{\lambda}\frac{(sin\theta_i + sin\theta_d)}{cos\theta_d}\frac{n_g}{n_{eff}} \tag{6.13}$$

It is desirable to design a PEG with a high LD to realize a high-resolution spectrometer. This can be achieved by further spacing the output waveguides apart from each other, i.e., by increasing the output waveguide pitch (ΔS), and/or reducing the PEG channel spacing ($\Delta\lambda$). However, it is obvious from Eq. 6.12 that LD is linearly proportional to *Rowland circle's* size, therefore the PEG size. Accordingly, there is a trade-off between the PEG resolving power and its size. The other important parameter used in quantifying PEG performance is the free spectral range (FSR), which describes the wavelength span between different diffraction orders. The FSR can be derived from the grating equation. According to Eq. 6.8, the diffracted mode $m + 1$ at wavelength λ must have the same diffraction angle given by the diffracted mode m at wavelength $\lambda + \Delta\lambda$, and therefore, the FSR is formulated as

$$FSR = \frac{\frac{\lambda_o}{m}}{(1 - \frac{m+1}{m}(1 - \frac{n_g}{n_{eff}}))} \tag{6.14}$$

Table 6.2 highlights the parametric values required to evaluate both LD and FSR of PEG [33]. Those values correspond to nanophotonic SOI platform with a slab thickness of 220 nm at a free-space center wavelength of $\lambda_o = 1.55\,\mu$m.

The PEG design procedure involves defining the desired channel spacing $\Delta\lambda$, the output waveguide pitch/spacing ΔS, the FSR, bandwidth, the central wavelength λ_o,

Table 6.2 Properties of a TE polarization slab mode for nanophotonic SOI platform with a 220 nm slab thickness [33]

Feature	Numerical value
Feature numerical value free-space central wavelength (λ_o)	1.55 μm
Slab mode effective index (n_{eff})	2.83
Ratio of group index to effective index ($\frac{n_g}{n_{eff}}$)	1.31
Rate of effective index change per wavelength shift ($\frac{dn_{eff}}{d\lambda}$)	−0.57 μ m^{-1}

and the desired number of grating facets. The PEG channel spacing $\Delta\lambda$ is determined by the application, and in the context of this work, it equals the FSR of the filter ring ($FSR_f = 5\,\text{nm}$). Although the PEG will be on this low resolution, it will have a small footprint. In order to minimize mode coupling between the adjacent output apertures (i.e., cross talk) as well as having a relatively large resolving power, the output waveguide pitch ΔS must be made sufficiently large, which would increase the radius of the Rowland circle, so a trade-off point must be found. The output waveguide pitch ΔS is selected to be $1.5\,\mu\text{m}$. The FSR has to be greater than the wavelength operation range (i.e., 40 nm), which is defined by the 3 dB bandwidth of the broadband light source. On the other hand, the central operating wavelength λ_o of the PEG is the same central wavelength of the Vernier-based optical sensor's spectrum at its output port, which is also the same as the light source central wavelength of $\lambda_o = 1530\,\text{nm}$.

The number of output waveguides (i.e., PEG channels) is set to equal the number of the spectral shifts within the operational bandwidth. This is in order to match the resonant tones in the filter micro-ring spectrum. At those resonant wavelengths, the output spectrum of the optical sensor comprises major tones. Therefore, separate output waveguides are needed to capture the diffracted light modes at different resonant wavelengths. The latter corresponds to different spectral shifts. Each diffracted light mode at each resonant wavelength is then guided to a unique photodetector. In our design, the spectrum at the output port comprises eight major tones within a 40 nm bandwidth, which implies the need for eight corresponding output waveguides as shown in Fig. 6.10. These waveguides are spatially distributed about the blazing point 'O'.

In order to position the input and output waveguides tangentially at the edge of Rowland circle, the incident and diffraction angles (θ_i and θ_d) have to yield a reasonably compact device with an acceptable cross-talk level. Furthermore, to visualize the effect of the incident angle on the PEG size (i.e., the Rowland circle radius), the

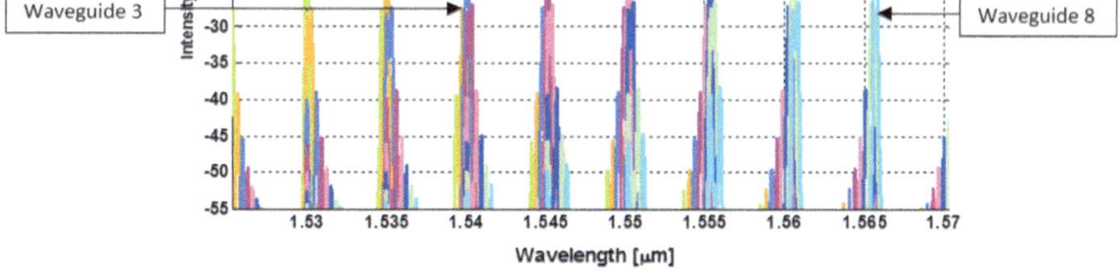

Fig. 6.10 Lumped spectrum at the drop port of the Vernier structure for specific effective index changes with the designated output waveguide per each resonance

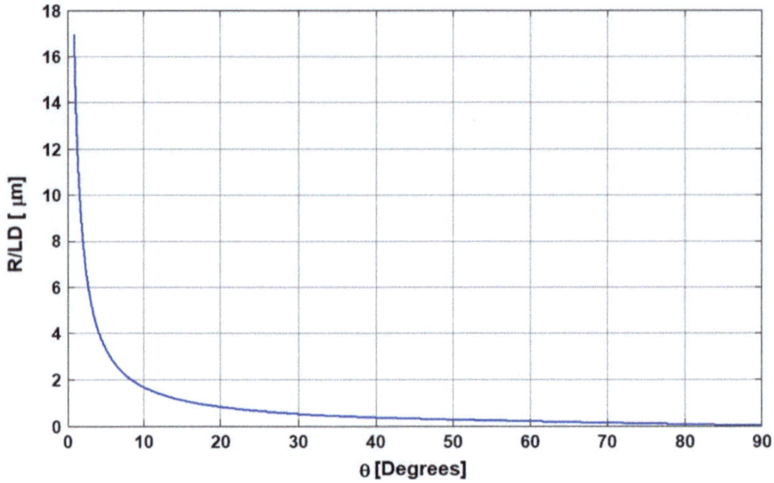

Fig. 6.11 Effect of incident angle on R/LD ratio

angle of incidence is set to equal the diffraction angle ($\theta_i = \theta_d = \theta$) as indicated in the LD expression (Eq. 6.13). Therefore, the Rowland circle radius can be expressed as

$$R = \frac{\lambda}{4}\frac{n_{eff}}{n_g}\frac{1}{tan\theta}LD \tag{6.15}$$

If the ratio n_g/n_{eff}, the channel spacing λ, and the output waveguide pitch ΔS are made constant, then increasing the angle θ will reduce the radius R and hence the PEG size. On the other hand, if the radius R is to be held fixed, then the effect of increasing θ needs to be compensated by increasing the output waveguide pitch ΔS; i.e., the output waveguides have to be distributed widely along the Rowland circle to avoid mode coupling between adjacent waveguides. To clarify the impact of θ on the PEG performance, for a constant $LD = 0.5\,\mu$m/nm, the Rowland circle radius R is calculated for different angles ($\theta = 10°, 30°, 50°$, and $70°$) at $\lambda = 1550$ nm, and $n_g/n_{eff} = 1.31$. The transmission profile of a single channel for different incident angles is shown in Fig. 6.11. It clearly demonstrates that at low incident angles the passband channel is narrower and a manageable level of cross talk is achieved, but the resulting device is quite large.

On the other hand, at high incident angles, the device is made more compact, but each channel becomes wider giving rise to more coupling between adjacent output waveguides. Hence, a compromise between moderate device size and good performance in terms of cross talk needs to be achieved (Fig. 6.12).

The angle of incidence and diffraction (θ_i and θ_d) are commonly selected to be within the range (30°–50°) [34–36], as this range provides a fair compromise between device size and performance in terms of cross talk. Hence, once θ_i is selected, the diffraction angle (θ_d) is calculated using the grating equation (Eq. 6.8), and the radius of Rowland circle R is calculated from the linear dispersion equation (Eq. 6.12).

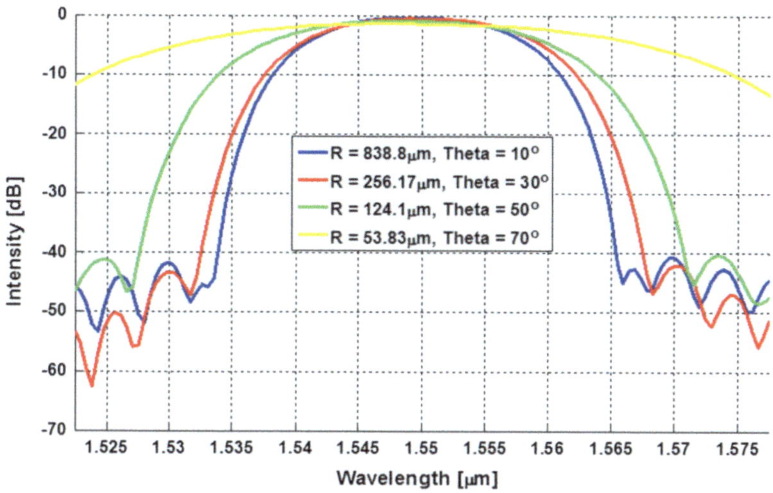

Fig. 6.12 Effect of incident angle on PEG cross-talk level

To improve PEG diffraction efficiency, grating facets must be illuminated perpendicularly that can be achieved using Eagle mounting. In this mounting configuration, grating facets are blazed individually to maximize phase coherence at the blazing point 'O' and achieve equal angles and close to zero ($\varphi_i \approx \varphi_d \approx 0$) between the incident and diffracted light beams with respect to the normal line on each illuminated facet.

The input and output waveguides are tapered to expand the mode size between narrow nanowire width and wide input/output apertures. Moreover, the emanating light from the input aperture diffracts widely as it propagates through the slab region, and the degree of this expansion is inversely proportional to the width of the input aperture. If a nanowire input waveguide is integrated directly into the slab region, the emanating light will expand greatly, which will accordingly demand etching more grating facets to capture most of the widely diffracted light and focus it at a specific output aperture. A more realistic approach is to widen the input aperture enough to minimize the degree of mode expansion, and therefore, fewer gratings are needed. Similarly, wider output apertures greatly improve the coupling of the reflected light modes from grating facets into those waveguides. In addition, larger widths excite higher-order modes, which when lumped together will result in flat passbands profile instead of a Gaussian one. This is also essential to overcome the misalignment between the resonant tones of the optical sensor and the central wavelengths of the PEG passband channels. However, wider apertures require a larger output waveguide pitch ΔS to maintain an acceptable cross-talk level. Both the input and output waveguides have their apertures optimized to 4 and 5.5 μm, respectively. They are tapered down to 480 nm using a 50 μm-long linear tapers. The rest of the design procedure determines the grating period d, and the number of grating facets N.

The grating period d is initially set equal to multiple wavelengths, and it is optimized to yield an acceptable insertion loss. Note that a small grating period

Table 6.3 PEG design parameters

Feature	Numerical value
Diffraction order (m)	8
Grating period (d)	6 μm
Rowland radius (R)	1500 μm
Number of grating facets (N)	300
Incident angle (θ_i)	45°
Diffraction angle (θ_d)	40.47°
Number of channels	8-*Channels*
Output waveguide pitch (ΔS)	1.5 μm
Channel spacing($\Delta\lambda$)	5 nm
Width of the input aperture	4 μm
Width of the output aperture	5.5 μm
Free spectral range (FSR_{PEG})	143.65 nm
Linear dispersion (LD)	1.4 μm/nm
Insertion loss	−0.5569 dB
Cross talk	−7.562 dB

yields smaller dimensions of both the illuminated and shadowed facets, making them susceptible to imperfection in the etching process, such as rounded corners resulting in higher levels of light scattering upon reflection. While larger grating period results in larger physical dimensions of grating facets, the wider dark facets contribute to significant light reflections that interfere destructively with those from the illuminated facets. Therefore, the diffraction efficiency will be degraded, leading to larger insertion losses. Finally, the number of grating facets N is chosen numerically to achieve an acceptable insertion loss (better than −2 dB is a good start). Due to the complexity of analytical PEG design, the above quantities are initially calculated from their respective formulate. Subsequently, numerical simulations are used to optimized those parameters necessary. Table 6.3 highlights the final values of the design parameters.

6.3 Sensor Performance

The sensor performance is evaluated based on the PEG and micro-rings channel spacing and insertion losses. A photograph of an 8-channel, 5 nm channel spacing PEG integrated with vernier micro-rings is shown in Fig. 6.13. The structure parameters were designed in silicon photonic platform with nanowire thickness of 220 nm, and a box oxide thickness of 2 μm. The upper cladding thickness was set to 1 μm. The width of the nanowires is 480 nm, and the gap between the ring and bus waveguides is 200 nm. The cladding on top of the sensor ring was fully etched to allow the optical field to interact with the analyte. The expected total design size is 1.0 mm^2.

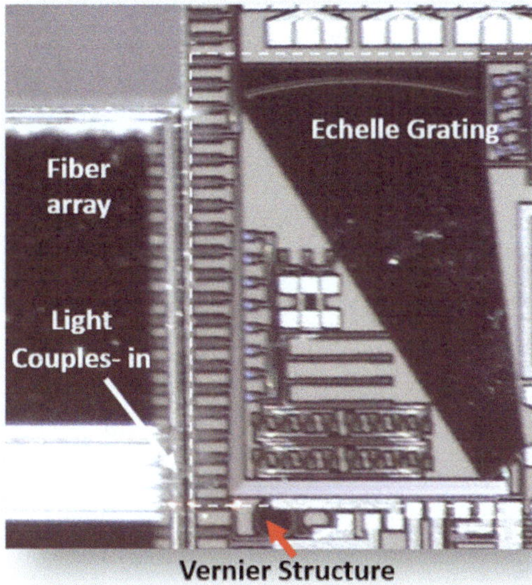

Fig. 6.13 A photograph of the fabricated 8-ch 5 nm echelle grating coupled with Vernier rings

The echelle grating passbands were optimized to have a box-shaped transmittance, which was achieved by widening the end facet of the output waveguide to 5.5 μm. Note that the channel cross-talk level was not critical as we only measured the transmittance at the channel center. This also helps to significantly reduce the PEG size. Furthermore, the grating facets were integrated with distributed Bragg reflectors (DBRs) in order to enhance power reflectance and minimize the insertion losses as noted in [20]. Figure 6.14a illustrates the transmission spectrum of the echelle grating, and Fig. 6.14b demonstrates the combined spectrum of the sensor chip with respect to different effective index variations.

Initially, the Vernier rings transmission was set at a wavelength of 1530 nm when no analyte was present. This corresponds to a central wavelength of 1550 nm (i.e., $Channel = 4$). In addition, the sensor and reference rings have FSRs of 587 and 625 GHz, respectively. The corresponding resonators waveguide widths and lengths are: 4.7 nm, $L_s = 111.1\,\mu\text{m}$ and 4.7 nm, $L_s = 111.1\,\mu\text{m}$. Note the effective index is ∼2.4. As a result, the minimum detectable frequency shift is 38 GHz (0.3 nm). Referring to (2), the limit of detection (LOD) to a minimum index change is 820×10^{-6} RIU. If we take the round-trip loss of each ring into consideration, an LOD of about 1500×0^{-6} RIU under the worst-case scenario ($\gamma = 0.92$). A more reasonable value is 1000×10^{-6} RIU at $\gamma = 0.98$. Furthermore, if we use a 1 nm channel spacing, the theoretical LOD is 160×10^{-6} RIU, whereas the typical value is 200×10^{-6} RIU at $\gamma = 0.98$. The worst-case scenario ($\gamma = 0.92$) is 300×10^{-6} RIU. For both examples, LOD can be enhanced to 80×10^{-6} if thermal tuning is used. For the de-multiplexer design, a planar echelle grating was considered. The operation principle of a planar echelle grating is highlighted in [37]. Our design

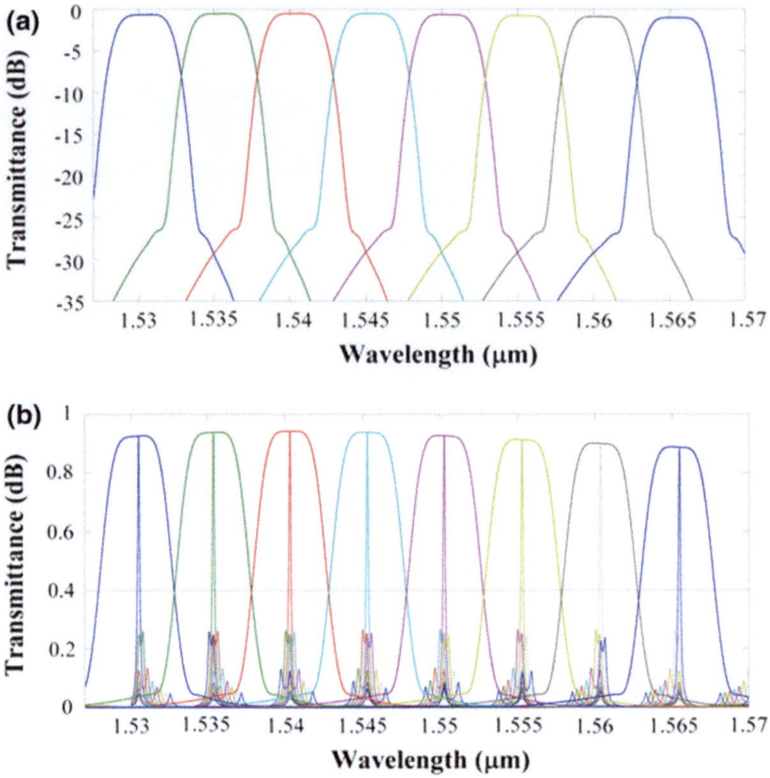

Fig. 6.14 **a** Combined theoretical spectrum of the Vernier-based structure at its drop port (in dashed black line) and transmission spectrum of (8×5 nm) PEG. **b** Final theoretical transmission spectrum for the optical sensor cascaded to a PEG spectrometer. The detection threshold is indicated by the dashed black line

is based on a Rowland configuration optimized to accomplish a trade-off between compactness and performance. The channel spacing of the PEG was selected to be 5 nm.

6.4 Conclusions

An optical sensor integrated with a spectrum analyzer has been proposed. The device is designed to detect changes in the optical index of contrast induced by the presence of target analyte. The sensor enables a considerable lower limit of detection when thermo-optic phase shifters are used, and it does not rely on the resonance state of the MRRs as in conventional designs. Furthermore, the proposed sensor provides large dynamic range detection, yet allows a high limit of detection (1.2×10^{-4} RIU) without the need of special optical components. For example, the PEG can be designed with low spectral resolution and no need for strict cross-talk performance as the detection only happens at the center of the passband. The sensor is compact and requires relatively small number of integrated PDs, which reduces the fabrication cost. Finally, the detection scheme has a discrete behavior that can greatly simplify the electronics readout circuit. The proposed device provides the benefits of reducing the detection limit as well as the required number of photodetectors, therefore, the overall sensor's footprint.

Acknowledgements This work has been supported by the Semiconductor Research Corporation (SRC) under the Abu Dhabi SRC Center of Excellence on Energy-Efficient Electronic Systems (ACE^4S), Contract 2013-HJ2440, with funding from the Mubadala Development Company, Abu Dhabi, UAE.

References

1. V.M. Passaro, M. La Notte, B. Troia, L. Passaquindici, F. De Leonardis, G. Giannoccaro, Photonic structures based on slot waveguides for nanosensors: state of the art and future developments. J. Res. Rev. Appl. Sci **11**, 402–418 (2012)
2. N.-A. Yebo, S.-P. Sree, E. Levrau, C. Detavernier, Z. Hens, J.-A. Martens et al., Selective and reversible ammonia gas detection with nanoporous film functionalized silicon photonic micro-ring resonator. Opt. Express **20**, 11855–11862 (2012)
3. K. Misiakos, A. Botsialas, I. Raptis, E. Makarona, G. Jobst, P. Petrou et al., Monolithically integrated frequency-resolved mach-zehnder interferometers for highly-sensitive multiplexed label-free bio/chemical sensing, in *Sensors, 2011 IEEE* (2011), pp. 1317–1320
4. Y. Liu, H. Salemink, Photonic crystal-based all-optical on-chip sensor, in *IEEE Sensors 2011*, vol. 20 (2012), pp. 19912–19920
5. G. Nemova, R. Kashyap, Theoretical model of a planar integrated refractive index sensor based on surface plasmon-polariton excitation. Opt. Commun. **275**, 76–82 (2007)
6. S.-Y. Cho, N.M. Jokerst, A polymer microdisk photonic sensor integrated onto silicon, Photonics Technology Letters. IEEE Photonics Technol. Lett. IEEE **18**, 2096–2098 (2006)
7. C.-Y. Chao, L.J. Guo, Biochemical sensors based on polymer microrings with sharp asymmetrical resonance. Appl. Phys. Lett. **83**, 1527–1529 (2003)
8. Q. Xu, V.R. Almeida, R.R. Panepucci, M. Lipson, Experimental demonstration of guiding and confining light in nanometer-size low-refractive-index material. Opt. Lett. **29**, 1626–1628 (2004)
9. D. Xu, A. Densmore, A. Delacge, P. Waldron, R. McKinnon, S. Janz et al., Folded cavity SOI microring sensors for high sensitivity and real time measurement of biomolecular binding. Opt. Express **16**, 15137–15148 (2008)
10. A.M. Armani, R.P. Kulkarni, S.E. Fraser, R.C. Flagan, K.J. Vahala, Label-free, single-molecule detection with optical microcavities. Science **317**, 783–787 (2007)
11. I.M. White, X. Fan, On the performance quantification of resonant refractive index sensors. Opt. Express **16**, 1020–1028 (2008)
12. J. Liu, X. Zhou, Z. Qiao, J. Zhang, C. Zhang, T. Xiang et al., Integrated optical chemical sensor based on an SOI ring resonator using phase-interrogation. IEEE Photonics J. **6**, 1–7 (2014)
13. L. Jin, M. Li, J.-J. He, Highly-sensitive silicon-on-insulator sensor based on two cascaded micro-ring resonators with vernier effect. Opt. Commun. **284**, 156–159 (2011)
14. D. Dai, Highly sensitive digital optical sensor based on cascaded high-Q ring-resonators. Opt. Express **17**, 23817–23822 (2009)
15. J. Hu, D. Dai, Cascaded-ring optical sensor with enhanced sensitivity by using suspended Si-nanowires. IEEE Photonics Technol. Lett. **23**, 842–844 (2011)

16. T. Claes, W. Bogaerts, P. Bienstman, Experimental characterization of a silicon photonic biosensor consisting of two cascaded ring resonators based on the Vernier-effect and introduction of a curve fitting method for an improved detection limit. Opt. Express **18**, 22747–22761 (2010)
17. O. Al Mrayat, M. Rasras, A digital-like on-chip photonics sensor, in Frontiers in Optics 2015, OSA Technical Digest (online) (Optical Society of America, 2015), paper JW2A.78
18. L. Chen, C.R. Doerr, P. Dong, Y.-K. Chen, Monolithic silicon chip with 10 modulator channels at 25 Gbps and 100-GHz spacing. Opt. Expres **19**, B946–B951 (2011)
19. M.S. Rasras, D.M. Gill, M.P. Earnshaw, C.R. Doerr, J.S. Weiner, C. Bolle et al., CMOS silicon receiver integrated with Ge detector and reconfigurable optical filter. IEEE Photonics Technol. Lett. **22**, 112–114 (2010)
20. M.S. Rasras, D.M. Gill, S.S. Patel, K.-Y. Tu, Y.-K. Chen, A.E. White et al., Demonstration of a fourth-order pole-zero optical filter integrated using CMOS processes. IEEE J. Lightwave Technol. **25**, 87–92 (2007)
21. J.B.D. Soole, A. Scherer, H.P. LeBlanc, N.C. Andreadakis, R. Bhat, M.A. Koza, Monolithic InP/InGaAsP/InP grating spectrometer for the 1.48–1.56 μm wavelength range. Appl. Phys. Lett. **58**(18), 1949–1951 (1991)
22. S.H. Kong, D.D.L. Wijngaards, R.F. Wolffenbuttel, Infrared micro-spectrometer based on a diffraction grating. Sens. Actuators, A **92**(1), 88–95 (2001)
23. K.C. Harvey, C.J. Myatt, External-cavity diode laser using a grazing-incidence diffraction grating. Opt. Lett. **16**(12), 910–912 (1991)
24. I.P. Kaminow, H.P. Weber, E.A. Chandross, Poly (Methyl Methacrylate) dye laser with internal diffraction grating resonator. Appl. Phys. Lett. **18**(11), 497–499 (1971)
25. I. Shoshan, U.P. Oppenheim, The use of a diffraction grating as a beam expander in a dye laser cavity. Opt. Commun. **25**(3), 375–378 (1978)
26. H. Fathallah, L.A. Rusch, S. LaRochelle, Passive optical fast frequency-hop CDMA communications system. IEEE J. Lightwave Technol. **17**(3), 397–405 (1999)
27. S. Pathak, P. Dumon, D. Thourhout, W. Bogaerts, Comparison of AWGs and Echelle gratings for wavelength division multiplexing on silicon-on-insulator. IEEE Photonics J. **6**(5) (2014)
28. K.A. McGreer, Theory of concave gratings based on a recursive definition of facet positions. Appl. Optics. **35**(30) (1996)
29. R. Marz, C. Cremer, On the theory of planar spectrographs. IEEE J. Lightwave Technol. **10**(12), 2017–2022 (1992)
30. H.A. Rowland et al., Preliminary notice of the results accomplished in the manufacture and theory of gratings for optical purpose. Philos. Mag. **13**, 469–474 (1882)
31. M. Born, E. Wolf, *Principles of Optic* (Pergamon, New York, 1980)
32. D. Chowdhury, Design of low-loss and polarization-insensitive reflection grating-based planar demultiplexers. IEEE J. Sel. Top. Q. Electron **6**(2), 233–239 (2000)
33. J. Brouckaert, W. Bogaerts, P. Dumon, D. Van Thourhout, R. Baets, Planar concave grating demultiplexer fabricated on a nanophotonic silicon-on-insulator platform. IEEE J. Lightwave Technol. **25**, 5 (2007)
34. E. Gini, W. Hunziker, H. Melchior, Polarization independent InP WDM multiplexer/demultiplexer module. IEEE J. Lightwave Technol. **16**(4), 625–630 (1998)
35. W.H. Wang, Y.Z. Tang, Y.X. Wang, H.C. Qu, Y.M. Wu, T. Li, J.Y. Yang, Y.L. Wang, M. Liu, Etched-diffraction-grating-based planar waveguide demultiplexer on silicon-on-insulator. Opt. Quant. Electron. **36**, 559–566 (2004)
36. Z.J. Sun, K.A. McGreer, J.N. Broughton, Demultiplexer with 120 channels and 0.29-nm channel spacing. IEEE Photonics Technol. Lett. **10**(1), 90–92 (1998)
37. J. Brouckaert, W. Bogaerts, S. Selvaraja, P. Dumon, R. Baets, D. Van Thourhout, Planar concave grating demultiplexer with high reflective Bragg reflector facets. IEEE Photonics Technol. Lett. **20**, 309–311 (2008)

Part II
Architectures and Circuits

"It is impossible to construct machinery occupying unlimited space, but it is possible to construct finite machinery and to use it through unlimited time. It is this substitution of the infinity of time for the infinity of space which I have made use of to limit the size of the engine and yet retain its unlimited power." Charles Babbage, *Life of a Philosopher*

Chapter 7
Design Challenges in Wireless Sensors for Dental Applications

Lutfi Albasha, Nasser Qaddoumi, Basil Hatahet, Nasir Quadir and Mansour Taghadosi

7.1 Introduction

In orthodontic treatment of teeth, classic bracket system is well known and has been extensively used [1]. The pressure of the wire brace forces the movement of the teeth. The classic system comprises three elements: the bracket, the wire and the bonding material. Figures 7.1 and 7.2 show examples. The pressure the brace system constantly applies forces the movement of the tooth to the desired location as predetermined by the dentist. Once in its new place, new bone grows to support the tooth. This process is called bone remodelling and bone deposition.

When a brace puts pressure on a tooth, it moves. Such movement is done slowly to avoid damaging the tooth. For this reason, the course of treatment usually lasts for many months, if not years. The patient must also regularly visit his dentist to check the movement of the teeth and apply any necessary adjustment. A tooth can move about a millimetre per month. The patient style of eating and the way his teeth move can affect the efficiency of the treatment, and this explains why people respond differently to orthodontic treatment [2].

Other type of treatment is dental retainers. Retention strategies usually include removable and fixed retainers, the latter having the advantage of minimal need for patient compliance. However, it does require good skills and regular checks [1]. Various types of fixed retainers have been described in [2] (Fig. 7.3).

Every year millions of people (mainly children) have their orthodontic appliances removed at the end of their treatment and have a bonded retainer fitted to their teeth

L. Albasha (✉) · N. Qaddoumi · N. Quadir · M. Taghadosi
American University of Sharjah, Sharjah, UAE
e-mail: lalbasha@aus.edu

N. Qaddoumi
e-mail: nqaddoumi@aus.edu

B. Hatahet
Denton Clinic, Denton, UK

Ibrahim (Abe) M. Elfadel and M. Ismail (eds.), *The IoT Physical Layer*,
https://doi.org/10.1007/978-3-319-93100-5_7

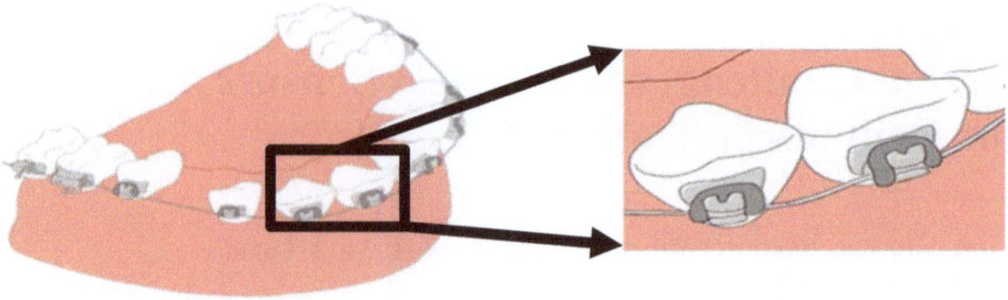

Fig. 7.1 System block diagram

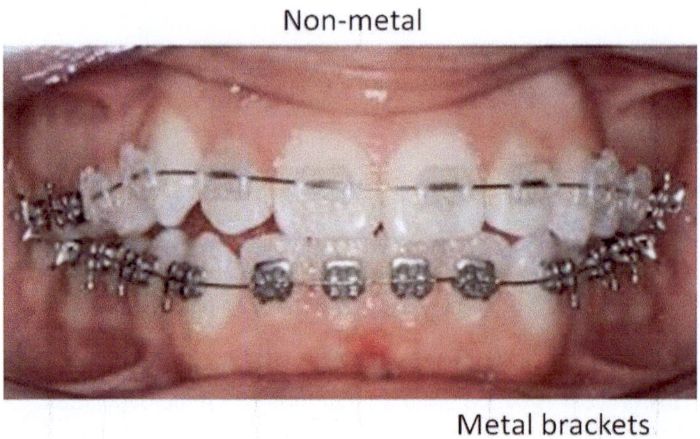

Fig. 7.2 Classic bracket system (reproduced with special permission from Denton Clinic, UK)

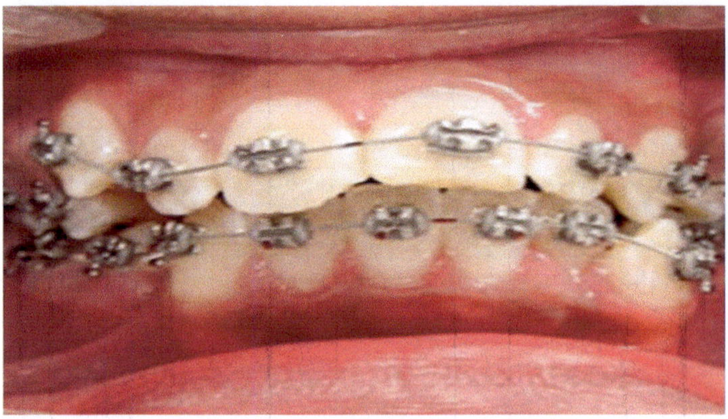

Fig. 7.3 Classic bracket system (reproduced with special permission from Denton Clinic, UK)

to prevent relapse of malocclusion. Up to 35% of these retainers fail prematurely when the bond breaks allowing one or more teeth to drift away from the retainer wire. Patients rarely discover the bond failure in time before this drift takes place. Following this, the patient either starts a new course of treatment costing thousands of US dollars and tens of hours or accepts treatment failure. Orthodontists usually review their patients regularly to check for any premature failure of retainers costing the patients money and the orthodontist time.

There needs to be an early-detection system that can determine if the placement of the bracket in the classic technique, for example, is still accurate and as planned. An embedded transmitter between the retainer and the tooth surface that can detect the presence of humidity at the site (indicating bond failure) and lets the patient or the orthodontist know that the retainer has failed will save patients and orthodontist money and time.

Bonded restorations such as filling, crowns and bridges on teeth fail when a small bond failure happens between the tooth surface and the restoration causing hidden decay of teeth that is usually noticed when the restoration completely fails and the tooth has lost significant amount of its substance. Again, the cost for repair to recover the situation runs into hundreds and sometimes thousands of US dollars. Here again, a transmitter detecting humidity level will alert the patients or their dentist to the bond failure and the situation could be recovered by simply re-cementing of the restoration at a small cost.

Millions of such restorations are fitted around the world DAILY! The average cost of these in the advanced world varies between hundreds to thousands of US dollars each. A small increase to the cost that can assist with the longevity of the restoration is negligible.

In the chapter, a wireless, self-powered (or battery-less) system that can power and drive an embedded chip for detecting bracket and bond failure is designed and discussed. The system operates a novel and dedicated protocol to power the circuitry, activate the sensor, read data and transmit back to the external unit. In the rest of the chapter, the system architecture will be discussed and circuits presented.

7.2 System Architecture

The overall system architecture is shown in Fig. 7.4.

The early-detection system comprises an internal unit placed on the bracket facing the tooth, means for wireless energy transfer and an external unit. The latter will transmit DC power and will also pick up received data from the implanted unit.

Figure 7.5 shows the proposed system block diagram of the complete wirelessly powered sensor-based system. The system consists of two main components: an external reader and the internal microsystem, both of which operate as receivers and transmitters of power and data accordingly. In particular, the external (receiver) reader is a hand-held device that performs the following two main functions:

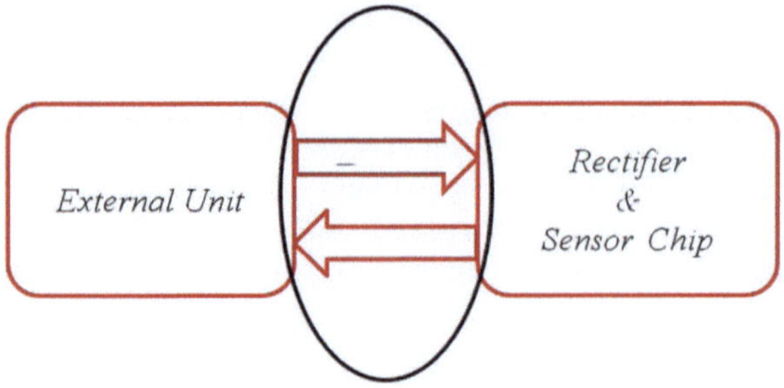

Fig. 7.4 Diagram showing the system sub-blocks

1. Transmit power to the implanted microchip.
2. Receive and analyse the data sent by the sensors.

On the other hand, the internal sensor microsystem has two main functions:

1. Receives power from the external device in order to power its on-board sensor.
2. Transmits the data received from the sensor to the external reader.

The humidity sensor (and the corresponding chip) is placed in between the bracket and the tooth. The system operates on modulating and transmitting readings from the humidity sensor through the internal coil to the external one and eventually to the data processing unit. This takes place as follows:

- When the brackets are properly affixed to the teeth, enough glue would be surrounding the chip to keep the humidity sensor readings constant. The moment the bracket is dislocated or shifted, the readings of the humidity sensor would change to reflect the dislocation.
- Data from the sensor and phase detector would be applied to the *load modulator unit* to be re-transmitted to the external unit.
- The load modulator is a fast-switching CMOS transistor to which output of the phase detector is applied to the gate (-source). This transistor would be placed such that the data is modulated onto the RF signal initially coupled to the secondary coil.
- Output data from the sensor is expected to take one of two forms:
  - No-fault data: Zero phase difference
    When no change in humidity occurs, the phase detector output is zero.
  - Fault-indicative data: Nonzero phase difference; data pulses output
    Any change in humidity level detected by the sensor is reflected into a series of data pulses that would be modulated onto the envelope of the RF waveform.
- The modulated RF signal would then be backscattered to the primary coil from which data will be extracted by the data acquisition and processing units.
- Since load modulation is, in principle, a form of amplitude modulation, both synchronous and asynchronous detection modes are possible. However, due to the low data rate requirement of this system and the need for reduced complexity at the external unit, asynchronous demodulation is selected.

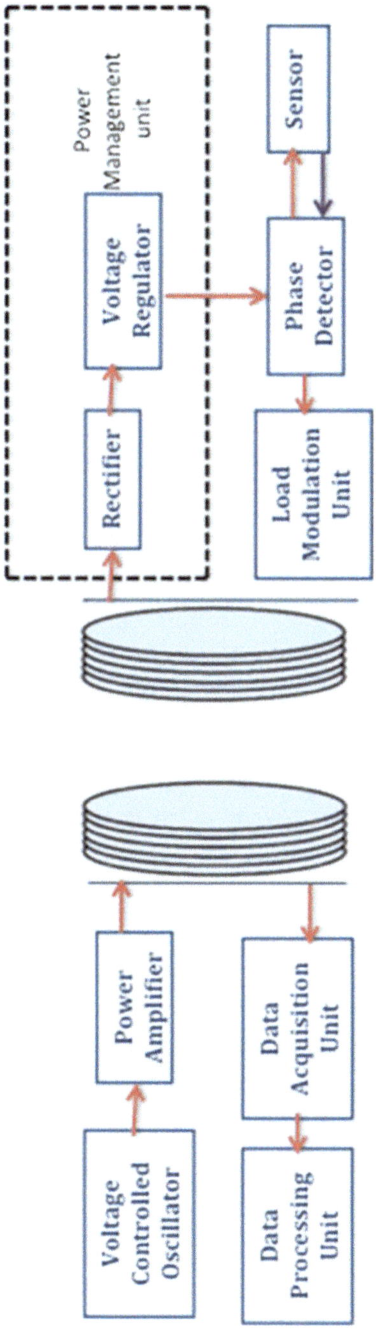

Fig. 7.5 System block diagram

- For asynchronous modulation, the data acquisition unit consists of a level shifter, followed by a peak detector/envelope detector that would capture the changes in signal envelope with respect to changes in humidity levels.
- This detected envelope would then simply be connected to an LED to demonstrate the humidity level.

An initial system simulation is conducted on ADS software to verify the principle of load modulation. The simulated circuit is shown in Fig. 7.6. For demonstration, a pulse waveform is applied to the load-modulating transistor representing sensor data. The obtained results are shown in Fig. 7.7. These initial results are further to be confirmed when simulated using the actual sensor circuitry [3–7].

7.3 Implanted Unit

The internally implanted unit is integrated on a single die. It consists of the receiving coil (RX), the rectifier circuit, data acquisition unit and the sensor blocks. In the following, the specifications of the rectifier and the RX coil are essential in designing the overall system such that it is capable of delivering the required power to the integrated sensor.

7.3.1 Integrated Rectifier

The integrated sensor and data processing units are powered up by wireless power transfer (WPT). Generally, WPT systems utilize an RF signal at a specified frequency in order to transfer the required power from the primary to the secondary side. Hence, the need of a rectifier circuit is essential in converting the received RF signal into a DC power suitable for running the integrated circuitry. This rectifier circuit should be capable of providing the total amount of current at the required voltage (the required power) for the integrated circuitry. The conventional Dickson charge pump rectifier is not suitable for biomedical applications utilizing ISM band, as the frequency of operation is lower compared to the GSM and higher standards. Zargham and Gulak present a rectifier circuit in [8] which is capable of delivering mW of power to a load at a high power efficiency. In this rectifier circuit shown in Fig. 7.8, a threshold cancellation technique is used partially to bias the transistors. The receiver coil along with its capacitor to form a resonance circuit is differentially connected to the circuit (between RF+ and RF- terminals). Only one side of the circuit (three transistors) conducts at each positive and negative cycles of the input RF voltage. The cross-coupled resistors along with the diode-connected transistors provide the gate bias voltage for NMOS transistors M3 and M4. The value of this gate voltage is critical as it is responsible to compensate for the threshold voltage of transistors M3 and M4. The higher the gate voltage, the less is the on-resistance of the transistor

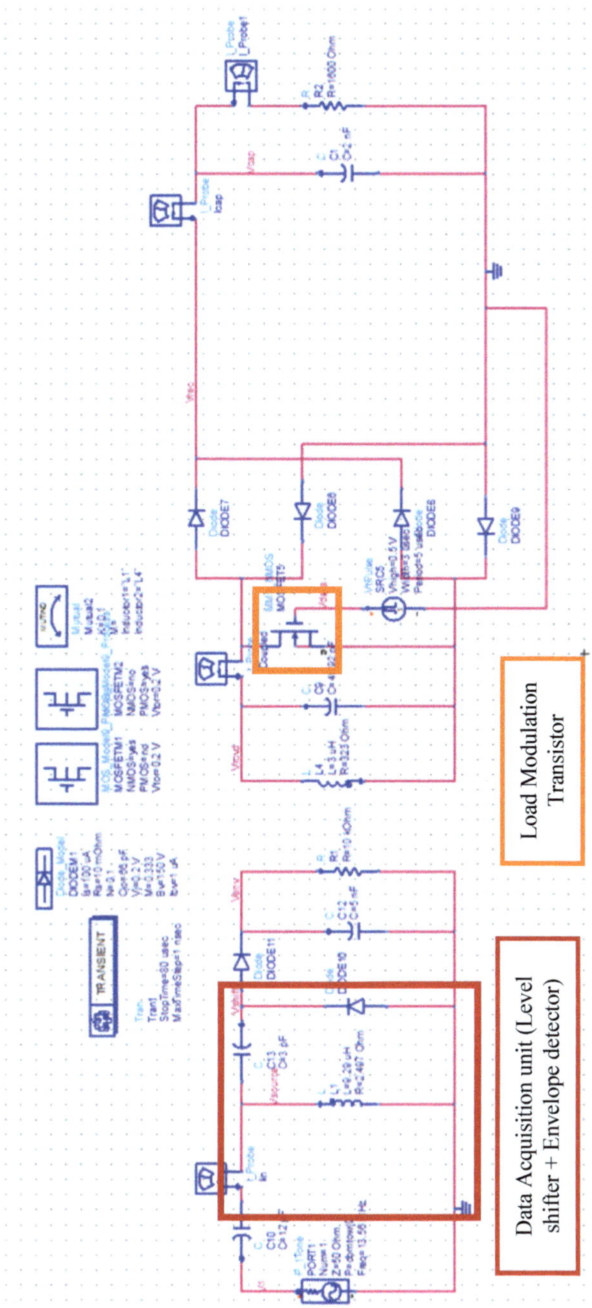

Fig. 7.6 System simulation in ADS

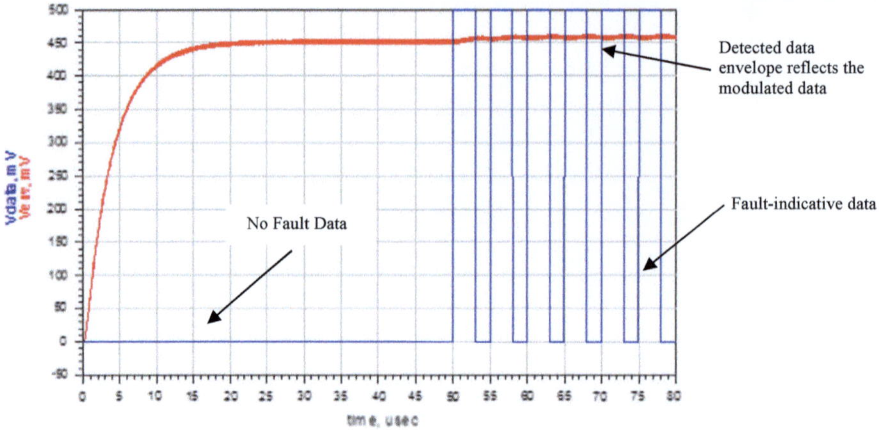

Fig. 7.7 System simulation result

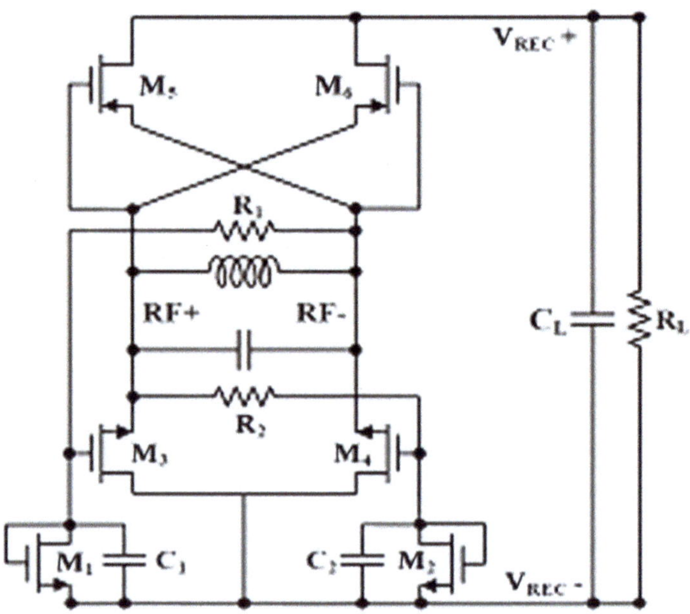

Fig. 7.8 Proposed rectifier [8]

during the conduction phase. On the other hand, the increase in this gate voltage would increase the leakage current of the transistor as the transistor does not turn off rapidly. Hence, an intermediate bias voltage is of importance in this partial threshold cancellation rectifier. Furthermore, the PMOS switches are cross-coupled to further reduce the voltage drop in the circuit and reduce the reverse current sinked from the load capacitor.

Table 7.1 Optimized sizes of transistors' rectifier in 180 nm RF process

Transistor	Channel type	Length (nm)	Width (μm)	No. of fingers	Total width (μm)
M1, M2	N-type	180	5	10	50
M3, M4	N-type	300	9.5	10	95
M5, M6	P-type	180	15	10	150

The rectifier circuit is aimed to deliver 600 μA of current at 1.5 V to the integrated sensor and data processing units. Table 7.1 presents the optimized sizes of the transistors in the rectifier, designed using GLOBALFOUNDRIES 180 nm RF process ensuring low cost and high performance. It was found that a peak RF voltage of 900 mV results in the required output DC voltage of 1.5 V at the specified 2 k load. Figures 7.9 and 7.10 show the rectifier's output voltage versus process and temperature variations at 120 MHz. The load capacitor in all simulations was set to 1 nF to clearly observe the output voltage with less ripples. As it is clear, the output voltage of the rectifier is stable around 1.45 V at 2 k load which results in 700 μA of current. Hence, the rectifier is capable of driving the sensor and the data processing units.

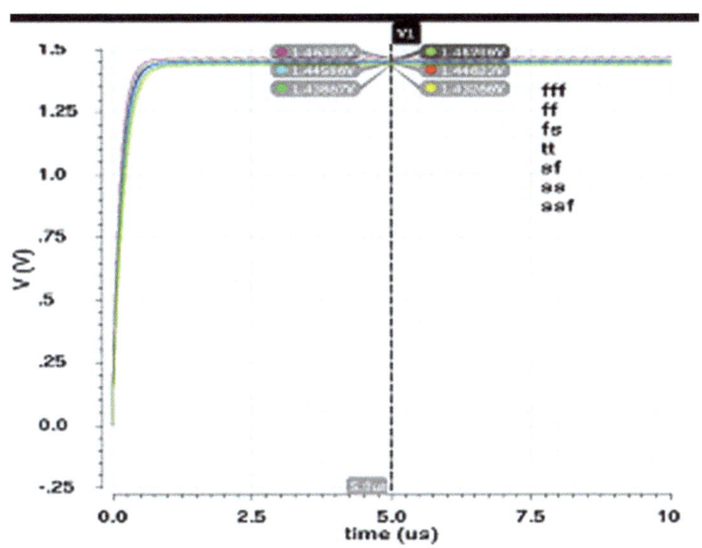

Fig. 7.9 Rectifier's output voltage versus process variations at 120 MHz, 900 mV peak input and 2 kΩ load

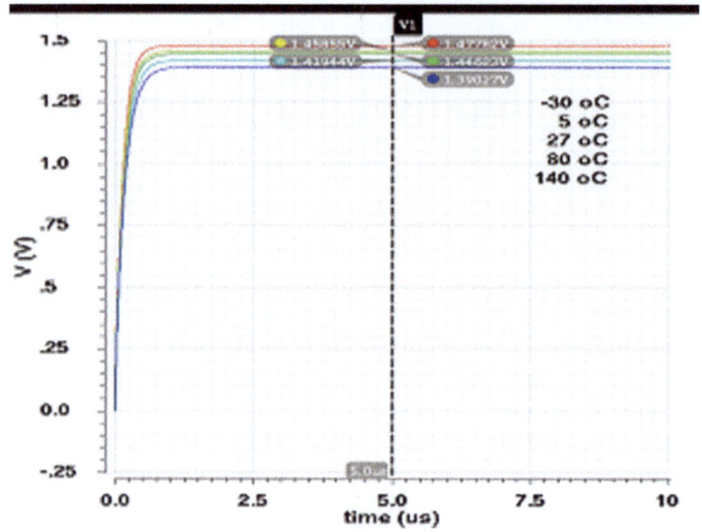

Fig. 7.10 Rectifier's output voltage versus temperature variations at 120 MHz, 900 mV peak input and 2 kΩ load

7.3.2 *RX Coil*

There is a restriction for the overall dimensions of the chip which is going to be placed on the braces [9]. Hence, it was decided to restrict the dimensions of the coil to be 2.5 by 2.5 mm^2 and optimize for the best achievable results. The square spiral inductor is the best structure to be integrated on chip. It was decided to use a planar spiral inductor due to their simple structure, ease of fabrication and integration. Figure 7.11 represents the substrate model of the 7-metal GLOBALFOUNDRIES 180 nm RF process in ADS momentum simulator. Different combinations including series and parallel configurations using available layers of the process were tested for the structure of the RX coil. However, the momentum simulation showed that similar multi-layer inductor value and Q-factor are achievable if a single-layer (using topmost metal layer AM) inductor is used. This is due to the fact that the thickness of this metal layer is 4 μm where all the other metal layers have 0.48 μm thickness. The optimized dimensions of the RX coil are presented in Table 7.2. Figure 7.12 shows the inductance and resistance of the RX coil in addition to its quality factor versus frequency. As it is pointed on the plot, the peak quality factor of 7.162 occurs at 74.5 MHz for this coil.

Each coil is considered as a one-port network which consists of real and imaginary impedances. The real part determines the resistive element, whereas the imaginary part is the reactive impedance at the frequency of operation. The following formulas are used to extract the coil information after S-parameter simulations.

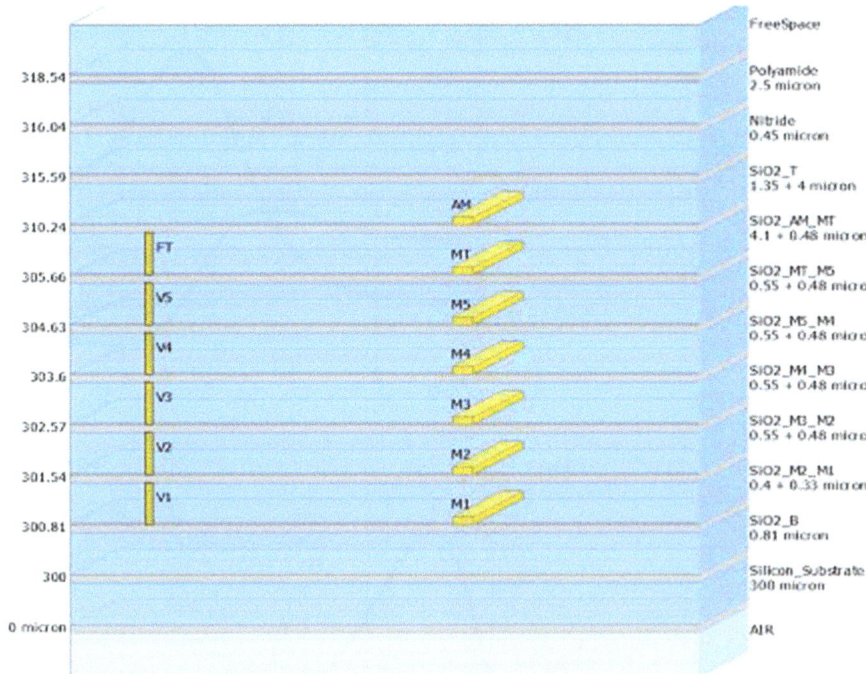

Fig. 7.11 Substrate model of the 180 nm CMOS process in ADS momentum (for RX coil)

Table 7.2 Optimized coils' dimensions

Parameter	TX1	TX2	RX
Number of turns	17	6	6
Overall length	40 mm	15 mm	2.5 mm
Overall width	40 mm	15 mm	2.5 mm
Conductor widths	750 μm	500 μm	100 μm
Conductor spacings	250 μm	350 μm	10 μm
Substrate	FR4	FR4	Silicon (on-chip)

$$L = \frac{Im(Z_{11})}{2\pi f} \tag{7.1}$$

$$R = Re(Z_{11}) \tag{7.2}$$

$$Q = \frac{Im(Z_{11})}{Re(Z_{11})} \tag{7.3}$$

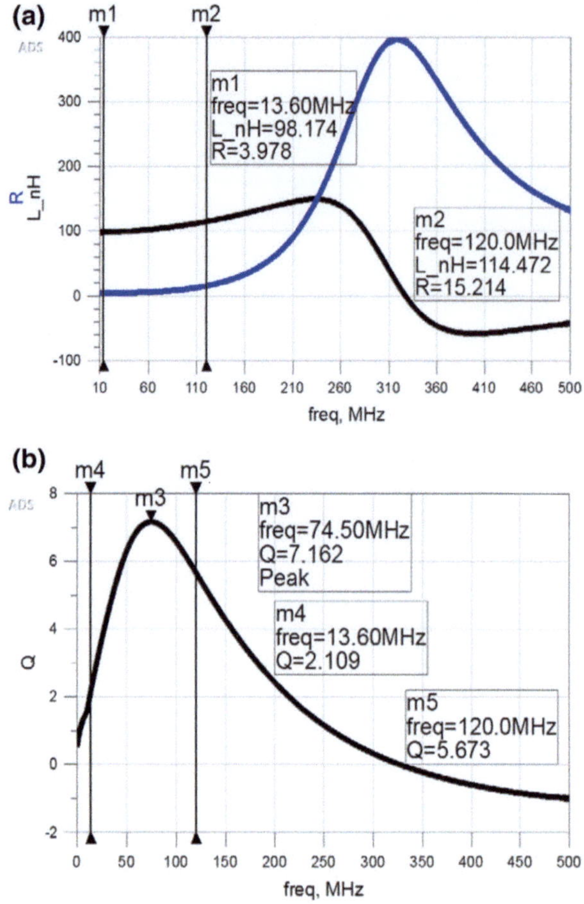

Fig. 7.12 Parameters of the RX coil: **a** inductance and resistance and **b** quality factor

It is noted that the second terminal of the coil is grounded in all simulations to have single-ended port. However, the differential impedance of the coils could be extracted from Eq. (4) when simulated in a differential configuration. Usually, the differential Q of the coil is 3–5% higher than the single-ended coil [10].

$$Z = R + jX = \frac{Y_{11} + Y_{22} + 2Y_{12}}{Y_{11}Y_{22} - Y_{12}^2} \tag{7.4}$$

7.4 External Unit

The external unit consists of two units which are the primary coil and the electronic circuitry. Figure 7.13 shows the proposed architecture of this unit. The peak of the acquired data from the data acquisition unit is extracted by the peak detector and is

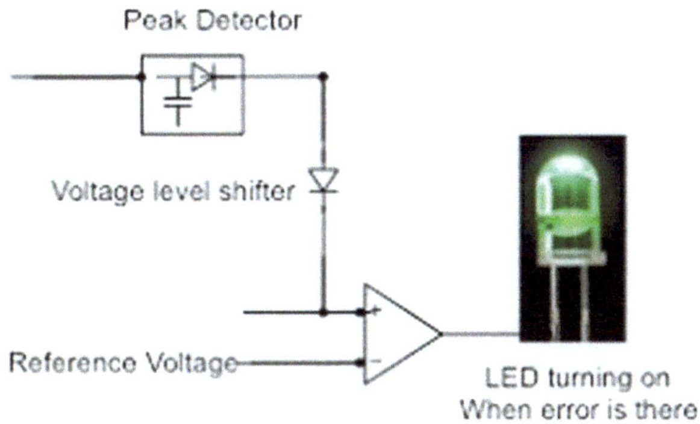

Fig. 7.13 Design concept of the external (receiver) unit

compared to a reference peak which is recorded when the bond/glue to the tooth of the braces is intact. The glowing of LED determines the error. The entire structure can be put into an electric toothbrush that could be passed over the teeth by the patient to check the condition of the braces.

7.4.1 TX Coil

The same planar square-shaped coil was chosen to be designed for the TX coil as well. This coil is to be fabricated on 1 oz FR4 substrate with 1.5 mm thickness as shown in Fig. 7.14.

7.4.2 TX1 Coil (13.56 MHz)

Initially, the frequency of operation was selected to be 13.56 MHz which is an ISM band allocated for biomedical applications. Figure 7.15 represents the optimized layout of TX1 coil with its dimensions presented in Table 7.2.

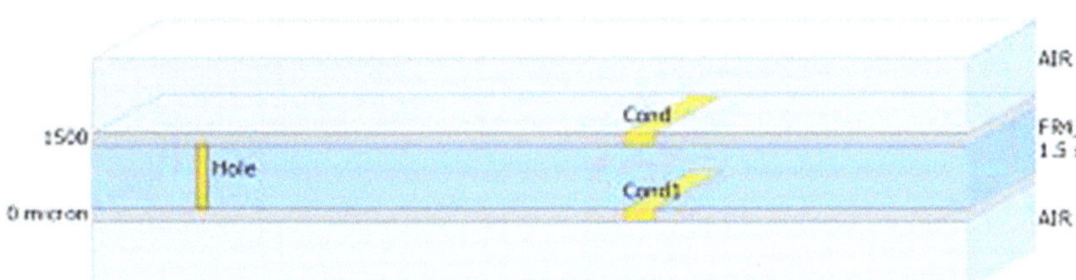

Fig. 7.14 Substrate model of FR4 in momentum for TX coil

Fig. 7.15 Layout of TX1 coil on FR4

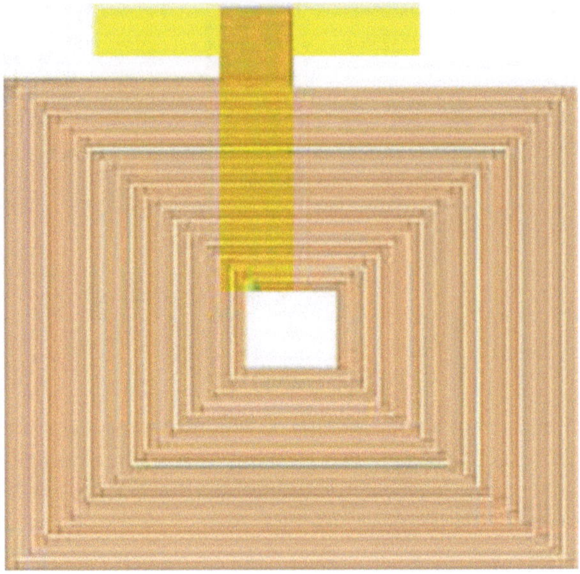

Figure 7.16 depicts the inductance and quality factor of the above coil versus frequency. As it is seen, the maximum quality factor of this structure occurs at 13 MHz where its inductance is 7.705 μH. The self-resonance frequency (SRF) of this structure, where it changes to be capacitive, is 33.3 MHz.

Fig. 7.16 Inductance and quality factor of TX1 coil versus frequency

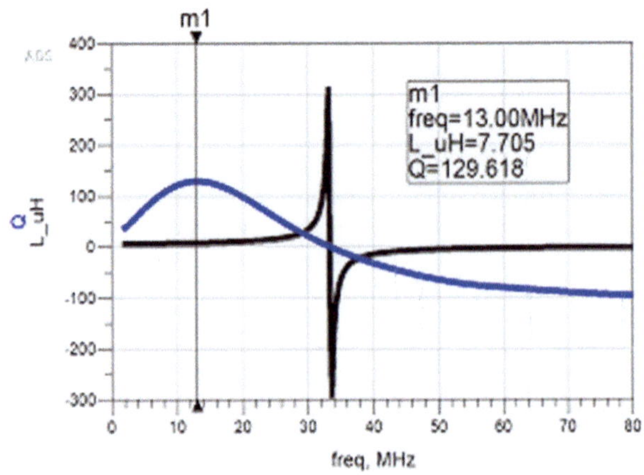

7.4.3 *TX2 Coil (40–120 MHz)*

It is a known fact that in order to operate at lower frequencies, the overall dimensions of the structure should be increased due to the increase in the wavelength. In addition, the required capacitor to provide a resonance at lower frequencies is large which is difficult to be integrated on chip. Hence, in order to have a small TX coil with high quality factor, the frequency of operation needs to be increased. The presented dimensions for the TX2 coil in Table 7.2 were found to be optimum to have the peak Q-factor of the structure on FR4 occurring at the peak Q of the RX coil. Figure 7.17a and (b) shows the inductance and quality factor of TX2, respectively. As it is pointed on the plot, the peak quality factor of 191.223 occurs at 66 MHz for this coil. At this frequency, the inductance of the structure is found to be 0.499 μH.

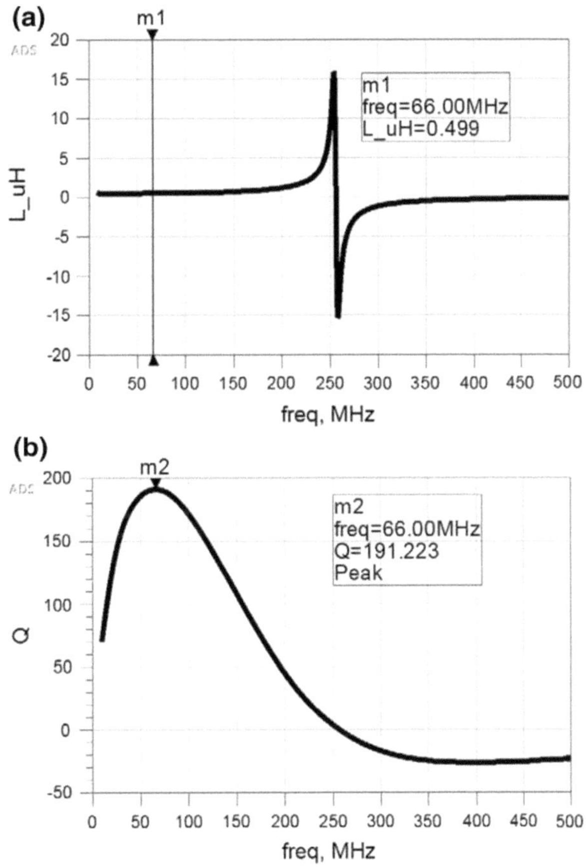

Fig. 7.17 TX2 frequency responses **a** Inductance **b** Quality factor

7.4.4 Inductive Coupling Simulation

The TX and RX coils could couple from inside or outside the mouth. For the case of the internal coupling, the media between the coils is the air whereas there is a restriction on the dimensions of the TX coil to be as small to enter the mouth. While the external coupling eliminates the size restriction, the media between the coils is changed from air to skin, fat and muscle which have high dielectric constants of 580, 38.72 and 308 at 5 MHz, respectively [11]. These materials are conductive at low frequencies. This increases the computation and simulation times accordingly.

Figure 7.18 represents a simplified model of an inductively coupled WPT link. As it is shown, both sides could be modelled as an inductor along with a resistive element which defines the quality factor (Q) of each inductor. The mutual coupling (M) between the two inductor is an important parameter in such systems to have maximum power transfer and coupling. The value of the mutual inductance between the two coils is dependent on how the turns of one of the coils interact with the magnetic field generated by the other coil.

The inductively coupled WPT system model is defined as a two-port network with the media in between. The efficiency of an inductively coupled system is defined as the ratio of the available power at the secondary and the provided power at the primary side ($\eta = P_L/P_S$). Zargham et al. express the efficiency of inductively coupled WPT system as [12]:

$$\eta_{max} = \frac{k^2 Q_1 Q_2}{\left(1+\sqrt{1+k^2 Q_1 Q_2}\right)^2} = \frac{\chi}{(1+\sqrt{1+\chi})^2} \tag{7.5}$$

where $\chi = \frac{|Z_{12}|^2}{rZ_{11}rZ_{22} - rZ_{12}^2} = \frac{|Y_{12}|^2}{rY_{11}rY_{22} - rY_{12}^2}$, and k is the coupling coefficient between the two coils in the specified media.

As inferred from the above equation, the efficiency of this two-port network is a function of the Q-factor of each coil along with their relative distance, coupling coefficient and the specified medium. In order to seek the coupling coefficient and the efficiency of the two coils when placed on top of each other (RX coil is perfectly aligned at the centre of TX coil), electromagnetic (EM) simulation using momentum simulator was performed. The maximum distance between the two coils was chosen to be 1 cm with air as the specified media. Figures 7.19 and 7.20 show the 2-D and

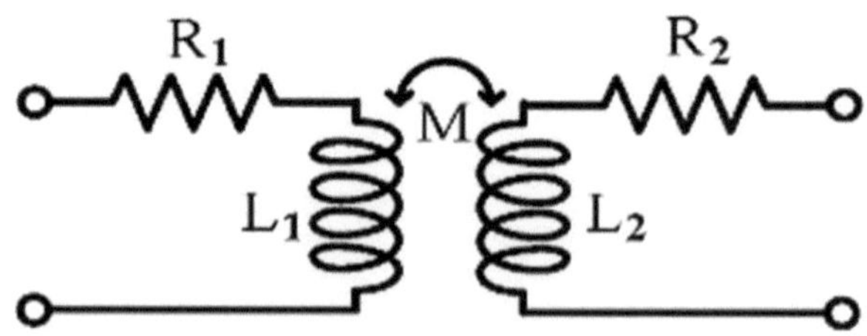

Fig. 7.18 Simple model for inductively coupled WPT system [12]

Fig. 7.19 2-D view of TX2-RX coils coupling in momentum simulation

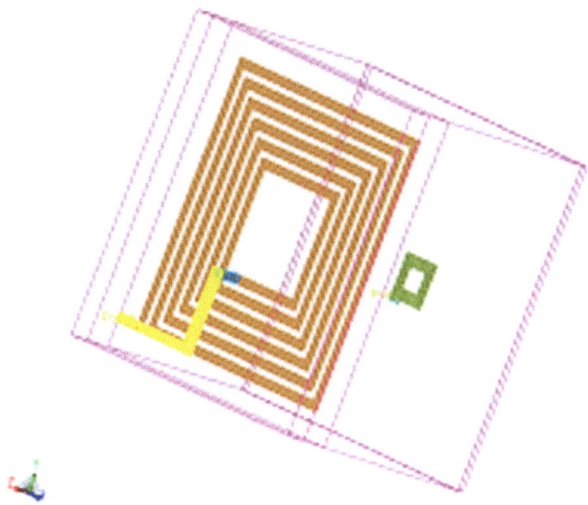

Fig. 7.20 3-D view of TX2-RX coils coupling in momentum simulation

3-D views of the simulation set-up in momentum, respectively. Equations (7.6) and (7.7) are used to calculate the mutual inductance (M) and the coupling coefficient (k) between the coils, respectively.

$$M = \frac{Im(Z_{21})}{2\pi f} \tag{7.6}$$

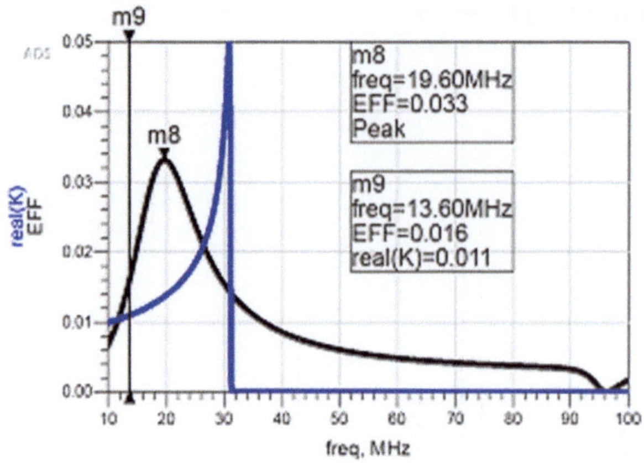

Fig. 7.21 Efficiency and coupling coefficient of TX1 and RX versus frequency at 1 cm distance

$$k = \frac{M}{\sqrt{L_1 L_2}} \tag{7.7}$$

Figure 7.21 represents the efficiency along with the coupling coefficient k of TX1 coil when placed on top of RX coil with 1 cm distance. As it is depicted on the plot, the maximum efficiency of 3.3% occurs at 19.6 MHz where k is 0.014. Both parameters are degraded at 13.6 mMHz and are found to be 1.6% and 0.011, respectively [12].

Figure 7.22 represents the efficiency along with the coupling coefficient k of TX2 coil when placed on top of RX coil with 1 cm distance. As it is depicted on the plot, the maximum efficiency of 13.3% occurs at 121.8 MHz where k is −0.036. Hence, better efficiency and coupling factor are achieved at higher frequencies as the maximum Q of the RX coil occurs at this range of frequencies. Furthermore, at ISM band of 40.6 MHz, an efficiency of 2.8% with a coupling coefficient of −0.01 is achieved as well.

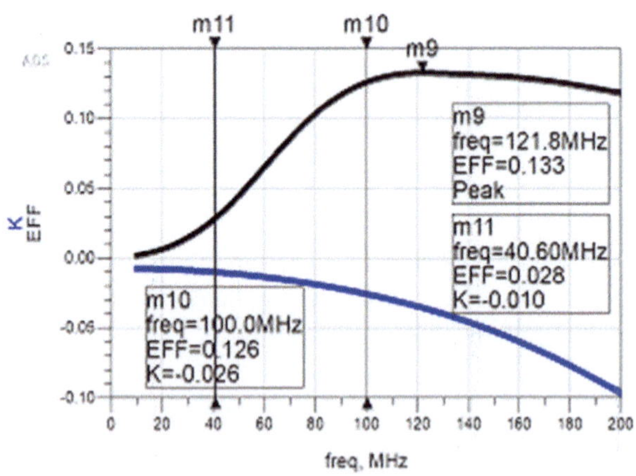

Fig. 7.22 Efficiency and coupling coefficient of TX2 and RX versus frequency at 1 cm distance

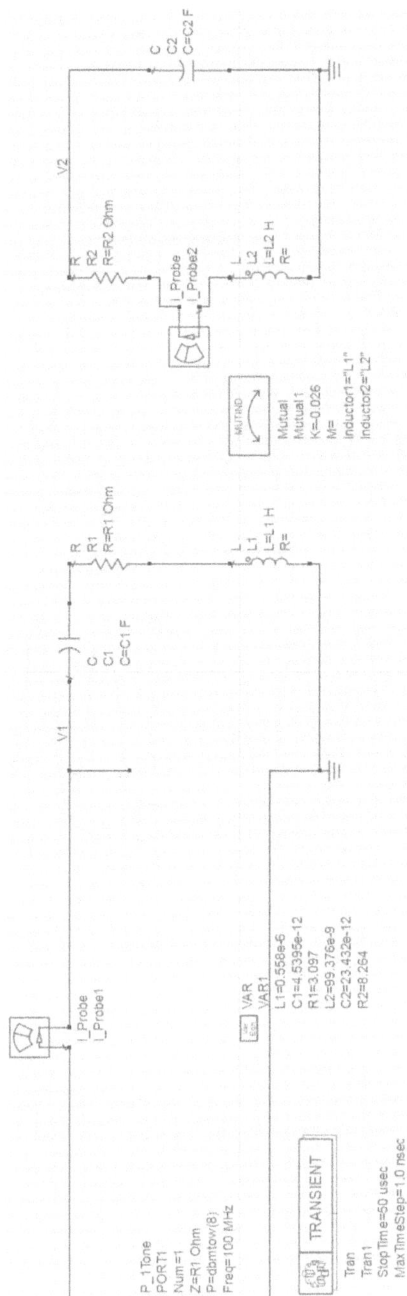

Fig. 7.23 Inductive coupling simulation with extracted parameters from EM simulation at 100 MHz

Table 7.3 Inductively coupled network parameters at 100 MHz

Parameter	TX2	RX
Q	113.269	8.219
L (nH)	558	108.099
R (Ω)	3.097	8.264
C (pF)	4.54	23.432

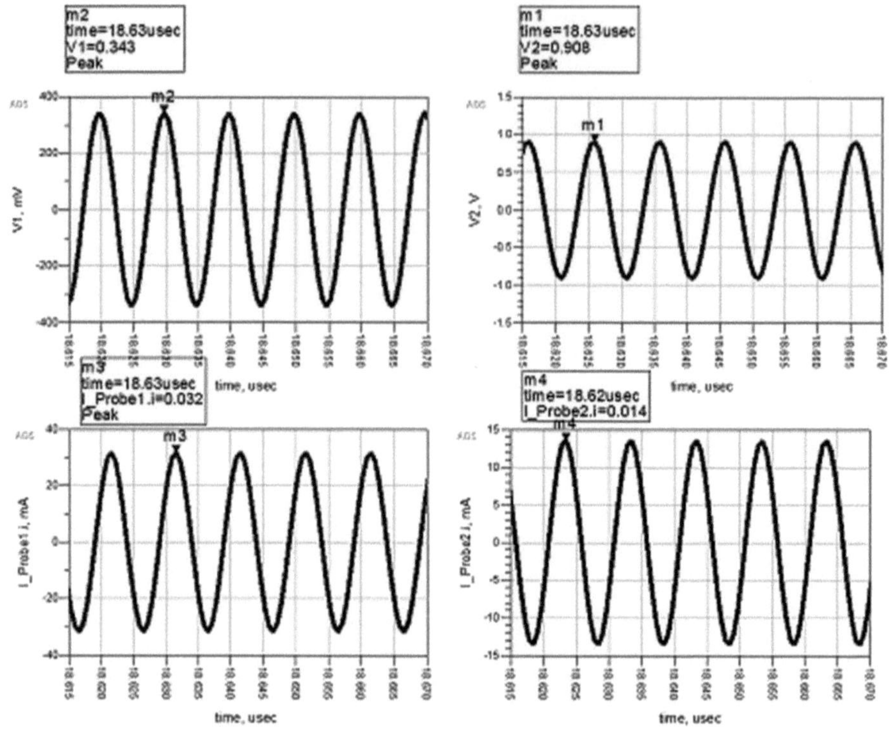

Fig. 7.24 Inductive coupling's primary and secondary voltages and currents waveforms at 100 MHz

In order to find out the value of the input power at the primary which results in the required peak-to-peak voltage at the input of the on-chip rectifier, the coupling coefficient, primary and secondary inductor values, and their corresponding resistances were extracted from EM simulation. Figure 7.23 shows the schematic view in ADS with transient simulation, and Table 7.3 summarizes the extracted parameters of the two-port network at 100 MHz. The required capacitance to achieve resonance at 100 MHz is also included in the table for each coil. As it is seen from the schematic, an input power of 8 dBm is required at the primary to generate a sinusoidal voltage with a peak value of 900 mV at the secondary (required by the rectifier to power up the sensor). The primary and secondary voltages and currents waveforms are shown in Fig. 7.24.

Table 7.4 Simulated power budgets of the various blocks

Simulated power requested by blocks			Simulated power supply by WPT to blocks		
Block	Voltage	Current	Block	Voltage	Current
Sensor	1.5 V	600 μA	Rectifier output to sensor input	1.5 V	700 μA
Rectifier input peak I-V	900 mV peak	4 mA peak	RX Coil by WPT to rectifier input	900 mV peak	14 mA peak
TX coil	343 mV	32 mA	External unit circuitry	NA	**NA**

7.5 Power Budget

As discussed in the previous sections, the rectifier circuit should be capable of delivering the required voltage and current to the sensor and other integrated circuits [13]. On the other hand, the input of the rectifier circuit is supplied by means of inductive coupling between the TX and RX coils. Table 7.4 represents the required and achieved voltage and current at each stage of the system. The simulated results validate the system operation.

7.6 Conclusions

The design of implanted electronic sensors into dental structures is a challenging task. This chapter has shown the principles of a new novel system that is powered wirelessly and is capable of detecting humidity leaks. It was shown that the simple task of delivering DC power requires extensive study of the environment and impairments associated. Factors such as changing impedances with mouth mechanics and movement dynamics and data acquisition points round the bracket system must be carefully considered and designed.

Acknowledgements This work has been supported by the Semiconductor Research Corporation (SRC) under the Abu Dhabi SRC Center of Excellence on Energy-Efficient Electronic Systems (ACE^4S), Contract 2013 HJ2440, with funding from the Mubadala Development Company, Abu Dhabi, UAE.

References

1. T.M. Graber, *Orthodontics? Principles and Practice*, 3rd edn. (Saunders, Philadelphia, 1972)
2. https://en.wikipedia.org/wiki/Dental_braces

3. Y.T. Liao, H. Yao, A. Lingley, B. Parviz, B.P. Otis, A 3- W CMOS glucose sensor for wireless contact-lens tear glucose monitoring. IEEE J. Solid-State Circuits **47**(1), 335–344 (2012)
4. M. Kuhl et al., A wireless stress mapping system for orthodontic brackets using CMOS integrated sensors. IEEE J. Solid-State Circuits **48**(9), 2191–2202 (2013)
5. M. Ahmadi, A wireless implantable microsystem for continuous blood glucose monitoring. PhD Thesis, University of Calgary, 2007
6. M. Kuhl et al., A telemetric stress-mapping CMOS chip with 24 FET-based stress sensors for smart orthodontic brackets, in *IEEE International Solid-State Circuits Conference (ISSCC)* (San Francisco, CA, 2011), pp. 108–110
7. K. Kotani, T. Ito, High efficiency CMOS rectifier circuit with self-Vth-cancellation and power regulation functions for UHF RFIDs, in *EEE Asian Solid-State Circuits Conference (ASSCC'07)* (Jeju, 2007), pp. 119–122
8. M. Zargham, P.G. Gulak, High-efficiency CMOS rectifier for fully integrated mW wireless power transfer, in *IEEE International Symposium on Circuits and Systems (ISCAS)* (Seoul, 2012), pp. 2869–2872
9. M. Taghadosi, L. Albasha, A. Zakaria, N.A. Quadir, N. Qaddoumi, Feasibility of employing orthodontic braces as radiating antenna for implantable devices, in *2017 IEEE Asia Pacific Microwave Conference (APMC)* (2017), pp. 499–502
10. R.L. Bunch, D.I. Sanderson, S. Raman, Quality factor and inductance in differential IC implementations. Appl. Notes IEEE Microw. Mag. **3**(2), 82–92 (2002)
11. S. Gabriel, R.W. Lau, C. Gabriel, The dielectric properties of biological tissues: III. Parametric models for the dielectric spectrum of tissues? Phys. Med. Biol. **41**(11), 2271–2293 (1996)
12. N.A. Quadir, L. Albasha, M. Taghadosi, N. Qaddoumi, B. Hatahet, Low-power implanted sensor for orthodontic bond failure diagnosis and detection. IEEE Sens. J. **PP**(99), 1–1 (2018). https://doi.org/10.1109/JSEN.2018.2791426
13. N.A. Quadir, M. Taghadosi, L. Albasha, N. Qaddoumi, Design of self powered chip integrated humidity sensor, in *IEEE 59th Midwest Symposium on Circuits and Systems* (Abu Dhabi, United Arab Emirates, 2016), 16–19 October 2016

Chapter 8
Energy-Efficient Body Area Network Transceiver Using Body-Coupled Communication

Jerald Yoo

8.1 Introduction

Body area network (BAN) refers to a network in and around the human body. There are three points to consider in BAN: 1) communication quality, 2) power consumption of the transceiver that forms the BAN, and 3) data rate.

Standard wireless communications such as wireless personal area network, Bluetooth or ZigBee can be used to support such network. However, it has been found that human body absorbs a significant portion of RF energy[1–3]; therefore, such RF-based communication is not an ideal solution for BAN. This absorption phenomenon is called "body shadowing effect", which requires a transmitter (TX) and a receiver (RX) to be in Line-of-Sight (LoS). Unfortunately, when it comes to BAN, having LoS is not always possible, since often human body is located in between the TX and the RX. In other words, the human body works as a blockage for the RF-based TX and RX. The body shadowing effect may attenuate the transmitted signal strength by more than 20 dB [4], and in addition, when RF antenna is placed in the vicinity of the human body, an additional 10–15 dB attenuation may occur [5].

In this chapter, we will explore the possibility of using the human body itself as a communication medium, which is called body-coupled communication (BCC) [3, 6–9]. By doing so, the human body is exploited as a transmission passage, rather than of blockage.

In BAN applications, power consumption of the transceiver (TRX) is also a concern (i.e., the lower power consumption, the better). Figure 8.1 shows the communication range (y axis to the right) and the power consumption (y axis to the left) for BAN versus data rate (x axis), data from recent publications [9].

As we can see, UWB/IRUWB has sufficient data rate, but power consumption is too high (>~10 s mW). Bluetooth Low Energy has lower power consumption than

J. Yoo (✉)
National University of Singapore, Singapore, Singapore
e-mail: jyoo@nus.edu.sg

Ibrahim (Abe) M. Elfadel and M. Ismail (eds.), *The IoT Physical Layer*,
https://doi.org/10.1007/978-3-319-93100-5_8

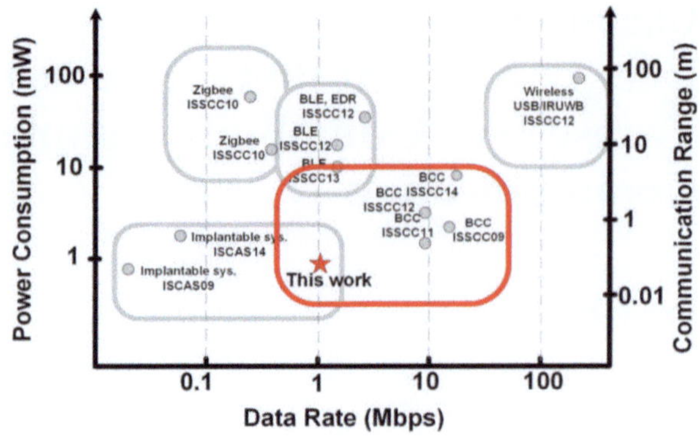

Fig. 8.1 Data rate versus power consumption of transceiver/communication range in BAN

the UWB/IRUWB, but still consumes high power (~10 mW). On the other hand, ZigBee / Implantable RF consumes much lower power (~mW) but at the cost of low data rate or low communication range, neither of which is suitable for BAN applications. Therefore, there is still an unmet gap which trades off the data rate and power consumption/communication range.

8.2 Body Channel Characteristics

8.2.1 Body-Coupled Communication Methods

The BCC can be done in three different ways: (a) magnetic coupling, (b) galvanic current, and (c) capacitive coupling, as shown in Fig. 8.2 [9]. In magnetic resonance coupling (Fig. 8.2a), resonant coils are utilized at the TX and RX to transmit and recover magnetic energy [10]; it is effective for coverage of up to 1 m distance, but with the existence of motion artifact, it is challenging to maintain constant resonance. Also, usage of coil limits on-chip integration, which results in a limitation on the form factor. In galvanic coupling (Fig. 8.2b), the TX and RX each has a pair of electrodes which subsequently transmits and receives a differential signal through the body [10]. The galvanic current is relatively independent with respect to the environmental variations; therefore, it is reliable. However, short-circuit current through the differential pairs at TX results in a high pathloss at frequencies over 1 MHz, which means it is preferred for short-distance communication of below 15 cm; hence, this is not an ideal BAN method.

In contrast, the capacitive coupling (Fig. 8.2) utilizes only a single electrode at both TX and RX [3, 11]. In this configuration, the TX and RX ground electrodes are floating; TX transmits the signal through the body, and the RX captures the signal through its single electrode. The floating ground electrodes at both ends create a return path through the parasitic capacitors between the floating ground and the earth ground. The fact that the return path is formed by the parasitic capacitor means

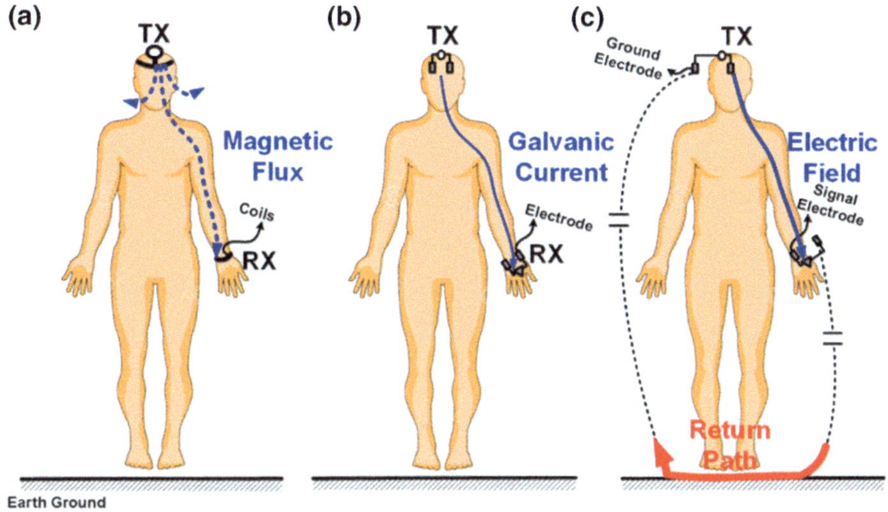

Fig. 8.2 BCC coupling methods: **a** Magnetic flux, **b** Galvanic current, and **c** Capacitive coupling

that this method is suitable for wearable applications, but at the cost of varying return path strength; for the capacitive-coupled BCC to become practical, the varying return path and multi-path issues should to be tackled. Nonetheless, this method enables higher data rates than the galvanic coupling, and it does not use any extra coils, so miniaturization is easier.

8.2.2 The Human Body Channel in Capacitive BCC

To see the effect of the body as a communication channel, we should first understand the pathloss of the channel.

Figure 8.3 shows the channel measurement setup for the capacitive BCC. It is important to note that we should avoid pathloss measurement by connecting TX (RF signal generator) and RX (spectrum analyzer) directly to body, since both instruments are firmly attached to the earth ground by the power cable. This means a strong return path will be explicitly formed between the two instruments, thereby resulting in less pathloss than in the wearable capacitive BCC case (where the return path is formed by the parasitic capacitance). To be more accurate, we can insert Balun as in Fig. 8.3 [3, 12, 13]. Figure 8.4 shows the frequency sweep channel pathloss measurement for different distances. Figure 8.4a shows the bandwidth up to 1 GHz, where the pathloss exhibits a band-pass shape over frequency. This trend is the same regardless of the location of the body, i.e., within head, head-to-arm, within arm, arm-to-arm, head-to-leg. Figure 8.4b is a closer look between 15–140 MHz,

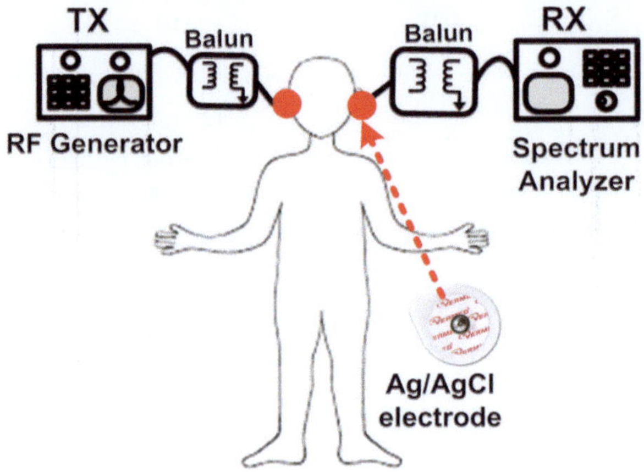

Fig. 8.3 Human body channel pathloss measurement setup for the capacitive coupled BCC. This measurement covers the whole human body area, from head to arm and legs

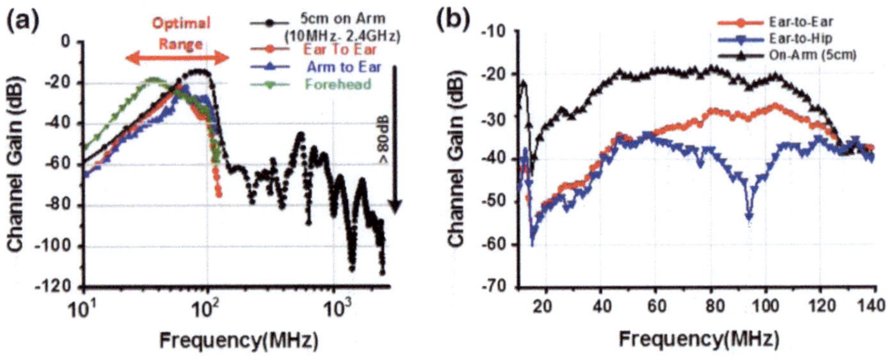

Fig. 8.4 Human body channel pathloss measurement result

where the different communication ranges result in different pathloss amounts. This measurement reveals an interesting phenomena: the smallest BCC pathloss in the human body is between 40–120 MHz. Hence, we can utilize the same bandwidth for the capacitive BCC. Note that RF-based radio under the same environment shows 20–40 dB more pathloss than the BCC, due to the body shadowing effect [14].

8.3 Non-Ideal Properties in BCC and Strategies to Overcome Such Issues

Varying Ground Connection and Return Path (Ground Effect): As discussed before, the return path in capacitive-coupled BCC relies on the parasitic ground of TX and RX; hence, the ground connection and the return path strength are dynamically changing. We call this phenomenon as the *Ground Effect*. It is therefore important to adapt the TRX to the varying channel condition.

Multi-Path: The signal may face multiple return paths with respect to the earth ground [8]. In addition, the signal may also experience multiple *forward paths* due to body compositions, where each one has a different conductivity and permittivity. When combined, these effects may steer the TX signal to follow different paths, depending on the location of the body where the transmission is initiated.

Varying Skin-Electrode Connection: The TX and RX electrode may suffer varying skin-electrode connection, due to misalignment or body movement. This may result in a weak connection that could affect the amount of the pathloss.

In-Band Interference: Human body may work as an antenna, so RF signals in the BCC in-band (e.g., FM radio or walkie-talkie) can interfere with the BCC signal, degrading the bit error rate (BER) or signal-to-noise ratio (SNR).

The aforementioned issues are the challenges capacitive-coupled BCC faces. These must be addressed by the TRX for a stable and reliable communication. Adoption of orthogonal frequency-division multiplexing (OFDM) can mitigate these issues, but at the cost of higher power consumption at TRX. Also, typical OFDM exhibits a high peak-to-average power ratio (PAPR), which makes the subsequent digital-to-analog converters (DACs) and power amplifier (PA) design challenging. This is because of the wide linear region and high dynamic range requirements resulting from high PAPR [8, 9].

8.4 Hybrid (Pseudo)-OFDM

To mitigate the issues mentioned in the previous section, we can adopt the Pseudo (or, Hybrid)-OFDM (P-OFDM) [6, 8]. Figure 8.5 shows the P-OFDM concept.

P-OFDM operates by transmitting the baseband OFDM symbols through an adaptive frequency hopping (AFH) FSK physical (PHY) layer TRX. Although the PHY

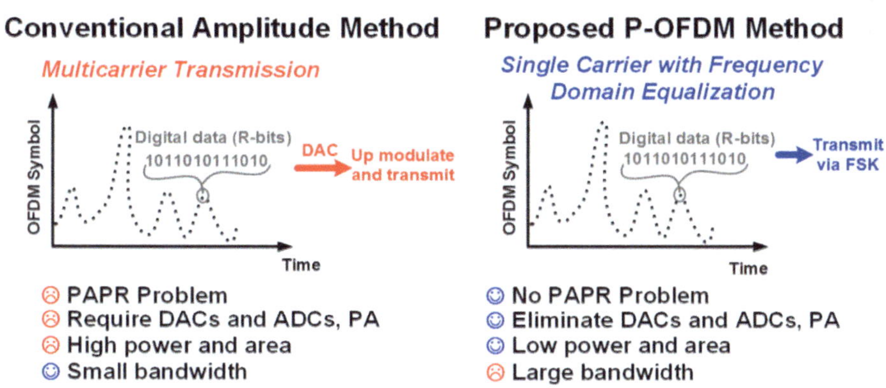

Fig. 8.5 Pseudo-OFDM concept

does not operate in the orthogonal multiplexing fashion, the OFDM symbols in baseband act as a pre-coded signal, which enables the channel estimation to happen in the frequency domain. This means the P-OFDM significantly reduces the channel estimation and equalization overhead, in contrast to FSK modulation with time domain equalization [6, 8]. Combination of the FFT processing and the cyclic prefix (CP) shows a similar performance to an OFDM system at lower complexity [6]. In the P-OFDM, the effective energy per bit will be given by the following equation [15]:

$$\bar{E}_b = \frac{1}{M} \cdot \sqrt{\frac{R}{2}} \cdot E_b \text{....................(1)}$$

where R is the number of bits for representing the OFDM symbol in the P-OFDM, and M is the number of constellation points adopted in the P-OFDM. E_b denotes the signal energy for each data bit. Therefore, the BER can be expressed as [15]:

$$BER_{P-OFDM} = \frac{1}{2} \cdot erfc\left(\sqrt{\frac{1}{M} \cdot \sqrt{\frac{R}{2}} \cdot \frac{E_b}{2N_0}}\right) \text{....................(2)}$$

where E_b/N_0 is the energy-per-bit-to-noise PSD ratio. Here, we can see that the BER_{P-OFDM} will be much lower than the BER_{FSK} [15, 16].

We should note that P-OFDM sends signal, through FSK with expanded bandwidth; the OFDM symbols are kept in the digital domain, and not converted to analog, so there is no need for the ADCs, DACs, and PA which are used in the conventional OFDM TRXs. Consequently, it will be free from the PAPR problem. Figure 8.5 shows the difference between the P-OFDM and the conventional OFDM. In the conventional OFDM, the OFDM symbol (R-bits) is converted to analog using DAC and then forwarded for up modulation. In contrast, the P-OFDM directly transmits these R-bits through the FSK.

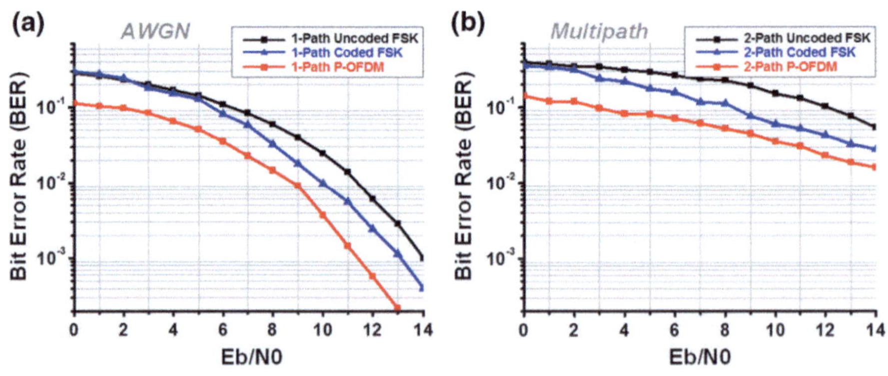

Fig. 8.6 Simulated BER performance versus SNR of an uncoded FSK, coded FSK, and P-OFDM, modeled under **a** AWGN and **b** 2-paths channel

Figure 8.6 shows the simulated BER waterfall curve of the P-OFDM modulation compared with an uncoded FSK and a coded FSK (Hamming (7-4)). The P-OFDM shows about 74 and 53% reduction in BER, when compared to a uncoded FSK and coded FSK, respectively. The multi-path scenario is also simulated; we assume a two-path case with the second path experiencing a 1ns delay plus a 0.5 dB attenuation. In both cases, we can observe that P-OFDM outperforms FSK-only case.

8.5 Pseudo-OFDM BCC Transceiver Design

Figure 8.7 shows the BCC transceiver exploiting the P-OFDM [7]. The input digital data is classified and grouped into lower data rate streams. Then, it is transmitted concurrently over 64 subcarriers. A 16-Quadrature Amplitude Modulation (QAM) having 64 subcarriers is utilized at the data rate of 1Mbps; the incoming data stream (1Mbps) is fed into the 16-QAM mapping block that generates 13 bits (I, Q) constellation points for each subcarrier input. The IFFT block in the TX side then modulates these constellation points. Subsequently, the cyclic prefix (CP) (which is a copy of the last 16 IFFT output subcarriers and 16 pilot subcarriers) is added for synchronization and channel estimation [7]. Each of these subcarriers' output is composed of 26 bits (13 imaginary + 13 real). Then, the output bits are streamed and passed into the miniaturized direct switching binary FSK modulator. In the PHY FSK modulation, the channel quality is constantly checked to identify a clean channel by employing adaptive frequency hopping (AFH). On the RX, the signal is demodulated by the miniaturized direct conversion FSK demodulator, with the CP being disregarded. The data stream is then forwarded to 64-point FFT, followed by the DFE for channel estimation using the pilot subcarriers [6, 7]. Finally, we demodulate the transmitted bits through the 16-QAM de-mapping block to recover the original transmitted data.

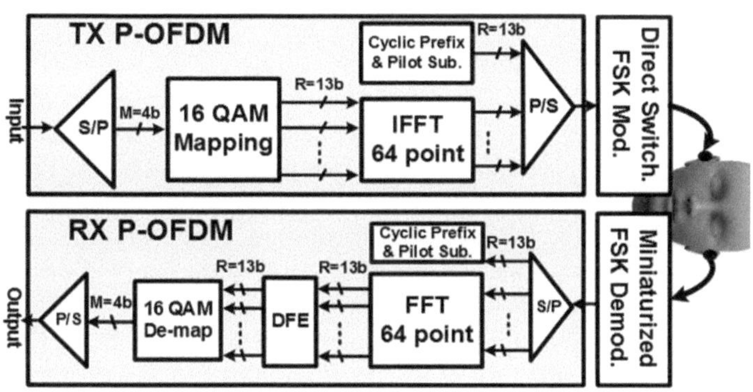

Fig. 8.7 Pseudo-OFDM BCC Transceiver architecture

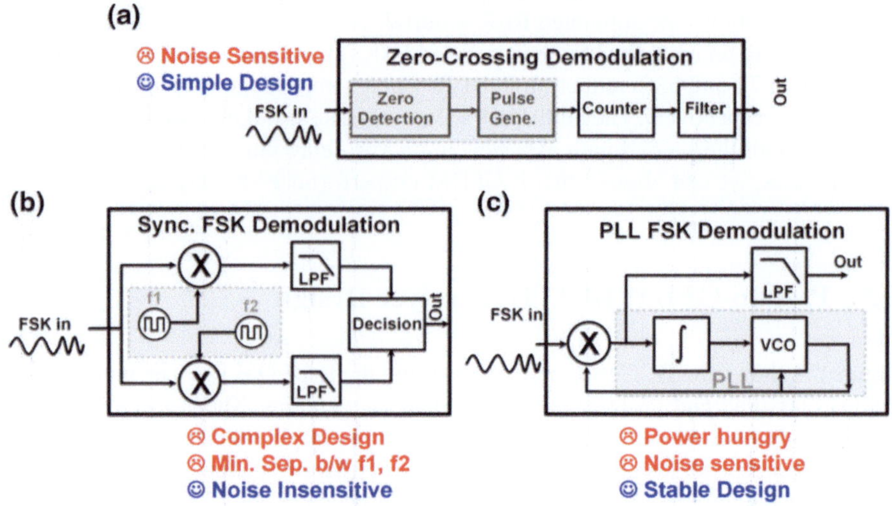

Fig. 8.8 Conventional FSK demodulation methods: **a** zero-crossing, **b** synchronous, and **c** PLL-based

PHY FSK: Asynchronous FSK demodulation employs an envelope detector for simple demodulation, but this is inappropriate for the BCC because the channel pathloss varies over the bandwidth. Figure 8.8 denotes the conventional FSK demodulation [7]. A zero-crossing demodulator (a) is area-efficient, but in BCC, due to the varying channel condition, it is not reliable. Synchronous FSK demodulator (b) uses two local carriers, with a decision circuit that decides the output bit. It is robust with respect to noise, but at the cost of a more complex design. Using a PLL (c) gives yet another design: with its closed loop, the feedback path adjusts the VCO to match the incoming signal frequency. However, in such design, a separate low-pass filter (LPF) is required to recover the original message. It consumes higher power than the other types.

Figure 8.9a highlights the FSK modulation and demodulation blocks within the presented P-OFDM BCC TRX. The AFH block seeks clean channel and drops the dirty channel to avoid hopping into a contaminated channel. The channel selector performs direct switching FSK modulation. The presented P-OFDM TRX is implemented, tested, and verified in 65nm 1P7M CMOS process. Figure 8.9b shows the LPF output (black) and the RX output (red) from the TRX. Figure 8.10 shows the measured eye diagram, with the RX input power of −74 dBm and the data rate of 1Mbps, at ear-to-ear communication. The eye diagram (a) shows the FSK RX-only result (both P-OFDM TX and RX are disconnected, measured at the FSK output); The eye diagram in the bottom (b) shows the P-OFDM TRX result (measured at the P-OFDM RX output). We can find that P-OFDM shows wider eye opening (960 ns) than the FSK TRX-only case (900 ns). The measured BER of the P-OFDM TRX and the FSK-only TRX is 10^{-7} and 10^{-6}, respectively.

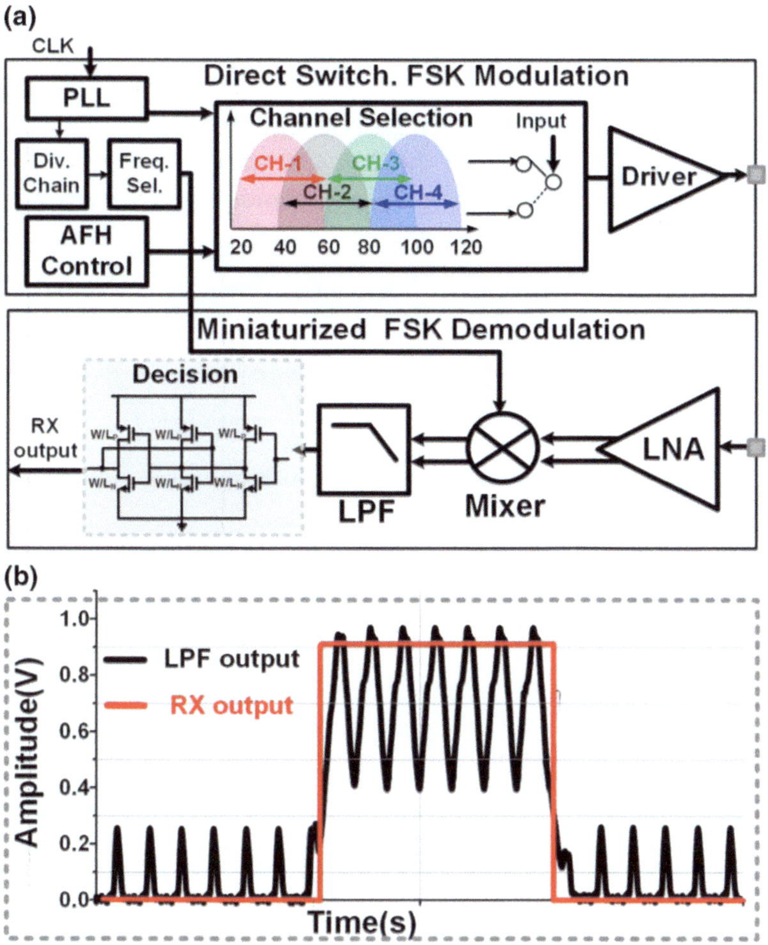

Fig. 8.9 **a** The architecture of the miniaturized zero-IF FSK demodulator within the implemented P-OFDM TRX, and **b** its measurement waveform

Figure 8.11 shows the measured PSD of the 64-point IFFT output, where the CP length of 16 subcarriers does not increase the variations in the desired band, and more importantly, the subcarriers do not overlap at their center frequencies; consequently, ISI is minimized. Figure 8.12 plots the constellation of the 16-QAM demodulated bits, while having an error vector magnitude (EVM) of −27.3 dB. In this measurement, the PRBS generator's random sequence is 212 bits.

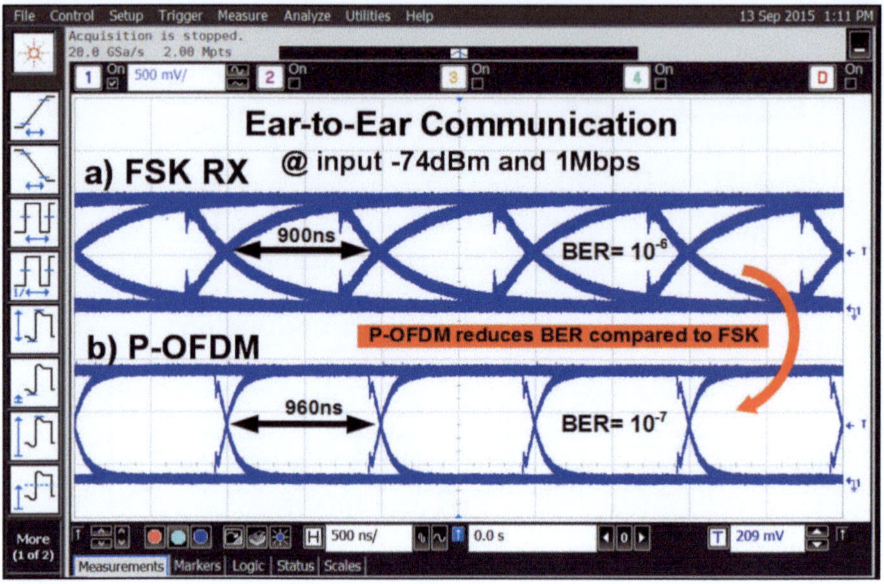

Fig. 8.10 Eye diagram from P-OFDM TRX

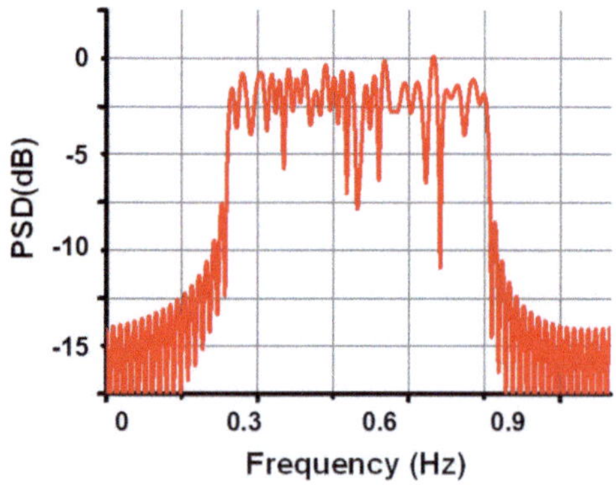

Fig. 8.11 Measured PSD of the 64-point IFFT output

Figure 8.13 shows the FSK demodulator block output of the designed P-OFDM TRX. We can see that the RX successfully recovers the original data. Figure 8.14 shows the power breakdown of the designed P-OFDM transceiver, with overall TX and RX power consumption of 1.1 and 1.4 mW, respectively. Figure 8.15 is the chip micrograph and the performance summary of the P-OFDM TRX.

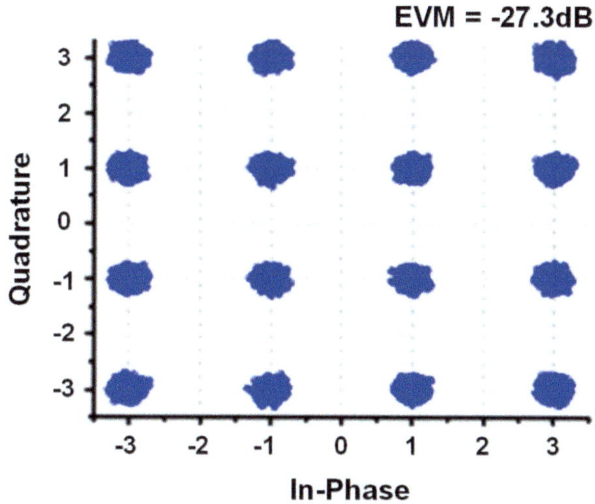

Fig. 8.12 Measured 16-QAM constellation plots of the received bits

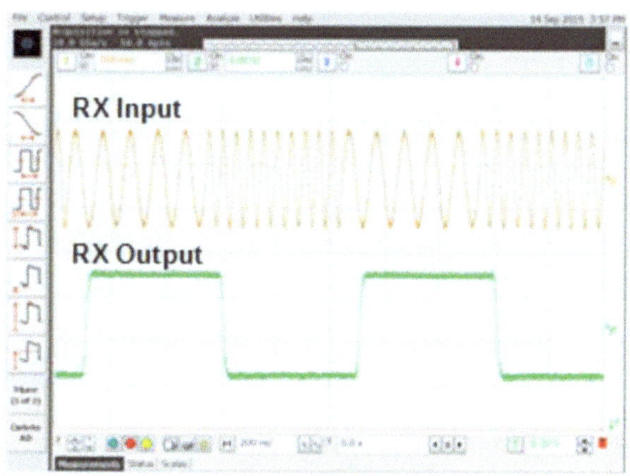

Fig. 8.13 Input and output of the FSK demodulation

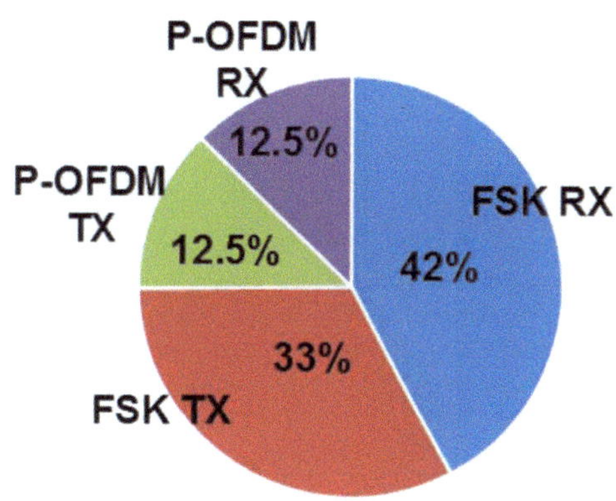

Fig. 8.14 Power breakdown of the P-OFDM TRX

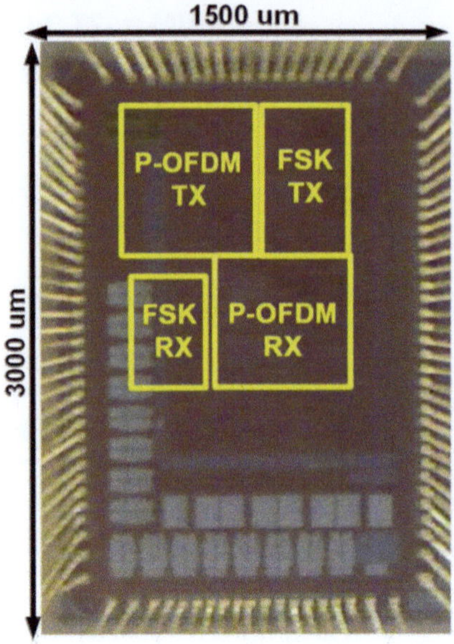

Implemented Results		
Process		65nm CMOS
Area		2.13mm^2
Supply Voltage		1.1V
Power	TX	1.1mW
	RX	1.4mW
Date Rate		1Mbps
Sensitivity		-78dBm
RX Noise Fig.		10.4dB
BER@ -74dBm		10^{-7}
Input Impedance		100Ω
1-dB Compression Point		-20.7dBm
PLL Phase Noise		-108 dBc/Hz @1MHz

Fig. 8.15 Chip micrograph of the P-OFDM TRX and the performance summary

8.6 Conclusions

Body-coupled communication is an attractive means to realizing an energy-efficient BAN. It overcomes the body shadowing effect, which is the biggest hurdle in RF TRX around the human body. BCC utilizes the human body itself as a communication medium. We have seen the channel pathloss characteristics of the body channel. In capacitive BCC, which shows the most energy-efficient approach among the BCC, return path is formed by parasitic ground (ground effect), and it varies over time, space, and posture. Therefore, in this chapter, we have presented a Pseudo (Hybrid)-OFDM technique to overcome the fluctuating channel pathloss. The implemented four-channel AFH P-OFDM BCC TRX is free from the PAPR problem as it eliminates the DACs, ADCs, and PA from the conventional OFDM design. The presented P-OFDM TRX in 65 nm 1P7M CMOS occupies an active area of 2.13 mm^2 at 1.1 V supply. It shows an average 74% BER improvement (compared to the FSK-only case) with a 1Mbps data rate.

Acknowledgements The author acknowledges Dr. Wala Saadeh and Mr. Yonatan Kifle for their research, design, and implementation of the BCC transceivers introduced in this chapter. He also thanks Mr. Owais Talaat Waheed for his technical help with the LaTeX typesetting of this chapter. This work has been supported by the Semiconductor Research Corporation (SRC) under the Abu Dhabi SRC Center of Excellence on Energy-Efficient Electronic Systems (ACE^4S), Contract 2013 HJ2440, with funding from the Mubadala Development Company, Abu Dhabi, UAE.

References

1. J. Yoo, H.-J. Yoo, Emerging low energy wearable body sensor networks using patch sensors for continuous healthcare applications, in *Proceedings of the IEEE Engineering in Medical and Biology Conference (EMBC)*, (2010), pp. 6381–6384
2. J. Yoo, Is RF really an answer for healthcare applications? - low power circuits for BAN, in *IEEE International Solid-State Circuits Conference (ISSCC) Evening Panel Discussion Session 3: What is the next RF frontier?*, (2012)
3. Y. Kifle, H.-S. Kim, J. Yoo, Human body and head characteristics as a communication medium for body area network, in *Proceedings of the IEEE Annual International Conference of the Engineering in Medicine and Biology Society (EMBC)*, (2015), pp. 1845–1849
4. J. Yoo, N. Cho, H.-J. Yoo, Analysis of body sensor network using human body as the channel. in *Proceedings of the ACM International Conference Body Area Network*, (2008)
5. A. Kara, H.L. Bertoni. Blockage/shadowing and polarization measurements at 2.45 GHz for interference evaluation between bluetooth and IEEE 802.11 WLAN, in *Proceedings of the IEEE Antennas Propagation Society International Symposium*, (2011), pp. 376–379
6. W. Saddeh, Y. Kifle and J. Yoo. A Hybrid OFDM body coupled communication transceiver for binaural hearing aids in 65 nm CMOS. in *Proceedings of the IEEE Symposium on Circuits and Systems (ISCAS)*, (2015), pp. 2620–2623
7. W. Saadeh, M.A.B. Altaf, H. Alsuradi, J. Yoo, A pseudo OFDM with miniaturized FSK demodulation body-coupled communication transceiver for binural hearing aids in 65 nm CMOS. IEEE J. Solid-State Circuits (JSSC) **52**(3), 757–768 (2017)
8. W. Saadeh, M.A.B. Altaf, H. Alsuradi, J. Yoo, A 1.1 mW ground effect-resilient body-coupled communication transceiver for head and body area network, in *Proceedings of the IEEE Asian Solid-State Circuits Conference (A-SSCC)*,(2016), pp. 201–204
9. W. Saadeh, M.A.B. Altaf, H. Alsuradi, J. Yoo, A 1.1 mW ground effect-resilient body-coupled communication transceiver with pseudo OFDM for head and body area network. IEEE J. Solid-State Circuits (JSSC) **52**(10), 2690–2702 (2017)
10. M.S. Wegmueller, M. Oberle, N. Felber, N. Kuster, W. Fichtner, Signal transmission by galvanic coupling through the human body. IEEE Trans. Instrum. Meas. **59**(4), 963–969 (2010)
11. M. Pereira, G. Alvarez-Botero, F. Rangel de Sousa, Characterization and modeling of the capacitive HBC channel. IEEE Trans. Instrum. Meas. **64**(10), 2626–2635 (2015)
12. J. Yoo, Body coupled communication: towards energy-efficient body area network applications, in*Proceedings of the IEEE International Symposium on Radio Frequency Integration*, (2017)
13. N. Cho, J. Yoo, S.-J. Song, J. Lee, S. Jeon, H.-J. Yoo, The human body characteristics as a signal transmission medium for intrabody communication. IEEE Trans. Microw. Theory Technol. **55**(5), 1080–1086 (2007)
14. A. Kara, H.L. Bertoni, Blockage/shadowing and polarization measurements at 2.45 GHz for interference evaluation between Bluetooth and IEEE 802.11 WLAN, in *Proceedings of the IEEE International Symposium Antennas Propagation*, (2011), pp. 376–379
15. J.G. Proakis, M. Saheli, *Digital Communication.*, pp. 95–148, (McGraw-Hill, Singapore, 2008)
16. S. Haykin, *Communication Systems*, pp. 368-417 (Wiley, USA, 2001)

Chapter 9
Ultra-Low-Power ECG Processor for IoT SOCs

Temesghen Tekeste, Yonatan Kifle, Hani Saleh, Baker Mohammad and Mohammed Ismail

9.1 Introduction

Among all global deaths, cardiovascular disease contributes the largest percentage of death (31% in the year 2012 [1]). Moreover, sudden cardiac arrest which is related to ventricular arrhythmia is the dominant cause of death of adults in the USA [2]. Monitoring of the heart activity and being able to predict heart anomaly would reduce the number of sudden cardiac deaths. Advances in health care devices and the emergence of the Internet of Things (IoT) have enabled wearable devices for continuous monitoring. Cardiac signals of a variety of vital signs are well monitored using electrocardiography (ECG), and it is vital to develop energy-efficient wearable systems to acquire and process ECG signals.

ECG devices provide real-time monitoring of cardiac activities, and prediction systems could alert patients prior to any critical situations. Prediction systems would aid in reducing deaths due to sudden heart failures such as sudden cardiac arrest (SCA). SCA originates from ventricular arrhythmia (VA), and VA detecting systems have been reported in [3, 4]. However, there are only few systems that are capable of prediction. A system that can predict VA up to 3 hours prior to its onset is reported in [5]. In order to perform detection or prediction, the first step is to extract ECG

T. Tekeste (✉) · Y. Kifle · H. Saleh · B. Mohammad · M. Ismail
Khalifa University, Abu Dhabi, UAE
e-mail: temesghen.habte@kustar.ac.ae

Y. Kifle
e-mail: yonatan.kifle@kustar.ac.ae

H. Saleh
e-mail: hani.saleh@kustar.ac.ae

B. Mohammad
e-mail: baker.mohammad@kustar.ac.ae

M. Ismail
e-mail: Ismail@kustar.ac.ae

Ibrahim (Abe) M. Elfadel and M. Ismail (eds.), *The IoT Physical Layer*,
https://doi.org/10.1007/978-3-319-93100-5_9

features and define parameters that could generate accurate classification results. ECG features include P, QRS and T waves that could be utilized to define factors and intervals for prediction or detection systems. Various techniques have been developed for extracting such features, include discrete wavelet transform (DWT) [6] and time domain analysis [7]. Efficient processing platforms are crucial to enable ultra-low-power operation for wearable devices, this in turn necessitates to design optimization at both algorithm and architecture levels.

9.2 ECG Processing Architecture

The proposed ECG system architecture is shown in Fig. 9.1. Its main components are the analog front end, ECG feature extraction and VA prediction. The first stage is the analog front end, and it forms the interface between the sensors and the digital processing. The second stage does the ECG feature extraction which delineates the ECG fiducial points. The last stage is a VA prediction system based on the Naive Bayes classifier. VA prediction is accomplished using a unique set of features. A custom-designed FSM controls all the parts of the architecture and coordinates the timing of each stage. A data path width of 16 bits is used, and scaling is done whenever necessary during the arithmetic operations. Clock gating is applied as a power reduction technique especially for the ECG feature extraction block.

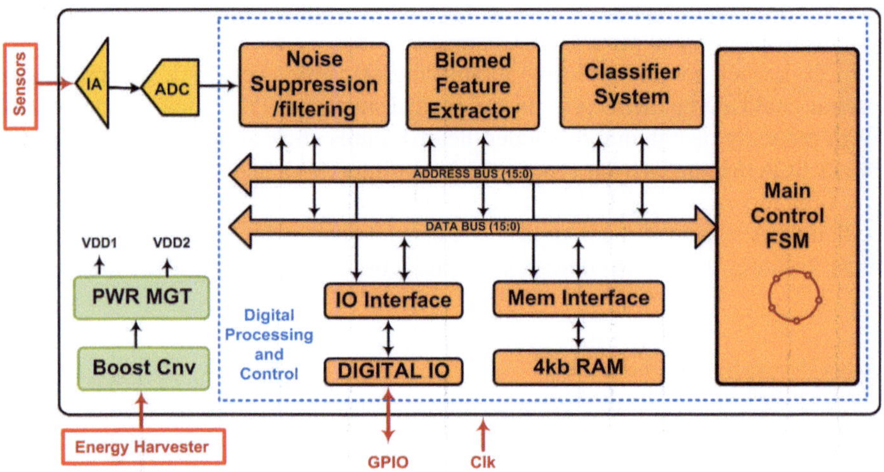

Fig. 9.1 Block diagram of proposed ECG processing architecture

9.2.1 Analog Front End

The analog frontend (AFE) needed for dealing with ECG signals should be able to deliver the ECG to the SoC with high signal-to-noise distortion ratio (SNDR). The amplitude of the ECG signal is in μV range, which sets stringent constraints on the system's noise requirement. It is also combined with $\pm 300\,\text{mV}$ DC offset that arises from the electrode−skin contact, hence the requirement for a high Common Mode Rejection Ratio (CMRR). Moreover, high power supply rejection ratio PSRR is essential as the ECG signal may be accompanied by the power−line interference which lies within the ECG band.

The complete AFE block diagram as part of the biomedical SoC is shown in Fig. 9.2. The AFE is composed of instrumentation amplifier with choppers to mitigate the dc offset and flicker noise, often followed by Gm-C filter to remove the chopping ripples. A programmable/variable gain amplifier (PGA/VGA) guarantees maximum signal-to-noise ratio (SNR) at all signal amplitudes. Finally the ADC digitizes the analog signal to be further processed by the SoC. As demonstrated in Fig. 9.2, there is a feedback path from the SoC to the VGA to control its gain digitally according to the input signal level. There is also a second feedback from the SoC to the instrumentation amplifier in order to cancel the inherent electrode dc offset.

Amplifier and Noise Cancellation Techniques

The whole system performs at a minimum dynamic range of 40 dB and SNDR of 32 dB so that it is able to deal with the ECG signal of as low as 10 μV. The gain of the first stage is limited to 26 dB in order not to saturate the following blocks at high-amplitude input. A second-order Gm-C filter is used to attenuate the chopping ripples.

The choppers modulate and demodulate the input signal at a frequency of 1 KHz which is fast enough as the highest ECG frequency is 150 Hz. The simulated gain bandwidth curve at 52 dB gain and 215 Hz bandwidth is shown in Fig. 9.3a 6 μV/sqrt(Hz) at the band of interest (0.5–150 Hz) is also achieved as shown in Fig. 9.3b. Figure 9.4 demonstrates the time domain simulation at several nodes of the system.

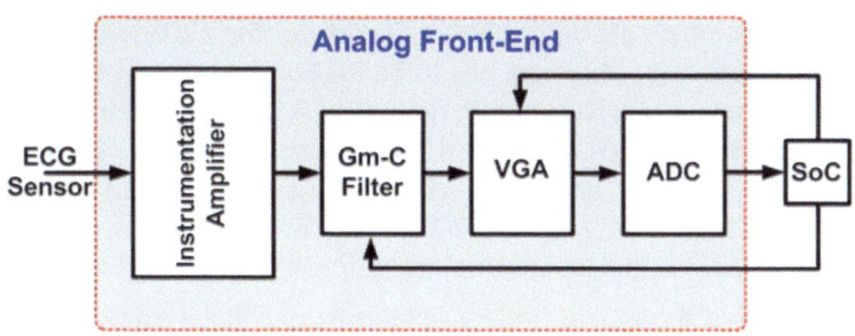

Fig. 9.2 The analog front-end block diagram

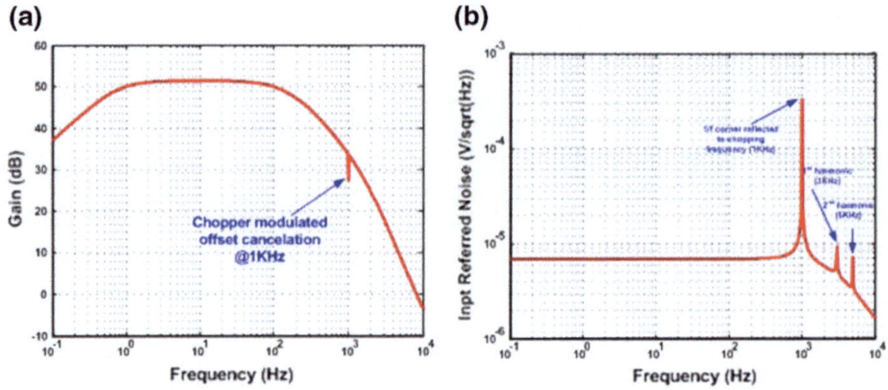

Fig. 9.3 **a** Gain bandwidth curve and **b** input-referred noise

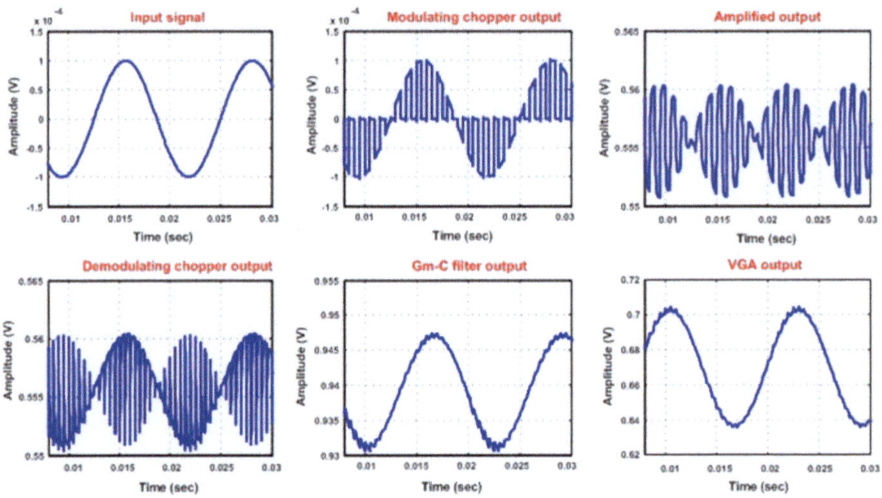

Fig. 9.4 Time domain simulation results

ADC

The AFE is followed by all-digital SoC. Therefore, the ECG where the noise is mitigated with the help of choppers has to be digitized for the SoC to perform the feature extraction and classification functions. The ADC does not need to operate at a fast clock rate which makes the SAR ADC an optimum choice. After performing several behavioural analysis, an 11-bit SAR ADC is implemented to digitize the ECG signal. The overall SAR ADC is shown in Fig. 9.5. The ADC operated at 1KS/s where the digitization is accomplished at sampling frequency multiplied by the number of bits (11 bits in this case).

During the sampling phase, the top plate of all the capacitors is connected to V_{CM} and the bottom plate to V_{IN}. During the hold phase, the top plate switch is opened

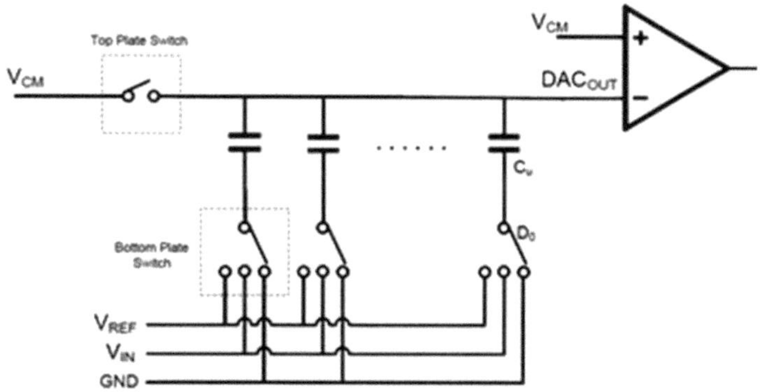

Fig. 9.5 Implemented SAR ADC

and the rest of the switches connect the bottom plates to ground therefore a charge of $-V_{IN}+ V_{CM}$ is stored in the capacitor arrays. The digitization is performed in the redistribution phase which can be summarized by the equation.

$$V_{DACout} = V_{IN} + V_{CM} + D_{10}\frac{V_{ref}}{2} + D_9\frac{V_{ref}}{4} + \cdots + D_0\frac{V_{ref}}{1024} \tag{9.1}$$

9.2.2 *ECG Feature Extraction Architecture*

The architecture for feature extraction is constructed based on an algorithm described in [8], and the architecture is described in [9]. Figure 9.6 illustrates the implementation architecture. The ECG processing flow starts with ECG preprocessing which is basically the filtering, followed by the ECG transformation and at last ECG delineation. Filtering enhances the SNR of the ECG wave and removes low frequency artefacts such as baseline wander and motion artefacts as well as high frequency interference. The QRS complex of an ECG wave has relatively higher amplitude than the other features, and its detection is a foundational step in feature extraction process. Robust and accurate detection of QRS complex is followed by refined determination of T and P waves. In the proposed architecture, curve length transform (CLT) is used for QRS detection and discrete wavelet transform (DWT) is utilized for T and P wave detection. CLT offers a computationally efficient QRS detection technique [10, 11]. The FSM in Fig. 9.7, controls the flow of the feature extraction. In addition, it provides the necessary signals required for clock gating, QRS detection and TP delineation.

The main features of the ECG are contained in the frequency range 0.5–50 Hz, hence the sampling frequency of our proposed architectures is only 250 Hz. Digitized ECG samples are fed to the system at the rate of 250 Hz and directly filtered. The filter is pipelined and operates at a clock rate of 250 Hz and is clock gated at other times.

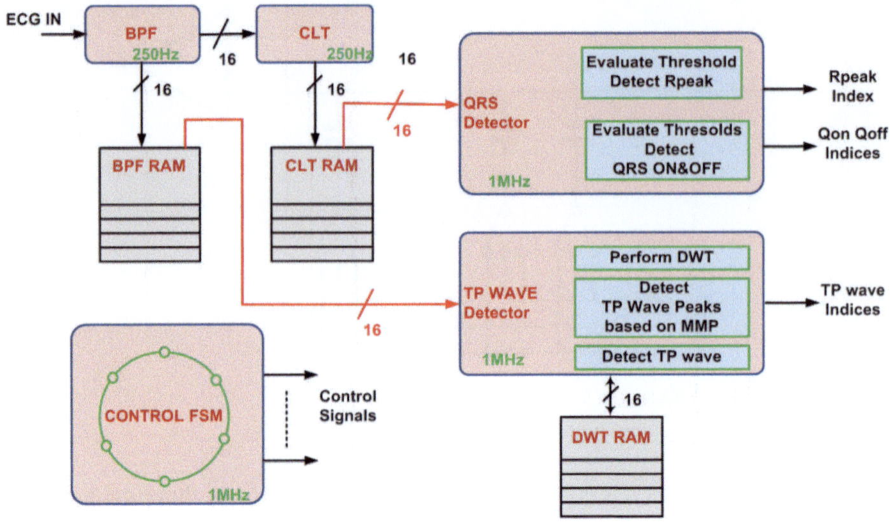

Fig. 9.6 ECG feature extraction architecture

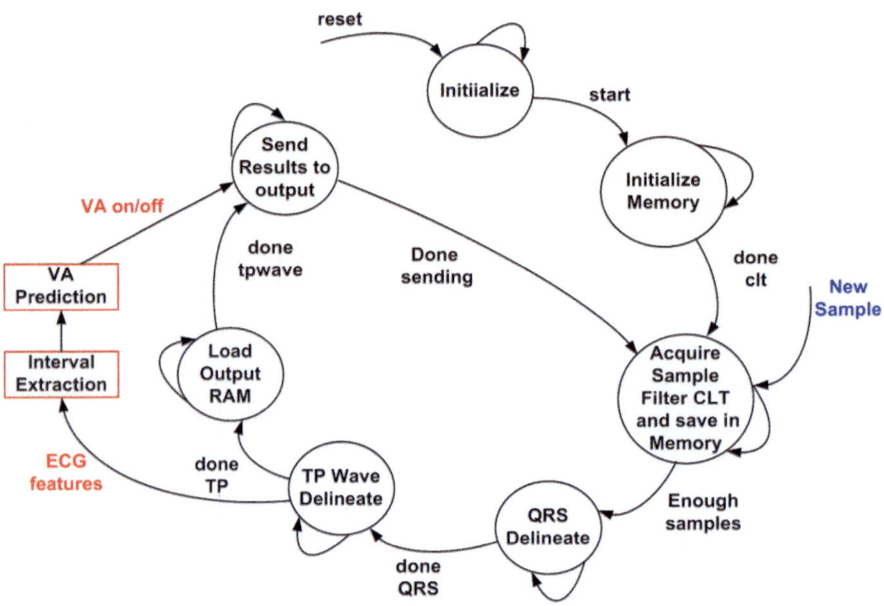

Fig. 9.7 Main control FSM

Filtered ECG is stored in memory since it is required for the TP wave delineation. Moreover, the CLT operates at rate of 250 Hz and its pipelined architecture will be described in the following subsection along with the QRS delineation. The output of the CLT is also stored in memory as it will be used by the QRS detection. The system buffers 512 samples and does the feature extraction. It requires 2 s in order to acquire 512 samples, and these samples are processed in a maximum of 2000 cycles depending on the morphology of the incoming ECG signal.

QRS Detection

QRS has the highest slope and relatively higher amplitude than the other ECG components. CLT extracts slopes and length of successive points of a wave as given on Eq. 9.2. When CLT is applied to ECG it enhances the QRS complex and relatively suppresses the TP waves (Fig. 9.8). In the proposed architecture, the CLT is implemented as on Eq. 9.3 in a pipelined fashion. During each stage of the pipeline, only one square root function is required which is the second term in Eq. 9.3, whereas the other terms are obtained from preceding transforms. In each stage of the pipeline, only one square root, one addition and one subtraction are enabled. The CLT signal which is saved in memory is used to evaluate the thresholds required for the detection of QRS peaks. The thresholds for the QRS detection are evaluated as in equation Eq. 9.4. After QRS_{peak} is detected, the system goes on to locate the Q_{on} and Q_{off} by defining a search window to the left and right of QRS_{peak}. Q_{on} and Q_{off} are also obtained by applying thresholds on the CLT signal.

$$L(\omega, i) = \sum_{k=i-\omega}^{i} \sqrt{C^2 + (\Delta y_k)^2} \tag{9.2}$$

where ω is the window size

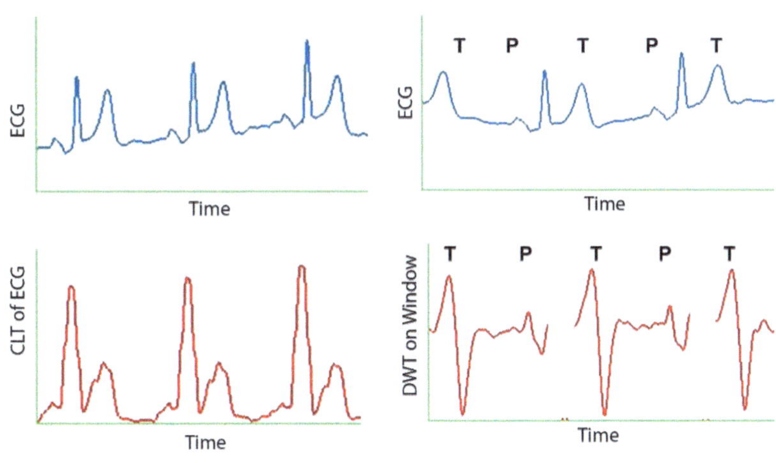

Fig. 9.8 CLT (left) and DWT on a window (right)

$$L(\omega, i) = L(\omega, i-1) + \sqrt{C^2 + (\Delta y_i)^2} - \sqrt{C^2 + (\Delta y_{i-1-\omega})^2} \tag{9.3}$$

$$Th_{up} = \frac{2}{3}\,[mean\ (CLT_{Pre_{beat}}) + Th_{pre}] \tag{9.4}$$

where $CLT_{Pre_{beat}}$ is CLT of previous beat, Th_{pre} is threshold of previous beat and Th_{up} is the updated threshold.

DWT-Based T and P Wave Delineation

T and P are obtained by applying DWT on a window after and before the QRS complex on the filtered ECG, respectively. The search windows are adaptive and are updated based on the previously detected beat-to-beat (RR) interval. DWT is implemented as a cascade of filter banks as illustrated in Fig. 9.9 and provides a multiscale decomposition. The filter h(n) and g(n) are lowpass and highpass filters respectively. Scale 2^3 is used for the TP wave delineation process. The wavelet decomposition forms a pair of maxima based on the concave or convex nature of the T and P waves around the baseline as shown in Fig. 9.8. These pair of maxima, known as maximum modulus pair (MMP), are detected in order to locate the positions of the T and P waves. The MMP steps are described in Fig. 9.10. First a threshold is set as mean of the absolute value of the DWT signal, and then a peak is detected followed by zero crossing and second peak. If the first peak is a maxima, then it is a concave signal around the isoelectric line, and if the first peak is minima, then it is a convex signal. The zero crossing point sets the location of the T or P wave peaks. Their respective onset and offset are obtained by extrapolating the distance between zero crossing of the DWT and the peaks to the left and right of the zero crossing.

$$[H]h[n] = \frac{1}{2}\,[\delta[n+2] + \delta[n+1] + \delta[n] + \delta[n-1]] \tag{9.5}$$

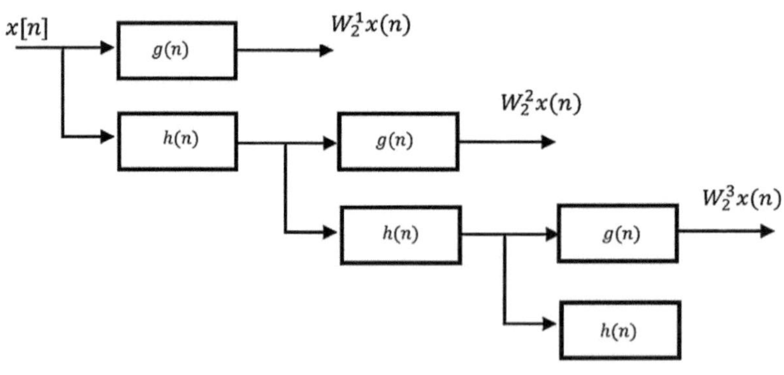

Fig. 9.9 DWT implementation as cascade of filter banks

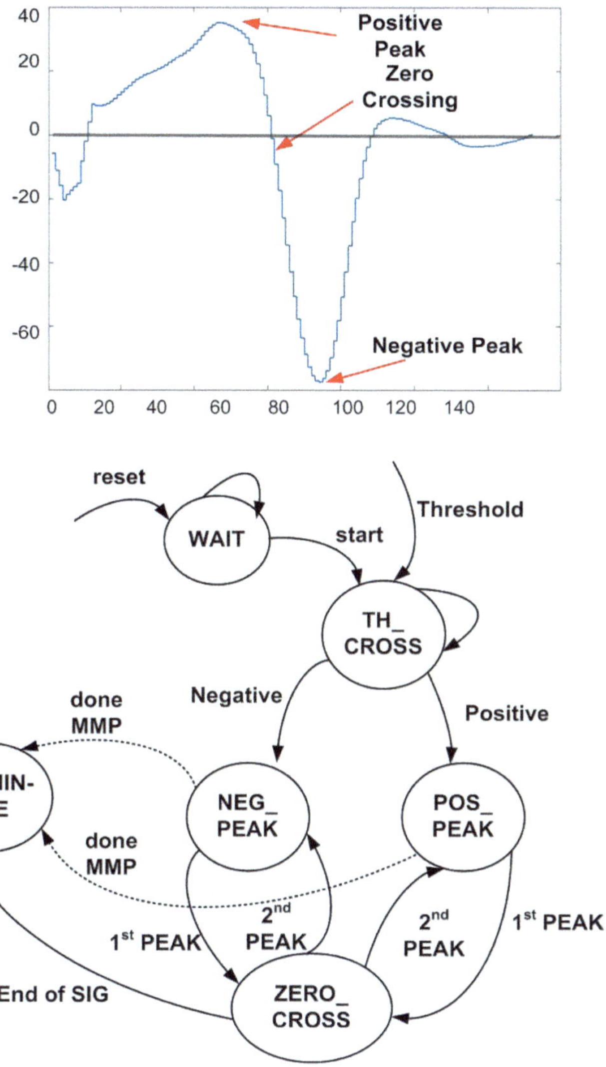

Fig. 9.10 MMP recognition and TP wave detection FSM

$$g[n] = 2\ [\delta[n+1] - \delta[n]] \tag{9.6}$$

$$[H]\Phi(\Omega) = j\ \Omega\ (\frac{sin(\frac{\Omega}{4})}{\frac{\Omega}{4}})^4 \tag{9.7}$$

9.3 VA Prediction Architecture

The first step in VA prediction is to extract the parameters or intervals that would give the highest accuracy. These intervals are selected based on their accuracy that is obtained during classification. Two consecutive ECG cycles are used to define seven intervals, namely RR, PQ, QP, RT, TR, PS and SP intervals. These intervals are obtained by taking the ECG fiducial points from the feature extraction. In this stage only additions and subtractions are required.

The VA prediction system is constructed using Naive Bayes classifier and is integrated along with the feature extraction and feature selection. The above-mentioned intervals are input to this system. The Bayesian classifier uses Bayes theorem to find out the probability of a data belonging to a particular class of observations. The Bayes theorem is given below where d is a set of feature vectors for class c_i.

$$P(c_i|d) = \frac{P(d|c_i)P(c_i)}{P(d)} \tag{9.8}$$

During classification, a data is assigned to a class which has the maximum conditional probability. This maximum conditional probability is represented by the following equation.

$$c = argmax P(c_i) \prod P(x|c_i) \tag{9.9}$$

Please refer to Chap. 13 for more details about the classification algorithm. For ease in hardware implementation, the product is changed to summation using logarithms and the same classification results are obtained by using the highest log score as given in the equation below:

$$c = argmax \left[log\left(P\left(c_i\right)\right) + \sum log P\left(x|c_i\right) \right] \tag{9.10}$$

In the classifier architecture, the main operations are finding the mean, standard deviation and conditional probability. The arithmetic operations required are addition, subtraction and division. For each of the features, used conditional probability is evaluated, and the commutative log score is utilized for classification.

9.4 Conclusions

The proposed architecture is computationally efficient. If we look at the feature extraction, it is performed using CLT and DWT. The CLT is pipelined and requires only one square, one square root and summation for each sample. And the DWT is implemented as cascade of filters. The filter coefficients are 2 and 1/2 which are implemented using shift and do not require multiplication or division. Besides, the DWT is performed on a small length of the signal on a window which is the prob-

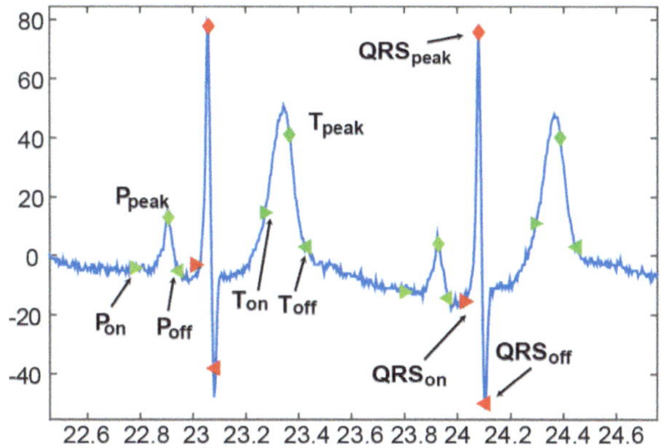

Fig. 9.11 ECF Features extracted using the chip

able location of TP waves. Only 256 bytes of memory are required for the DWT since only a single scale is used for the delineation process and the maximum window size does not exceed 256. The feature selection and feature classification were implemented heavily using combinatorial circuits which aid in reducing the overall power consumption. The VA prediction requires finding conditional probabilities, for which logarithmic scores were used to avoid the need for complex multiplication and division operations. Results shown in Fig. 9.11 demonstrate ECG features extracted using the test chip. Moreover Fig. 9.12 demonstrates the VA prediction process by initializing a high pulse when a VA is predicted.

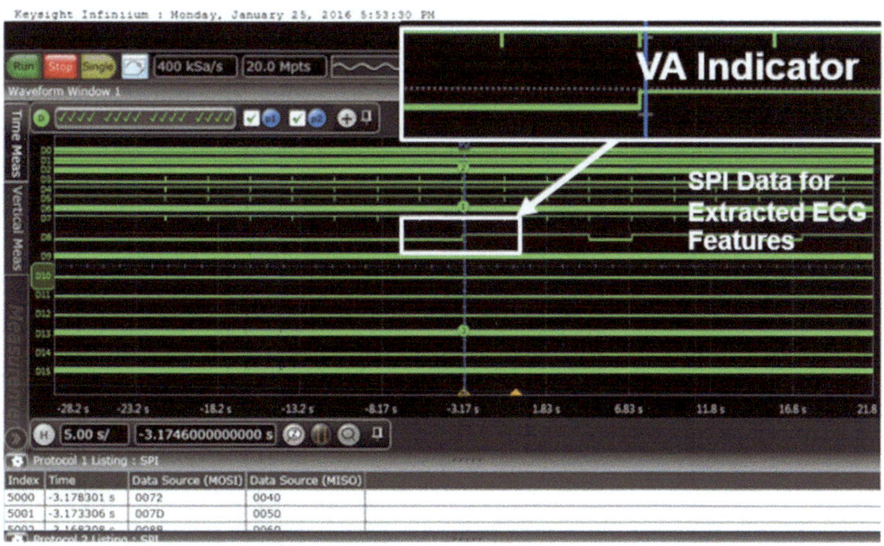

Fig. 9.12 Test chip VA prediction example

Acknowledgements This work has been supported by the Semiconductor Research Corporation (SRC) under the Abu Dhabi SRC Center of Excellence on Energy-Efficient Electronic Systems (ACE^4S), Contract 2013 HJ2440, with funding from the Mubadala Development Company, Abu Dhabi, UAE.

References

1. WHO, (2015) Cardiovascular diseases (cvds). [Online]. Available: http://www.who.int/mediacentre/factsheets/fs317/en/
2. Sudden cardiac arrest: A healthcare crisis (2015) [Online]. Available: http://www.sca-aware.org/about-sca
3. C.-Y. Chiang, H.-H. Chen, T.-C. Chen, C.-S. Liu, Y.-J. Huang, S.-S. Lu, C.-W. Lin, and L.-G. Chen, Analysis and design of on-sensor ECG processors for realtime detection of VF, VT, and PVC, in *2010 IEEE Workshop on Signal Processing Systems (SIPS)*, (2010), pp. 42–45
4. Q. Li, C. Rajagopalan, G. Clifford, Ventricular fibrillation and tachycardia classification using a machine learning approach. IEEE Trans. Biomed. Eng. **61**(6), 1607–1613 (2014)
5. N. Bayasi, T. Tekeste, H. Saleh, B. Mohammad, A. Khandoker, Low-power ECG-based processor for predicting ventricular arrhythmia. IEEE Trans. Very Large Scale Integr. (VLSI) Syst. **PP**(99), 1–13 (2015)
6. J. Martinez, R. Almeida et al., A wavelet-based ECG delineator: evaluation on standard database. IEEE Trans. Biomed. Eng. 348–570 (2004)
7. N. Bayasi, T. Tekeste, H. Saleh, A. Khandoker, B. Mohammad, M. Ismail, Adaptive technique for P and T wave delineation in electrocardiogram signals, in *2014 36th Annual International Conference of the IEEE Engineering in Medicine and Biology Society (EMBC)* (2014), pp. 90–93
8. T. Tekeste, N. Bayasi, H. Saleh, A. Khandoker, B. Mohammad, M. Al-Qutayri, M. Ismail, Adaptive ECG interval extraction, in *IEEE International Symposium on Circuits and Systems (ISCAS)* (IEEE, 2015), pp. 998–1001
9. T. Tekeste, H. Saleh, B. Mohammad, A. Khandoker, M. Ismail, A biomedical SoC architecture for predicting ventricular arrhythmia, in *IEEE International Symposium on Circuits and Systems (ISCAS)* (IEEE, 2016), pp. 2262–2265
10. W. Zong, The studies on ECG feature extraction and arrhythmia detection and classification by fuzzy reasoning, Ph.D. dissertation(1993)
11. W. Zong, G. Moody, D. Jiang, A robust open-source algorithm to detect onset and duration of QRS complexes. Comput. Card. 737–740 (2003)

Chapter 10
Time-Delay Array Beamforming for Millimeter-Wave IoT Systems

Owais Talaat Waheed, Ahmed Elian, Ibrahim (Abe) M. Elfadel and Ayman Shabra

10.1 Introduction

Millimeter-wave frequencies have emerged as an essential enabler for high-speed wireless data systems. The abundance of bandwidth at these frequencies was initially adopted by short-range indoor standards such as wireless HD and WPAN 802.15.3c that demonstrated 4 Gbps data rates [1]. More recently, the FCC released multiple mm-wave bands for 5G systems [2]. Systems with target data rates as high as 10 Gbps will appear by utilizing the available bandwidth in addition to complex modulation schemes with spectral efficiencies as high as 8bits/s/Hz. Field trials have demonstrated the feasibility of such a system at 28 GHz [3, 4] with investigations targeting frequencies as high as the 73.5 GHz band [4].

Beamforming is essential at millimeter wave frequencies to address issues related to multi-path interference while creating multiple simultaneous spatial channels to increase data rate or range. Moreover, the use of arrays increases the effective antenna aperture and hence improves the received signal-to-noise ratio (SNR). In the world of IoT, one important application domain of millimeter-wave beamforming is in autonomous vehicles and the V2X technologies at the intersection between the automotive industry and the upcoming 5G telecommunication technologies [5].

O. T. Waheed · Ibrahim (Abe) M. Elfadel (✉)
Masdar Institute at Khalifa University, Abu Dhabi, UAE
e-mail: ielfadel@masdar.ac.ae

O. T. Waheed
e-mail: owaheed@masdar.ac.ae

A. Elian
Department of Electrical and Computer Engineering, University of Toronto, Toronto, Canada
e-mail: ahmed.elian@mail.utoronto.ca

A. Shabra
MediaTek, Woburn, MA, USA
e-mail: aushabra@yahoo.com

Ibrahim (Abe) M. Elfadel and M. Ismail (eds.), *The IoT Physical Layer*,
https://doi.org/10.1007/978-3-319-93100-5_10

The current state of the art of millimeter-wave beamforming [1, 3, 6, 7] relies on phase-array techniques in various embodiments and has demonstrated data rates up to 4 Gbps at a receiver energy efficiency of 0.4 nJ/bit [1]. The prospects of scaling these architectures as data rates increase seem limited due to a combination of large fractional bandwidth and higher-order modulation, which results in array-induced intersymbol interference (ISI) [8]. A well-established approach to address this problem is through the use of time-delay arrays [8–10], but it requires the implementation of time-delay component in the path of a wideband signal on each channel.

In this chapter, we present an approach to design a time-delay array without the need for delaying the received signal. We will demonstrate how the delays needed can be realized by delaying the sampling clock in the receiver's analog-to-digital converters.

In Sect. 10.2 of this chapter, we discuss the fundamental limits of phase-array architectures and some of the outstanding issues with existing implementations of time-delay arrays. In Sect. 10.3, we introduce a novel time-delay array architecture that addresses these limits, and in Sect. 10.4, we discuss its timing requirements. In Sect. 10.5, we analyze and compare the performance of the various beamforming architectures. In Sect. 10.6, we illustrate the benefits of antenna grouping in relaxing the stringent timing requirements of time-delay arrays and in realizing hybrid architectures combining the advantages of both types of beamforming arrays.

10.2 Conventional Beamforming

Figure 10.1 shows a simplified block diagram of a phase-array receiver, where the signal arrives at an angle θ, such that each antenna receives the wavefront at a slightly different instance in time. The phase shifter blocks ϕ_i in the figure correct for the phase differences in the RF carrier received at each antenna, such that the summation is constructive.

It is important to note that this correction does not impact the modulated signal at all, and as such this approach is likely to run into important limitations for higher data rates. In Fig. 10.2, we illustrate this with an edge-fire array receiving its signal from a broadside transmitter. We can see that the raw data bits at each of the antennas are not aligned in time even after the carrier phase is corrected. This results in an array-induced intersymbol interference (ISI), which increasingly limits the performance of wideband single-carrier modulation as the bandwidth of modulated signal becomes a larger fraction of the RF carrier frequency [8]. This effect is plotted in Fig. 10.3 for a linear array, where we can also see that the ISI increases with the number elements in the array N. Even with moderate number of elements in the linear array, it is easy to exceed the typical rule of thumb ISI budget of 10%.

A possible solution to these problems is shown in Fig. 10.4 and is commonly used in radar systems where active or passive RF delay elements are used [9, 10]. For a linear array of size N with element spacing $\lambda/2$, and incident angle θ, the delays are

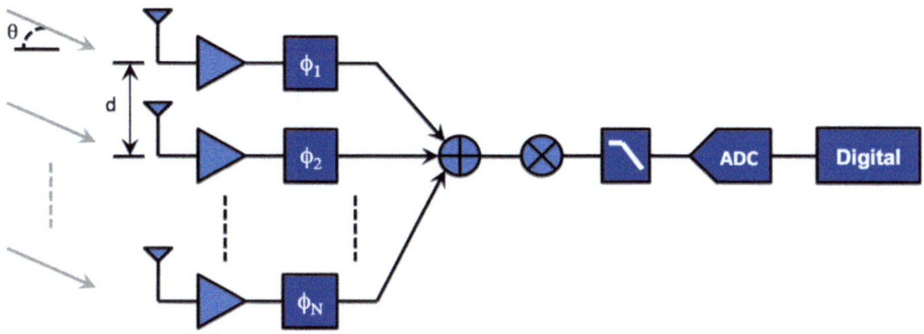

Fig. 10.1 Conventional phase-array beamforming receiver

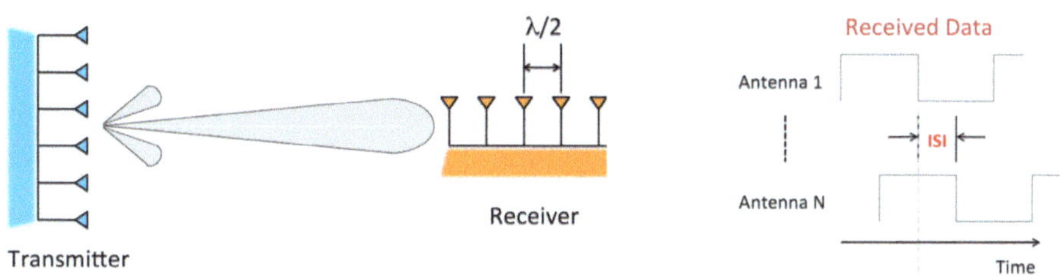

Fig. 10.2 Array-induced intersymbol interference

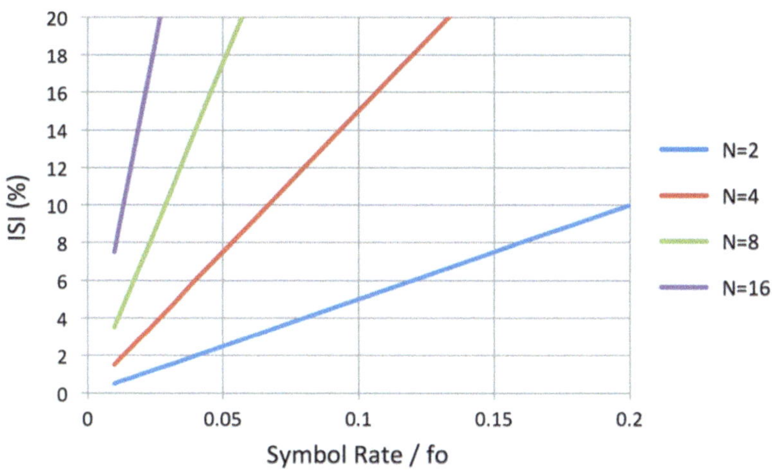

Fig. 10.3 ISI increases with both fraction bandwidth and array size

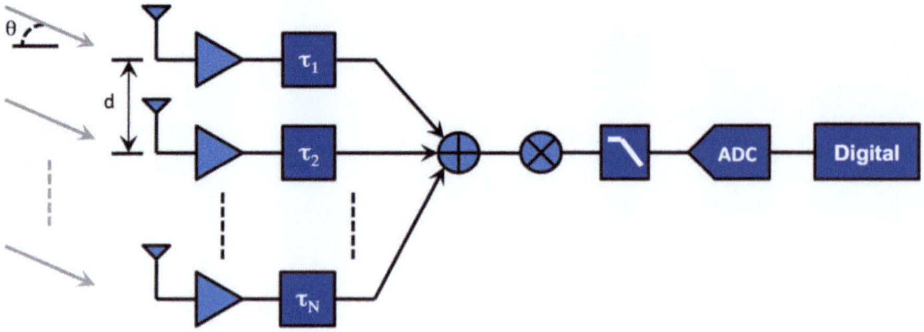

Fig. 10.4 Delay array receiver architecture

$$\tau_i = (i - 1)\frac{sin\theta}{2f_{RF}} \tag{10.1}$$

where i is the element number, and f_{RF} is the RF frequency.

Naturally, these techniques occupy an area proportional to the wavelength, which could be significant, but more importantly it does not scale with process shrinks. Furthermore, a variety of RF mismatches and parasitics will impose increasingly important limitations as data rates increase since small phase angle mismatches will impact the quality of the beams with limited degrees of freedom available for calibration. Another consideration is the mutual coupling between antenna array elements [11] that results in the antenna element impedance being a function of frequency and the location in the array. Here again, we have very little recourse to the compensation of these effects to enable very high data rates. It is worth mentioning that these RF delay techniques are also limited in their ability to rapidly adapt to the obstruction of the link line of sight (LOS). Since these solutions are RF in nature, they require significant power and area overhead to support multiple input sources with different spatial origins. One might think that an ideal solution to all these problems would be to use digital beamforming based on oversampling or digital interpolation [12], as shown in Fig. 10.5.

To illustrate the timing correction needed in a delay array, let us consider the case of an eight-antenna system operating at 120GHz. Figure 10.6 illustrates (10.1) for the timing correction required for each antenna as the beam angle is swept from −90 to +90. Although the correction range is in the 10's of picoseconds, the minimum correction required for a commonly used five beam steering step is 300 fs and will be a small fraction of a clock period. To achieve this level of correction in the digital domain will require a fractional delay digital filter, which in turn requires a long-duration digital filter with coefficients realized using full multipliers operating at a very high rate for each antenna. This will come at the cost of area and power.

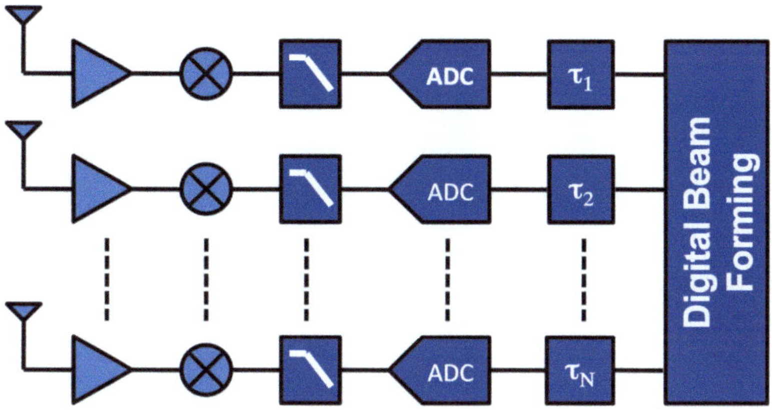

Fig. 10.5 Digital beamforming

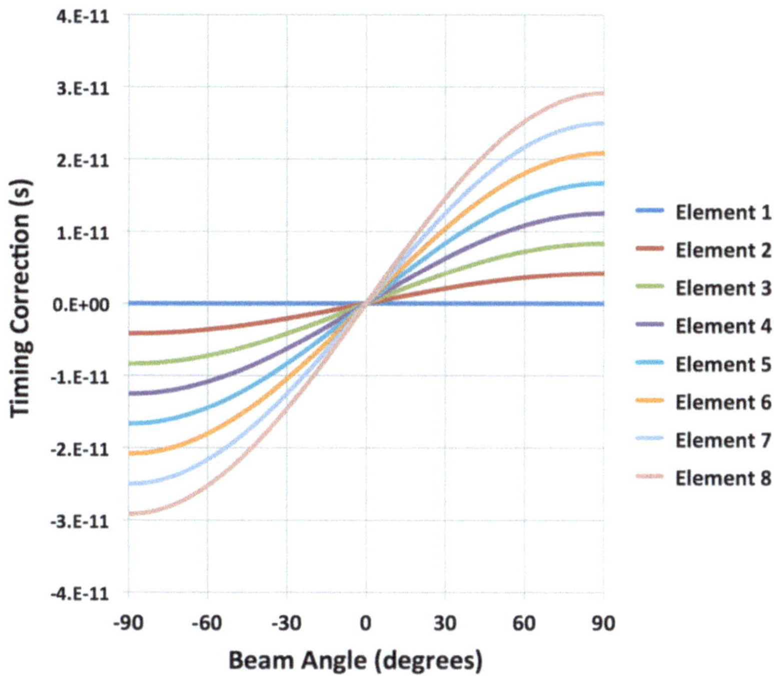

Fig. 10.6 Timing correction required for delay array beamforming

10.3 Beamforming Using ADC Sample Control

The delay operations needed in beamforming arrays can be performed much more simply in the mixed-signal domain. In this section, we describe how a number of techniques can be used to achieve very fine grain timing adjustments. These techniques open the door for the calibration of many RF impairments and will improve receiver performance and reduce power.

Techniques for time skew adjustments have been used successfully for many years in time-interleaved ADCs and achieve sub-picosecond timing correction [13]. These techniques can be used to adjust the clock timing of an ADC, as illustrated in Fig. 10.7, and can achieve exactly the same function as the RF or digital delays described in the previous section. This approach will be more area compact compared to the other approaches because it will eliminate long RF delay lines or complex high-order digital filters. This approach will also benefit from the decrease in gate delay with scaling by allowing finer delay adjustments. The phase differences between the antenna paths are corrected using digital blocks that multiply the signal with complex phase rotators $W_i, i = 1, \ldots, N$, where N is the number of antennas.

Note that this approach does require an ADC for each antenna, but these ADCs will have thermal and quantization noise specifications that are relaxed by a factor $\sqrt{N}$. The impact of all these ADCs combined will therefore be to a first order equal to the ADC used in the RF delay approach in terms of area and power. This is similar to what is seen in split ADC architectures [14]. Interferers will need to be processed by this ADC, which is not the case in a RF delay approach since the interferers are attenuated before the ADC. The impact on the ADC dynamic range (and area) can

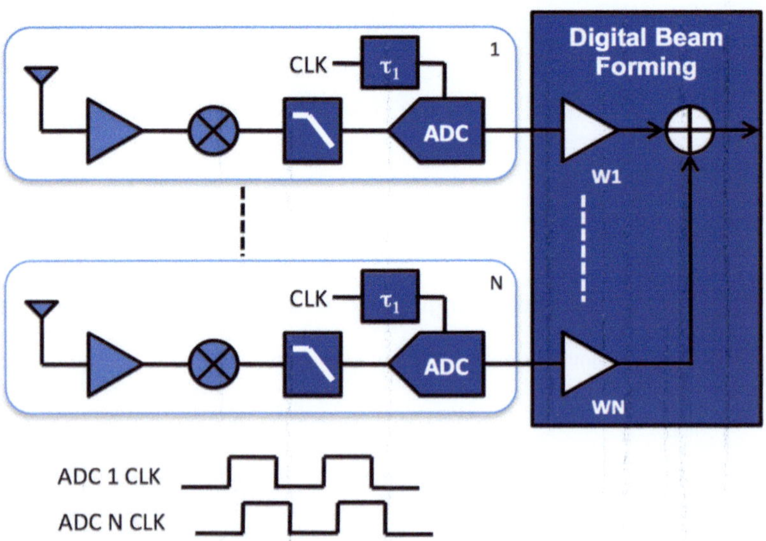

Fig. 10.7 Mixed-signal beamforming

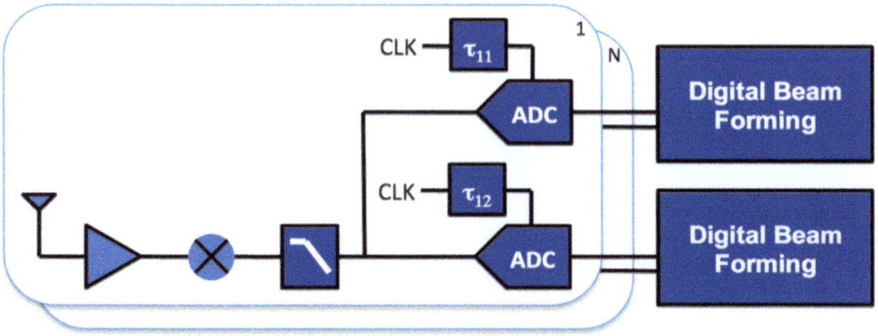

Fig. 10.8 Mixed-signal beamforming support for multiple simultaneous beams

be inferred from indoor experimental results reported in [15] to be in the order of a 1-bit increase in resolution. In an outdoor environment with stronger interferes, the overhead is likely to be larger.

Nonetheless, having a dedicated ADC per antenna will provide many of the benefits of digital beamforming such as precise beam steering, interference nulling, and RF impairment calibration. It also allows for the creation of multiple independent antenna beams. All that needs to be done is to add an additional ADC, and its associated delay correction, to each antenna path for each additional beam that needs to be supported as shown in Fig. 10.8.

The proposed beamforming approach falls naturally within the design considerations of an ADC needed in a high data rate system. CMOS implementations of a high-speed ADC will very likely utilize time-interleaving techniques which require ADC sample time control. In an effort to reduce power, time-interleaved ADC now commonly uses SAR ADC cells since they have demonstrated extremely low power and low area [16].

10.4 Clock Delay Requirements

The techniques presented in the last section rely on the ability to adjust the clock phase. From (10.1), it is possible to determine the delay correction required as

$$\Delta\tau_i = \tau_i - \tau_1 = (i-1)\frac{cos\theta}{2f_{RF}}\Delta\theta \tag{10.2}$$

Note that the timing correction range increases with the array size and the maximum beam sweep angle. It decreases with increasing RF frequency because we assume a $\lambda/2$ antenna separation. The minimum correction step, on the other hand, does not depend on the array size, but otherwise has similar dependence as the range. At millimeter-wave frequencies, these values are realizable in CMOS implementation. For example, if we consider a 120 GHz system with an eight-element linear array, then

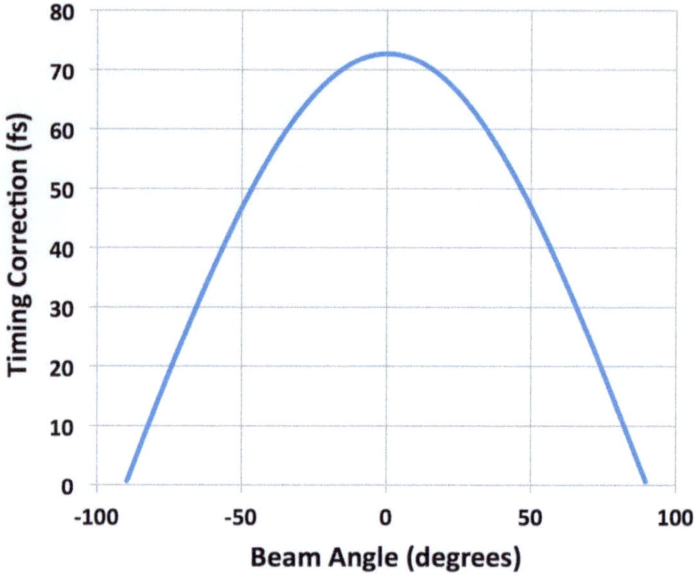

Fig. 10.9 Timing correction for a 120 GHz system requiring 1° beam steering step

the minimum correction step required for a 1° beam steering is shown in Fig. 10.9. Typically, systems steer the beam in larger steps such as 5°. For a ±45° beam range this translates to a 250 fs delay step.

A classical approach to generating multiple clock phases exploits the multiple phases generated by a ring oscillator. This limits the delay resolution of a CMOS buffer, which is at least two orders of magnitude larger than 250 fs even in the most advanced technologies. An alternative is to utilize a phase interpolation buffer to generate a delay that is a fraction of buffer delay step [17]. This approach provides high linearity in the delay control but is sensitive to variations in the circuit's power supply, temperature, and process variation. For robust and programmable delay, closed-loop control based on delay locked loops as reported in [18] can be used to provide sub-picosecond resolution in a mature 0.13 μm CMOS process. Open-loop voltage-controlled delay lines reported in [19] achieve a time resolution as low as 10 fs.

The impact of the quantization of the delay control on the array factor is shown in Fig. 10.10 for an array nominally pointing in the −30° direction. A coarse delay step of 10 ps results in a deformation of the array factor such that the antenna array is no longer pointing in the desired direction. For 1 ps delay quantization, the array gain is slightly reduced and the side lobes increase in magnitude. Although the quantization is larger than the smallest step determined from (10.2), the overall impact on the array factor is not significant.

In the next two sections, we will investigate this issue further by considering its impact on the communication system.

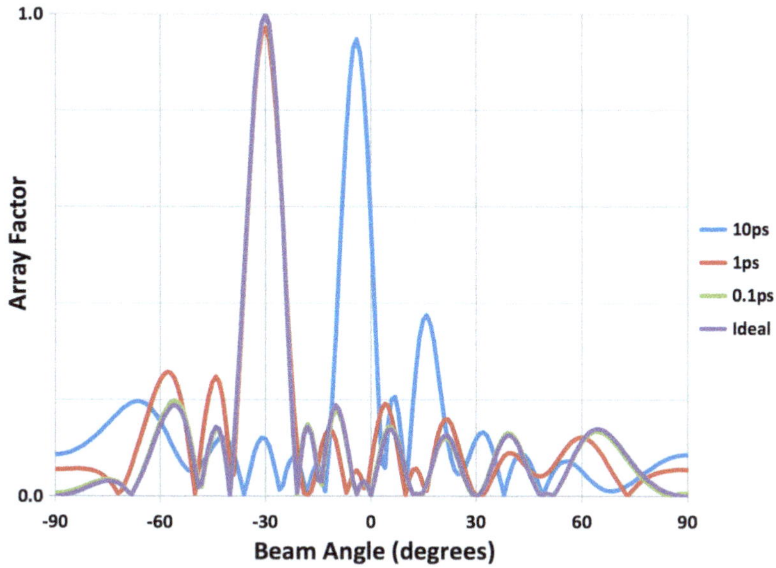

Fig. 10.10 Impact of delay quantization on the array factor of an eight-element linear array operating at 120 GHz

10.5 Bit Error Rate Performance

The metric we have adopted to evaluate the beamforming system performance is the bit error rate (BER) for the bit stream received over a communication channel. The bit errors occur due to factors associated with the transmitter, the channel, and the receiver. Since our focus is on the receiver design, factors that degrade performance include noise, interference, distortion, synchronization, and correction precision in each channel. Of these factors, we will consider the ones associated with the array, and the phase and time corrections.

We have performed the BER analysis using MATLAB models for phase and time arrays. We have evaluated the BER as a function of the fractional bandwidth, number of antennas, and per-bit signal-to-noise E_b/N_0, under ISI conditions. The fractional bandwidth is defined as the ratio f_s/f_{RF} of the signal bandwidth f_s to the center frequency of the signal f_{RF} [20]. Ideally, the nulls of the pulse-shaped symbols from all the antennas in the array should be aligned and no interferer should occur with time-adjacent symbols at the sampling instant. In a wideband context, where the fractional bandwidth is high, this may not practically be the case. The side lobes of pulse-shaped symbols interfere constructively or destructively, resulting in an error at the receiving end. This phenomenon of intersymbol interference (ISI) is detrimental to the performance of antenna arrays not only for large arrays but also for large fractional bandwidth applications.

Table 10.1 BER experiment specifications

Array type	Uniform linear array
Modulation scheme	16-QAM
RF frequency (f_{RF})	120 GHz
No. of antennas (N_a)	2 to 15 (variable)
E_b/N_0 (dB)	– 10 to 6 (variable)
Symbol rate (f_s) (GSy/s)	5, 10, 15, 20, 25 (variable)
Total symbols	10^6
Angle of arrival	Azimuth = 40, to observe significant ISI. Elevation has no impact in a ULA.

Table 10.1 summarizes the assumptions made to analyze the impact of large fractional bandwidth on BER. The system parameters we have selected are intended to illustrate the issues that might be encountered in realistic use cases in which high performance is required.

Noise in the channel impacts the BER through the E_b/N_0 of the received signal, which is defined as the ratio of energy per bit to noise spectral density [21]. The channel model used for this analysis is the additive white Gaussian noise (AWGN).

To quantify the impact of noise, we plot BER vs. the number of antennas N_a in a linear array for a given value of the fractional bandwidth, using E_b/N_0 as a parameter. For small arrays, when the symbol waveforms from across the array are well aligned, the BER improves with larger E_b/N_0 and N_a for both time and phase arrays. However, for large arrays the time- and phase-array cases do not follow the same trend. We observe a degradation of BER in phase arrays with a large number of antennas N_a. This degradation is due to ISI.

Figure 10.11 shows this behavior for a fractional bandwidth of 10GHz/120 GHz. The performance of phase and time arrays is only similar for values of N_a up to at

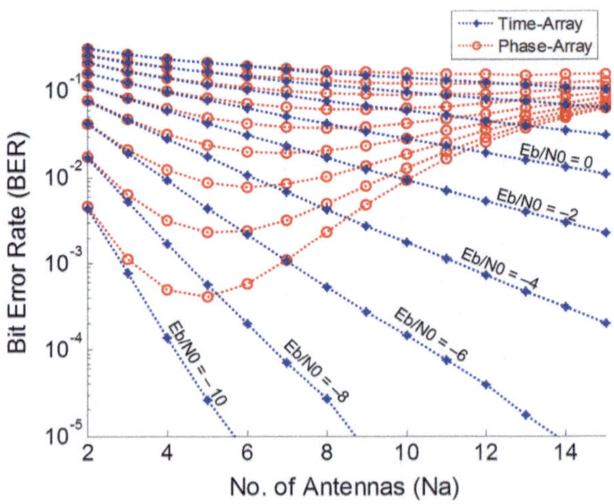

Fig. 10.11 BER vs. N_a for varying E_b/N_0 at f_s/f_{RF}=(10/120)

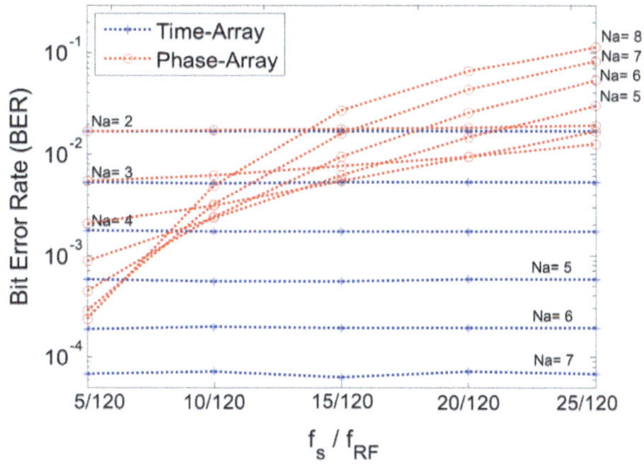

Fig. 10.12 BER vs. fractional bandwidth at $E_b/N_0 = 4$dB for varying array sizes

most 6 for a linear array, after which phase-array performance degrades due to ISI. The same trend is observed if we scale f_s and f_c with the same factor. For example, a fractional bandwidth of 5GHz/60 GHz will give the same results as 10GHz/120 GHz. This feature of time-array beamforming is achieved only when the time correction for the ADCs is within the required range as was explained in Sect. 10.4. The results in Fig. 10.11 do not include the impact of some practical system artifacts associated with large arrays such as the dependence of antenna impedance on the location of the antenna in the array.

The impact of the fractional bandwidth is illustrated in Fig. 10.12, where BER vs. f_s/f_{RF} is plotted using N_a as a parameter for a value of E_b/N_0 of 4dB. The performance degradation for phase arrays is visible even for a fractional bandwidth of 10%. The performance of time array does not change with fractional bandwidth at all in this analysis, but this does not account for the impact of some practical system limitations associated with very wide bandwidth circuits such as the frequency dependence of noise, gain, and impedance matching.

To ascertain the BER trends in these plots, the number symbols were selected to provide an acceptable level of accuracy. Depending on the BER, we used between 10^6 to 10^7 symbols per BER computation for each pair of E_b/N_0 and N_a values.

For a given BER target, one can optimize a phase-array system by trading-off N_a and fractional bandwidth. Such a trade-off is not necessary for an ideal time-delay array. On the other hand, practical time-delay arrays are constrained by the granularity of the achievable time correction. Such granularity can be relaxed using antenna grouping as will be explained in the next section.

A detailed study on the ISI induced by higher fractional bandwidth and its impact on BER is presented in [22]. The same reference addresses the choice of the modulation scheme and its impact on the BER. The impact of both BPSK and 16-QAM is quantified in [22].

10.6 Antenna Grouping

To relax the stringent requirements for time correction in the proposed mixed-signal beamforming approach, it is possible to combine phase-shift and time-delay arrays. This may be achieved by grouping antennas and correcting the fixed time delay for each group while correcting the phase within each group. The array still has N_a antennas which are divided into N_a/N_{grp} groups, each having N_{grp} antennas. The ideal case is 'group of 1', where time correction is made at each antenna in the array. In 'group of 2', time correction is performed for each pair of antennas.

In antenna grouping, the true time correction requirements scale with the number of grouped antennas, N_{grp}. The time correction required between adjacent antenna groups may be derived from (10.1) as

$$\Delta t_g = N_{grp}\frac{sin\theta}{2f_{RF}} \tag{10.3}$$

The number of antennas that can be grouped together is very much driven by the impact of ISI on BER. The larger the ISI, the smaller the number of antennas that can be grouped together. We can observe this in Fig. 10.13, where the BER is plotted against array size, using E_b/N_0, as in Fig. 10.11, and group sizes as parameters.

Although antenna grouping has the benefit of relaxing the strict time correction requirements, higher fractional bandwidth (e.g. at 20/120) leads to BER degradation with increasing antenna group sizes. The effect of ISI would be minimal for lower fractional bandwidths, giving us the choice to select a larger group. For example, if we use a grouping of four antennas, the time correction requirement is relaxed to about 0.6 ps (as compared to about 0.15 ps when corrected at each antenna) while achieving the same accuracy of 1° at the same carrier frequency of 60 GHz. Note

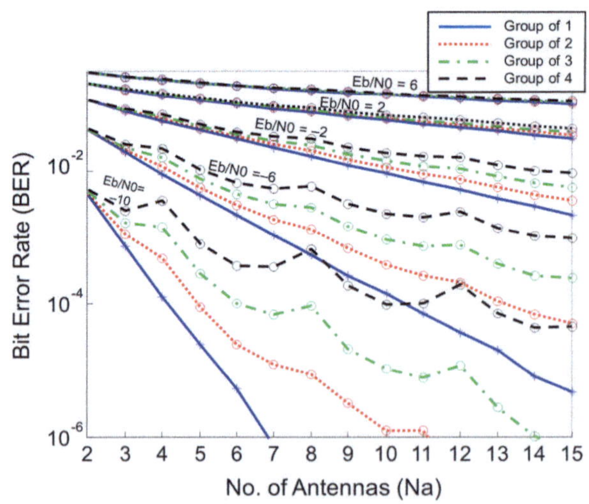

Fig. 10.13 BER vs. N_a (f_s/f_{RF} = 20/120 GHz) for varying E_b/N_0 and antenna group size

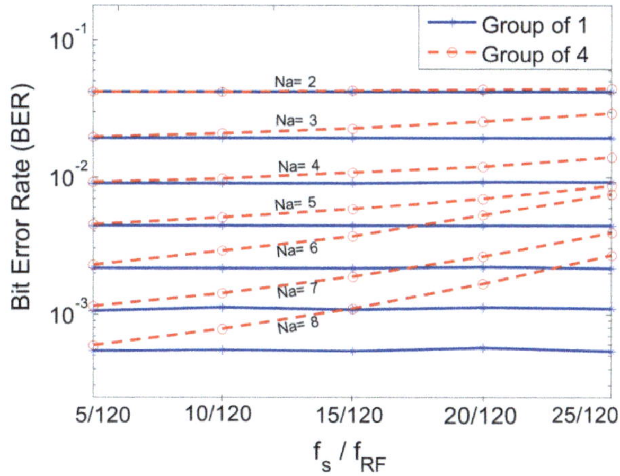

Fig. 10.14 BER vs. fractional bandwidth at $E_b/N_0 = 2$dB for varying antenna sizes, with four antennas grouped within the array

that this conclusion is valid for lower fractional bandwidths, where ISI's impact on BER is manageable.

To better evaluate the BER trends in the presence of antenna grouping, we plot Fig. 10.14, where the BER of $N_{grp} = 4$ is compared with the ideal case $N_{grp} = 1$. We notice that at higher fractional bandwidths, there is degradation in BER just as in Fig. 10.12. However, antenna grouping reduces the BER as compared to the ungrouped case of Fig. 10.12.

Based on the BER analysis presented in Fig. 10.14, the number of antenna groups should be added as a design parameter on a par with the number of antennas. The design task is to decide how many antennas to group together while maximizing the fractional bandwidth under a BER target. The selection of these design parameters is application-specific and very much depends on the allowable BER limits. To quantify these trade-offs, we use the 'BER percentage degradation' to evaluate the relative degradation in BER with respect to the ideal case. Note that this relative metric only shows how much is the degradation relative to best case. The best case for any f_s/f_{RF} is the case where we correct time at each antenna. For example, if there is a deterioration in BER from 10^{-2} to $(1.01 * 10^{-2})$ at a given E_b/N_0 then there is relative degradation of 1%. On the other hand, if there is a degradation in BER at some other value of E_b/N_0 from 10^{-5} to 10^{-3} then the relative degradation is 100 times larger. Although 10^{-3} is a better BER than $(1.01 * 10^{-2})$, the ISI impact has is different. This relative degradation is an important metric to quantify the impact of ISI. We fix the maximum allowable BER percentage degradation at 10%.

Figure 10.15 shows a plot of BER percentage degradation vs. fractional bandwidth for two antennas $N_{grp} = 2$ grouped together and for two different E_b/N_0 values. We observe that for $E_b/N_0 = 2dB$, the BER degradation is below 10% for an array size of up to eight antennas even for the high fractional bandwidth of 25/120. On the other hand, for lower E_b/N_0 values, e.g., down to $-6dB$, the maximum allowable array size is three antennas.

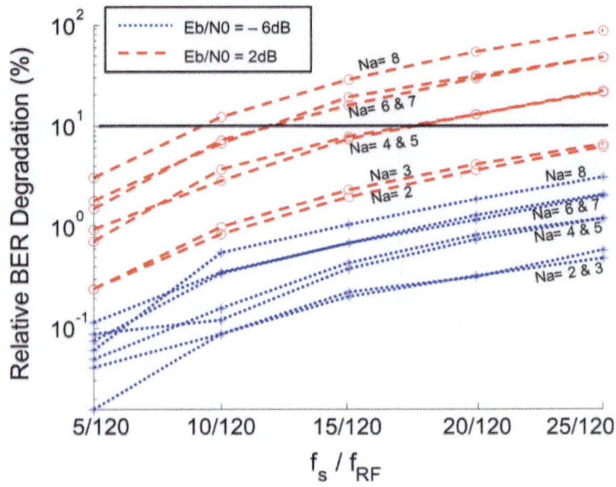

Fig. 10.15 BER degradation (percentage) vs. fractional bandwidth for varying antenna size, with two antennas grouped within the array

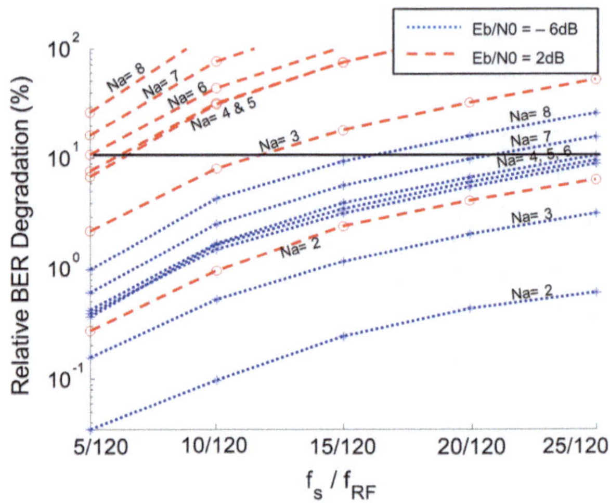

Fig. 10.16 BER degradation (percentage) vs. fractional bandwidth for varying antenna size, with four antennas grouped within the array

For a large antenna group size of $N_{grp} = 4$(Fig. 10.16), we observe that for $E_b/N_0 = 2\,\text{dB}$, the BER degradation is below 10% for an array size of less than six antennas at higher fractional bandwidth. On the other hand, for lower $E_b/N_0 = -6\,\text{dB}$, the maximum allowable array size is two antennas for a fractional bandwidth of 25/120.

We can of course achieve smaller BER percentage degradation for large array sizes but at the expense of having higher signal power and/or lower fractional bandwidth. It is worth noting that Figs. 10.15 and 10.16 show us partial information on relative percentage degradation of BER. Such information is not sufficient to select the design parameters and should be augmented with Figs. 10.13 and 10.14 for the selection of the array parameters that achieve the target BER performance while maximizing the fractional bandwidth.

10.7 Conclusions

In this chapter, we have presented a scalable, mixed-signal approach to digital beamforming using antenna time-delay arrays and numerically evaluated its performance as function of the array design parameters and the fractional bandwidth of the millimeter-wave signals. One important conclusion of our extensive numerical studies is that the BER performance of time-delay arrays improves with antenna array size while for phase arrays a large size will result in BER degradation due to intersymbol interference. Another important conclusion of the numerical studies is that for a given array size, the fractional bandwidth of the signal has no impact on the BER of the time-delay array but results in significant degradation of the phase-array BER. In order to relax the design specifications on the time-delay element, we have also conducted an extensive numerical analysis of antenna groupings and have determined that loss of beamforming performance can be minimized by appropriately choosing the group size relative to the array size. The numerical studies are all based on Monte Carlo simulations that are time-consuming and so far restricted to linear arrays. While their extension to 2D arrays is possible, we believe that a more fruitful direction would be to develop mathematical time-delay array models that can help in quickly evaluating time-delay antenna array performance in the context of a millimeter-wave beamforming system design flow. In terms of applications, it is very much worth exploring the implementation our scalable, mixed-signal beamforming approach in the context of the connected IoT vehicle and the emerging V2X technologies.

Acknowledgements This work was supported by the Semiconductor Research Corporation (SRC) with customized funding from Mubadala Development Company, Abu Dhabi, UAE, under the 2011 Program on Minimum Energy Electronic Systems (MEES I).

References

1. S. Emami et al. A 60GHz CMOS phased-array transceiver pair for multi-Gb/s wireless communications. *2011 IEEE International Solid-State Circuits Conference (ISSCC)*, pp. 164–166, San Francisco, CA, USA, Feb. 2011
2. Federal Communications Commission. Spectrum Frontiers Rules Identify, Open Up Vast Amounts of New High-Band Spectrum for Next Generation (5G) Wireless Broadband. https://apps.fcc.gov/edocs_public/attachmatch/DOC-340310A1.pdf
3. T. Obara, T. Okuyama, Y. Aoki, S. Suyama, J. Lee and Y. Okumura. Indoor and outdoor experimental trials in 28-GHz band for 5G wireless communication systems. *IEEE 26th Annual International Symposium on Personal, Indoor, and Mobile Radio Communications (PIMRC)*, pp. 846–850, Hong Kong, China, Sep. 2015
4. A. Harada, Y. Inoue, D. Kurita, T. Obara, 5G Trials with Major Global Vendors. NTT DOCOMO Technical Journal **17**(4), 60–69 (2016)
5. J. Choi, V. Va, N. Gonzalez-Prelcic, R. Daniels, C.R. Bhat, R.W. Heath, Millimeter-Wave Vehicular Communication to Support Massive Automotive Sensing. IEEE Communications Magazine **54**(12), 160–167 (2016)

6. K. Takinami et al. A 60GHz wireless transceiver employing hybrid analog/digital beamforming with interference suppression for multiuser gigabit/s radio access. *2015 Symposium on VLSI Circuits (VLSI Circuits)*, pp. C306-C307, Kyoto, Japan, Jun. 2015
7. M. Tabesh et al. A 65nm CMOS 4-element Sub-34mW/element 60GHz phased-array transceiver. *2011 IEEE International Solid-State Circuits Conference (ISSCC)*, pp. 166–168, San Francisco, CA, USA, Feb. 2011
8. A. Niknejad, H. Hashemi. *mm-Wave Silicon Technology 60GHz and Beyond.* Chapter 7, pp. 250, Springer 2010
9. J. Roderick, H. Krishnaswamy, K. Newton, H. Hashemi, Silicon-Based Ultra-Wideband Beam-Forming. IEEE Journal of Solid-State Circuits (JSSC) **41**(8), 1726–1739 (2006)
10. T. S. Chu, J. Roderick and H. Hashemi. A 4-Channel UWB Beam-Former in 0.13 μm CMOS using a Path-Sharing True-Time-Delay Architecture. *IEEE International Solid-State Circuits Conference. (ISSCC) Digest of Technical Papers*, pp. 426–613, San Francisco, CA, USA, 2007
11. P. Hannan, The element-gain paradox for a phased-array antenna. IEEE Transactions on Antennas and Propagation **12**(4), 423–433 (1964)
12. L.C. Godara, Application of antenna arrays to mobile communications. II. Beam-forming and direction-of-arrival considerations. Proceedings of the IEEE **85**(8), 1195–1245 (1997)
13. K. Poulton, M Flynn. GHz ADC: from exotic to mainstream. *IEEE Custom Integrated Circuits Conference (CICC)*, San Jose, CA, USA, Sep. 2010
14. J. McNeill, M.C.W. Coln, B.J. Larivee, "Split ADC" architecture for deterministic digital background calibration of a 16-bit 1-MS/s ADC. IEEE Journal of Solid-State Circuits (JSSC) **40**(12), 2437–2445 (2005)
15. K. Hassan, T. S. Rappaport and J. G. Andrews. Analog Equalization for Low Power 60 GHz Receivers in Realistic Multipath Channels. *IEEE Global Telecommunications Conference GLOBECOM 2010*, pp. 1–5, Miami, FL, USA, Dec. 2010
16. I. Dedic. 56Gs/s ADC : Enabling 100GbE. *Conference on Optical Fiber Communication (OFC/NFOEC), collocated National Fiber Optic Engineers Conference*, pp. 1–3, San Diego, CA, USA, 2010
17. S. Sievert et al. 2.9 A 2GHz 244fs-resolution 1.2ps-Peak-INL edge-interpolator-based digital-to-time converter in 28 nm CMOS. *IEEE International Solid-State Circuits Conference (ISSCC)*, pp. 52–54, San Francisco, CA, USA, Feb. 2016
18. S. Callender and A. M. Niknejad. A phase-adjustable Delay-Locked Loop utilizing embedded phase interpolation. *IEEE Radio Frequency Integrated Circuits Symposium*, pp. 1–4, Baltimore, MD, USA, Jun. 2011
19. M. Straayer et al. 27.5 A 4GS/s time-interleaved RF ADC in 65nm CMOS with 4GHz input bandwidth. *IEEE International Solid-State Circuits Conference (ISSCC)*, pp. 464–465, San Francisco, CA, USA, Feb. 2016
20. S. Emami, C. Corral and G. Rasor. Peak-to-average power ratio (PAPR), fractional bandwidth and processing gain of UWB schemes. *Eighth IEEE International Symposium on Spread Spectrum Techniques and Applications - Programme and Book of Abstracts (IEEE Cat. No.04TH8738)*, pp. 929–933, Sydney, Australia, Sep. 2004
21. F.P. Fontan, P.M. Espineira, *Modelling the Wireless Propagation Channel: A simulation approach with Matlab.* Chapter 5, pp. 105, Wiley, 2008
22. O. T. Waheed, A. Shabra and I. M. Elfadel. Impact of Fractional Bandwidth on the Bit Error Rate of a Beamforming System. *IEEE 59th International Midwest Symposium on Circuits and Systems (MWSCAS)*, pp. 225–228, Abu Dhabi, UAE, Oct. 2016

Part III
Algorithms and Protocols

"All the same," said the Scarecrow, "I shall ask for brains instead of a heart; for a fool would not know what to do with a heart if he had one." "I shall take the heart," returned the Tin Woodman; "for brains do not make one happy, and happiness is the best thing in the world." L. Frank Baum, The Wonderful Wizard of Oz

Chapter 11
Nature-Inspired? Optimization in the Era of IoT: Particle Swarm Optimization (PSO) Applied to Indoor-Distributed Antenna Systems (I-DAS)

H. Hildmann, D. Y. Atia, D. Ruta, K. Poon and A. F. Isakovic

11.1 Introduction and Motivation

We live in a brave new world where half of the global population is living in urban areas [1] often growing at a rate that outpaces the growth of their infrastructure. For the year 2050, the UN predicts the existence of more than two dozen so-called *megacities*, metropolitan areas spanning multiple cities and exceeding a population of 10 million each, forecasting that by that date 70% of the global population will be living in cities [1]. For the younger readers, this will happen in their lifetime. Yet, while the excessive over-consumption of resources and the devastating effect of climate change are known to the general public, surveys have shown that 60% of those recognizing the problem (here: climate change) are not acting upon it [2].

Without a single 'culprit' to blame, a fundamental change of how we do things is required. In this chapter, we foster such rethinking for the field where we are qualified to make a difference: Information–Communication Technologies (ICT) are becoming pervasive and are impacting almost all areas of our lives. In 2013, the global (ICT) ecosystem constituted an aggregated demand rivalling the combined power

H. Hildmann
Universidad Carlos III de Madrid (UC3M) Dep. de Ingenieria de Sistemas y Automatica - Av. Universidad, 30, 28911 Leganes, Spain
e-mail: hanno.hildmann@uc3m.es

D. Y. Atia · A. F. Isakovic (✉)
Khalifa University, 127788 Abu Dhabi, UAE
e-mail: iregx137@gmail.com; abdel.isakovic@kustar.ac.ae

D. Y. Atia
e-mail: dina.atia@kustar.ac.ae

D. Ruta · K. Poon
EBTIC, Khalifa University, 127788 Abu Dhabi, UAE
e-mail: dymitr.ruta@kustar.ac.ae

K. Poon
e-mail: kin.poon@kustar.ac.ae

Ibrahim (Abe) M. Elfadel and M. Ismail (eds.), *The IoT Physical Layer*,
https://doi.org/10.1007/978-3-319-93100-5_11

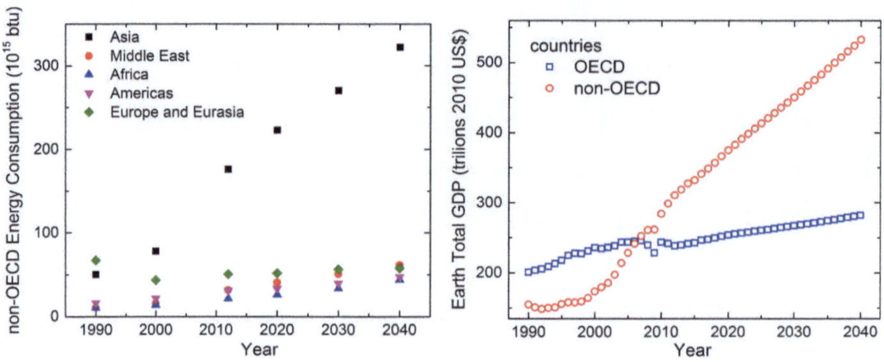

Fig. 11.1 The projected global electricity consumption (left) and the projected global resource (all types) consumption as expressed in world GDP (right) Source: U.S. Energy Information Agency http://www.eia.gov/forecasts/ieo/world.cfm (accessed October 2016)

generation of Germany and Japan, equivalent to approximately 150% of the demand of the global aviation industry. To put this into perspective, ICT today accounts for roughly one-tenth of the entire global power consumption [3].

The global power consumption is rising rapidly, and its production is increasingly falling behind as the widening gap between supply and demand can no longer be mitigated by consuming more and more natural resources [4]. The World Energy Outlook[1] projects a 75% increase between 2007 and 2030 [5] (cf. Fig. 11.1).

A future *smart-grid* will make the distribution networks more efficient [6] while controlling flexible demand wherever possible [7], but this is a loosing battle when focusing only on power-hungry installations such as data centres, server farms or telecommunication networks [8, 9] and other main electricity consumers in ICT.

It is expected [10] that wire*less* (as opposed to *wired* or physically *connected*) communication technologies will become pervasive and with the rapid decrease in size and cost for wireless devices (specifically passive and active sensors [11]) the number of connected devices in the so-called *Internet of Things* (IoT) is predicted to increase to 50 billion by the year 2020 [12]. The forecasted compound annual growth rate (CAGR) of the IoT is 24.42% for the period 2014–2019 [13]. Examples of connected objects range from large physical structures like factories, plants, bridges and high rises to the smallest devices currently available like RFID tags, and microchips, which will all be connected and able to communicate with one another. The global IoT / machine-to-machine (M2M) communication market is expected to grow *'from $128.7 billion in 2014 to $498.92 billion in 2019'* [13].

Using intelligent control mechanisms [14] and even projected savings of up to 20% through correcting faults in buildings [15] alone won't be enough. Change cannot just come at the level of some types of installations such as the national electricity grid; we have to improve, holistically, the efficiency of the underlying infrastructures as well as the generation and distribution paradigms [16].

[1] http://www.world-nuclear.org/info/inf16.html.

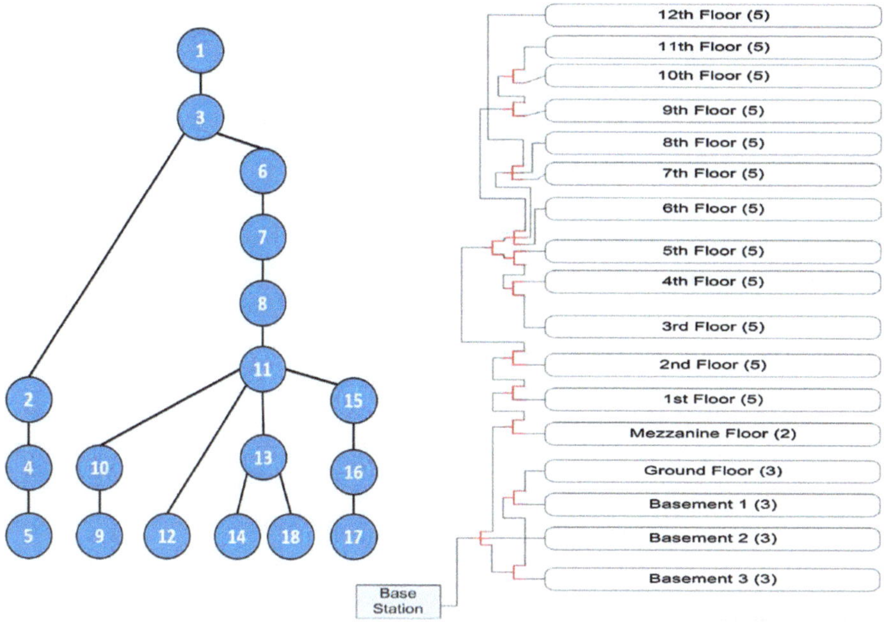

Fig. 11.2 This chapter proposes the use of Particle Swarm Optimization (PSO) techniques to optimize large indoor wireless access infrastructure: In-building Distributed Antenna Systems (I-DAS). The image shows a solution for a 17−floor I-DAS created by our implementation

In this chapter, we focus on the optimization of energy- and cost-efficient wireless access network infrastructure for wireless sensors and actuators. Designing such networks in a cost-aware fashion is a complex and computationally demanding task that will have to consider the usual trade-off between initially investing into the hardware and the installation (Capital Expense, CapEx) and the resulting cost for operating and maintaining the installation (Operational Expenses, OpEx).

With regard to the traffic handled by wireless access networks, indoor users are the main contributors to traffic volume and the demand for bandwidth and access points is only expected to increase [17]. To facilitate this in an energy-aware fashion, multiple access points and antennas are used to cover an area instead of installing a single node [18]. Unfortunately, due to quality-of-service constraints regarding signal strength and area coverage [19], the complexity of Distributed Antenna Systems (DAS) networks hinders the calculation of the best possible distribution of splitters and cabling for all but the small networks (Fig. 11.2).

There is evidence that—especially for problems of high complexity —nature-inspired approaches (e.g. [20]) have the potential to be extremely efficient [21]; principles and approaches found in nature have successfully been applied to increase the accuracy of models and to improve algorithms (cf. [22]). Collective behaviour such as clustering and collaboration based on self-organization [23] has been shown in group-living animals from insects to vertebrates and at the cell level [24]. These insights have been used in the coordination of (semi-)autonomous multi-agent systems based on self-organization [25]. We will use Particle Swarm Optimization (PSO) [26, 27] to optimize design decisions for Distributed Antenna Systems.

Summary

Future buildings will require larger and denser Distributed Antenna Systems. This is partly due to the increasing data consumption (and generation) by the individual user as well as the trend towards megacities and urbanization. Improving connection and cabling designs of such larger networks has the potential to reduce both, CapEx as well as OpEx. Our aim is to (i) reduce the combined cost for all hardware (splitters and cable) as well as (ii) optimize the operational power requirements.

11.2 Background

Offering wireless access means providing coverage using antenna-based sending/receiving approach. This used to be realized using a small number of high powered and spatially antennas. Recently, however, Distributed Antenna Systems (DAS) (using multiple antennas of lower power) have been found to be a promising alternative. They can provide good wireless access coverage [28] in areas otherwise suffering from reduced service due to signal blocking by physical objects such as walls. It furthermore offers high spectral efficiency for wireless communication [29] and—since the antennas are spatially separated—reduced correlations [30]. Each of these antennas is connected by cable to a central device [31] acting as the backbone.

11.2.1 (Indoor) Distributed Antenna Systems (I-DAS)

In-building deployment of DAS (henceforth I-DAS) has been studied analytically in [18] which concluded that dedicated in-building DAS can deliver service to a large number of indoor users while providing good signal quality. In [32], the author discusses the importance of using a dedicated I-DAS to solve the issue of providing indoor wireless access network coverage and emphasizes the concern for network throughput degradation due to performance bottlenecks caused by users in the building. The author found that in-building DAS would be a dominant solution for offering in-building coverage, especially for 3G networks.

11.2.1.1 DAS Deployment Optimization

Due to its potential to significantly reduced cost and improve the delivered indoor coverage, various approaches to DAS deployment optimization have already been investigated: e.g. [17] considers Mixed Integer Linear Programming (MILP) with

Table 11.1 Unique trees with n nodes. A problem with n antennas has (n^{n-2}) different solutions. Evaluating all of them becomes computationally unfeasible for moderate values for n

n	n^{n-2}	If it takes 1 s to evaluate 1000 trees
5	125	< 1 s
7	16807	≈ 17 s
8	262144	≈ 4.4 min
10	1.0×10^8	≈ 28 days
15	1.94×10^{15}	$\approx 61,713$ years
20	2.62×10^{22}	$\approx 8.3 \times 10^{12}$ years

topology constraints on the building, the choice of the splitters and antenna power requirements. The main goal is the reduction of cable length and the resulting minimization of deployment cost. These investigations show a dependency on the deviation between the required and supplied power allowed at the antennas. Unfortunately, the MILP-based approach does not scale well.

Alternatively, [19] focuses on (i) selecting the splitters / tappers and (ii) assigning equipment and antennas to specific ports. For individual antennas, the provided exact pseudo-polynomial time algorithms and the fully polynomial time approximation schemes minimize power deviation, outperforming greedy algorithms.

In [33], the authors solved multi-objective DAS design problems using a genetic algorithm (GA). Using adapted crossover and mutation methods, they optimized the number of antennas and their locations as well as the power transmitted from each antenna. Their simulation results—carried out on a relatively small DAS network—showed that varying the transmitted power of each antenna, instead of operating them homogeneously, leads to fewer antennas and reduced power leakage.

The examples above describe some different approaches, their aims and possible shortcomings. For a more detailed literature study, the reader is referred to [34].

11.2.1.2 DAS Complexity

A DAS topology can be represented as a connected graph without cycles, commonly called a *tree* [35]: all n antennas in an area are represented by the n nodes in a graph, in which any two nodes are connected to each other (connected graph) and where the number of edges is $n - 1$ (i.e. there are no cycles) (Table 11.1).

Given a set of n nodes, [36] showed that the family of different trees that can be constructed over this set has n^{n-2} members. This result is commonly known as *Cayley's Theorem* due to [37] (cf. [38]). This, in turn means that in order to find the best possible tree-like topology for a DAS of n antennas, a total of n^{n-2} trees have to be evaluated in order to identify the best one. The table below illustrates this:

11.2.2 Nature-Inspired Optimization and Swarm Intelligence

11.2.2.1 Nature-Inspired Optimization

The distributed nature of self-organization is one of its defining characteristics and maybe its most advantageous feature. Distributed methods [39] are used in nature to determine paths for individuals in large populations. Global goals, such as the social optimum, can be found using only local information [40]. Furthermore, behavioural features can induce positive feedback, potentially leading to collective decision-making in a social context [41]. Reliable information is a crucial factor influencing decision-making. The ability to communicate information to other members of the population is considered as a key factor and the ability of individuals to work together has almost certainly contributed to insect societies' great ecological success [42]. The social insect metaphor for problem-solving emphasizes distributedness, the interaction of members in a population, flexibility and robustness [43]. The growing understanding of complex collective behaviour offers the prospect of creating artificial systems which are controlled by emergent collective behaviour [44].

The field of swarm intelligence (SI), often operating in the domain of fragmented knowledge [45], offers an alternative way of designing intelligent systems in which autonomy, emergence and distributed functioning replace control, pre-programming and centralization. Individual members of a swarm typically use only local (onboard) sensing and coordinate their activity via the shared environment [46].

11.2.2.2 Swarm Intelligence (SI)

The collective behaviour of the swarms that is governed by the interactions of individuals with each other and their environment forms the basis for their emergent intelligence. Swarm intelligence principles are best applied to dynamic systems that require self-organization, robustness and scalability but where ensuring the absolute optimality of the solution is not a critical factor.

Swarm intelligence approaches typically strike a balance between random exploration of the local solution space and sharing distributed knowledge through simple interactions within the swarm. As a result, these processes constitute an emergent, flexible and adaptable soft optimization or search methodology, able to quickly find good solutions for very complex problems for which other methods are intractable.

Swarm intelligence (SI) models [27, 47] and genetic algorithms (GA) [48] can help optimize a complex network design. In a wider network optimization context, GA and SI were used for a variety of applications. Self-organizing behaviour of ant colonies was utilized for adaptive routing on a telecom network [47], as well as to solve a traffic signal control problem in [49]. Particle Swarm Intelligence model was used in [50] to optimize wireless sensor networks and perform load balancing.

Particle Swarm Optimization (PSO) was also successfully applied to minimizing the cost of heating in the system planning context [51]. GA and PSO were applied to classification model design, especially in the context of the classifier fusion [26]. PSO was used for efficient energy clustering for wireless sensor networks by minimizing the distance and number of hops between two transmission nodes [50].

11.2.2.3 Particle Swarm Optimization (PSO)

Particle Swarm Optimization (PSO) is a meta-heuristic where particles embodying the candidate solutions move through the solution space (in various directions) at different velocities, thereby imitating natural swarms of birds, insects or fish [27].

In PSO, all particles initially *fly* along random directions and with random velocities (cf. Algorithm 4 on p. 185). Gradually, through interaction with other particles, their exploration directions are adjusted and eventually the swarm converges on an area in the solution space that is likely to contain the best or very good solutions.

The increasingly coordinated movement is the key feature of PSO which distinguishes it from other evolutionary algorithms. Adding randomness to the velocity can further improve performance by avoiding getting trapped in local optima [51].

Orientation and speed, with respect to moving through the solution space, translate into the direction of a vector and the length. For any particle, given these two values and its current position, we can calculate the location where it will be next. This results in guided exploration of the solution space and convergence on good solutions. Obviously, the parameters that control the swarming and the speed of convergence are crucial in avoiding either searching too long or converging too quickly (thereby zooming right past the best solution without noticing it).

11.2.3 Prüfer Code Representation

In Sect. 11.2.1.2, we briefly discussed the complexity of the DAS problem on the basis of the assumption that we can basically represent a solution to the DAS problem as a tree structure. Since *Cayley's Theorem* [37, 38] states that there are exactly n^{n-2} different trees with n nodes, this gives us the complexity of the DAS problem. Without going into the details, the first combinatorial proof for this theorem was provided by Püfer [52] in 1918 [35] using a mapping that represented trees with n nodes as strings of length $n-2$. By showing that the set of strings of that length had n^{n-2} members, Püfer proved the *Cayley's Theorem*.

Since any tree can be uniquely represented by its Prüfer code [53], all possible spanning trees of n nodes can be represented by all $n-2$-element Prüfer codes from the set $\{1, .., n\}$. Although Prüfer code has been shown to be a suboptimal choice as encoding for certain problems [54], we have shown it to be applicable to the I-DAS problem [55], a claim supported by the good results obtained using our implementation [56]. Encoding and decoding to and from Prüfer code follows a simple linear algorithm (cf. Algorithms 1 and 2, respectively), details of which can be found in [53]. The complexity of both translations (tree → Prüfer strings; Prüfer

Algorithm 1 Encoding a (DAS-)tree to a unique Prüfer code

```
L ← leaves of T
for i ← 1 to (n − 2) do do:
    v ← node removed from the head of L
    PC[i] ← neighbour of v
    delete v from T
    if deg(PC[i]) = 1 then
        add PC[i] to L
```

strings → tree) increases linearly with n, i.e. their complexity is $O(n)$ (the algorithms translate trees to PC, and vice versa, in linear time).

Algorithm 2 Decoding a Prüfer code to a unique (DAS-)tree

```
L ← nodes that do not appear in the Prüfer code PC
for i ← 1 to (n − 2) do do:
    v ← node removed from the head of L
    add edge {v, PC[i]} to T
    if i is the rightmost position of v in C then
        add v to L
    v ← node removed from the head of L
    add edge {v, PC[n − 2]} to T
```

We adopted Prüfer string encoding to represent the search space for the I-DAS problem. The key advantage of using Prüfer code is its unique and efficient encoding (and decoding), which maps any valid tree structure into $n - 2$-dimensional space (and vice versa). Such $n - 2$-dimensional Euclidean spaces are known to work well with swarm and evolutionary search algorithms.

11.3 Particle Swarm Optimization to Optimize DAS

The $O(n^{n-2})$ complexity of the DAS design problem makes it unfeasible for traditional algorithms even for moderate n [17]. As motivated in Sects. 11.2.2.1 and 11.2.2.2, nature-inspired approaches such as PSO are likely to offer a good alternative [26, 27, 51]. Similar to other optimization problems, I-DAS has functional constraints which make solving it challenging. Among these are the requirement to provide a high quality of service in terms of full indoor coverage and minimal power leakage while, at the same time, coming at a minimal deployment cost (i.e. minimizing the combined cost incurred from cable length and splitters). This cost, however, depends on the configuration and the organization of the system components. Other important cost factors, such as the number of antennas used, their types and positions, are in our case assumed to be given and optimized before the task of providing power commences.

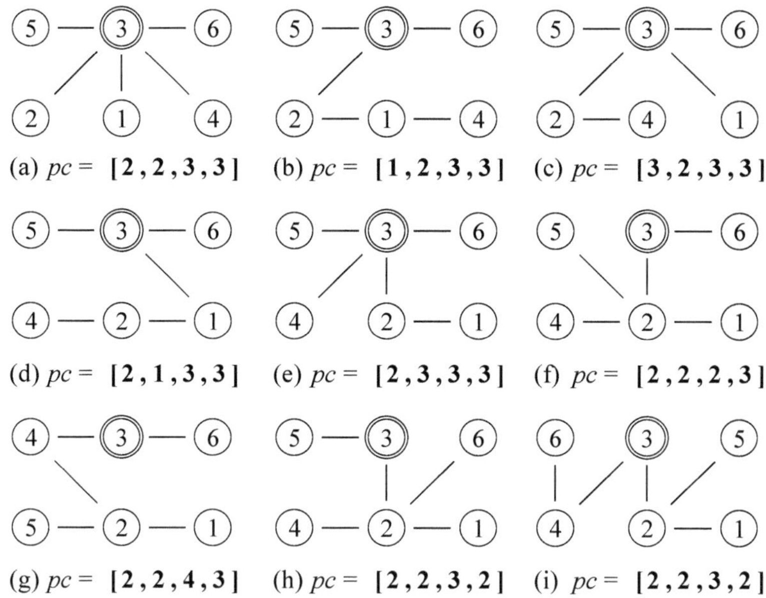

Fig. 11.3 The Prüfer codes similar to [2,2,3,3] (six-floors building). All variations from **b** to (i) differ from the original string **a** in only one digit and by only 1. Furthermore, in 6 out of 8 cases similar Prüfer numbers also map to similar graphs differing from the original by just a single vertex. In the representation above, we have fixed node 3 as the root of the tree to enhance readability

Therefore, the goal of our PSO-based I-DAS optimization is to determine the best (or at least very good) configuration of cabling and splitters (minimizing the total cable length as well as the number of splitters), while at the same time meeting all functional requirements and satisfying power constraints.

11.3.1 Modelling Distributed Antenna Systems

The objective of the In-Building Distributed Antennas System (I-DAS) is to provide cellular coverage to all usable areas of the building, ideally at a minimum cost of equipment and power. An I-DAS consists of a centre base station supplying power to all the floors using coaxial cables routed through each floor's telephone room. It also contains passive splitters to split the incoming signal into multiple output ports. Figure 11.4 below illustrates this and shows the components of an I-DAS problem.

In our case, all splitters on a floor are located in one location (which we refer to as *the telephone room*) and only four types of splitters are used (with regard to the number of their output ports): 2-way, 3-way, 4-way and 6-way splitters, with each incurring a power loss of 3.0, 4.8, 6 and 7.8 dBm, respectively. This means that both the splitters connecting the antennas on a specific floor as well as the splitters

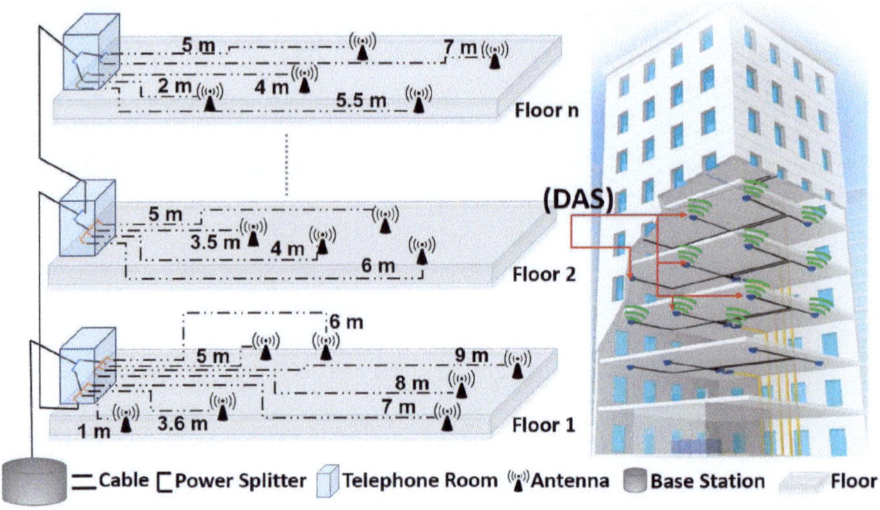

Fig. 11.4 An illustration of an In-building DAS and the components typically used

that connect all floors to the base station (cf. Fig. 11.4) are physically located in the telephone rooms. The antennas distributed over the floors are assumed to be already placed in optimal locations, resulting in fixed cabling requirements to connect them all to the telephone room on each floor and thus in a fixed cabling cost per floor. What is not fixed, however, is the connection between the antennas and the splitters on each floor, and the choice of splitter type. All antennas have minimum power requirements which have to be satisfied in order for the antenna to be operational. Therefore, the decision of which antenna to connect to which splitter (and the decision of which type of splitter to be used and to which other splitters it will be connected) will impact the splitters' power requirements (Fig. 11.3).

Each floor has a single connection to the rest of the building, i.e. the telephone rooms on all floors can be seen as a single input, the power requirement of which are equal to the requirements of the one splitter from which the entire floor is powered, be it through further nestings of splitters or through cables connecting directly to antennas. This means that we have a second level of optimization to consider, namely the connection of all floors to the base station in the building. Due to this, we can treat the I-DAS problem as two independent I-DAS optimization problems:

- **Intra-floor optimization** assumes fixed antennas locations and tries to find an optimal spanning tree between the telephone room and all antennas on the floor.
- **Inter-floor optimization** is dedicated to find an optimal connection spanning tree from the base station to all the floors (telephone rooms).

11.3.1.1 Intra-Floor Optimization

As discussed, the number and locations of antennas on any given floor are fixed, as are the lengths of cable required to connect them to the telephone room. The splitters are physically co-located so that the minimal cabling required to connect them can be ignored. The hardware cost $Cost^i_{CapEx}$ for a single floor F_i is thus given by:

$$Cost^i_{CapEx} = \sum_{\substack{t \\ s \in S_t}} c_{splitter_t} \times \sigma^i_s + \sum_{a \in A} dist_a \times c_{cable} \times \sigma^i_a \tag{11.1}$$

With $t \in \{2, 3, 4, 6\}$, S_t the set of all splitters of type t (i.e. the set of 't-*way splitters*'), σ^i_s a boolean that has the value 1 *iff* splitter s is located on floor F_i, A the set of all antennas, $dist_a$ the distance of the antenna a to the telephone room and σ^i_a a boolean that has the value 1 *iff* antenna a is located at floor F_i. The constants c_{cable} and $c_{splitter_t}$ are the cost per meter cable and for a splitter of type t.

In order to meet the power requirements for all antennas on a floor F_i (each with their own requirement $power_a$), the power input to the floor has to be at least:

$$min_i = \max_{a \in A}(\sum_{a \in A} dist_a \times c_{cable} \times \sigma^i_a + \sum_{\substack{t \\ s \in S_t}} s \times c_{splitter_t} \times \sigma^i_s \times \rho^a_s + power_a) \tag{11.2}$$

This is the combined power attenuation incurred in (a) the cable (assumed to be a constant per meter), (b) the splitters used in connecting this antenna to main input splitter of this floor and (c) the actual power requirements of the antenna ($power_a$). We assume that the attenuation of the cables as well as the individual splitter types is constant (c_{cable} and $c_{splitter_t}$) and use another boolean ρ^a_s which has the value 1 *iff* splitter s is along the path between the base station and the antenna a.

The intra-floor optimization problem for a floor F_i can be described as finding an assignment to all $\sigma^i_a \in \Sigma$ and $\rho^a_s \in P$ such that $Cost^i_{CapEx}$ is minimized (Fig. 11.5).

After an initial test of all four types of splitters (2-way, 3-way, 4-way and 6-way) two 'extreme' end approaches were considered: the connection was primarily built (i) from either the 6-way splitter (the largest available splitter) or (ii) from 2-way splitters such that all outputs of the splitters were connected (cf. Fig. 11.6).

In both cases only a few antennas will remain in the end and those can be connected by using the appropriate 4-, 3- or 2- way splitter (when using 6-way splitters) or by adding a single 3-way splitter (in the 2-way approach). The tree is to be constructed such that the maximum depth is minimized, i.e. splitters connect only to splitters until the combined available outputs of the connected splitters are equal or greater than the number of antennas on the floor (i.e. $\geq |A|$). This is motivated from the maintenance point of view: normally building operators impose a limit on the number of splitters in the connection path to an antenna (typically set to at most 3). It is straightforward to see that the algorithm will minimize the connection depth.

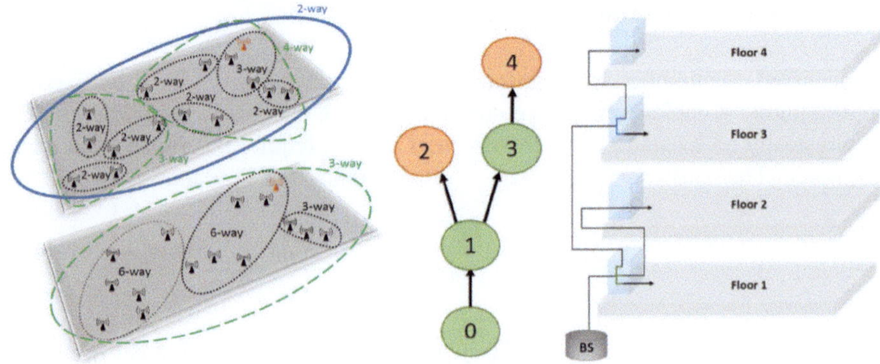

Fig. 11.5 A visual representation of the 2- and 6-way splitter tree algorithm for the **intra-floor problem** (left) and an example of a solution to an **inter-floor problem** with the spanning tree (middle) and the actual splitter structure (right) connecting floors in a four-level building

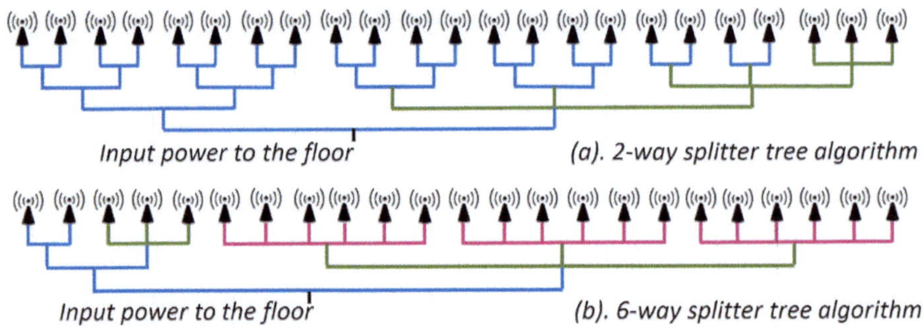

Fig. 11.6 A visual representation of the 2-way (top) and 6-way (bottom) splitter tree algorithm for 23 antennas. The blue, green and pink tree branches correspond to the 2-, 3- and 6- way splitters, respectively. The order of connections follows the logic of connecting first the antennas that require more power to the ports of splitters with smaller power loss

Table 11.2 A cost comparison between 2-, 3-, 4- and 6-way splitter algorithm applied to 15 antennas

Splitter type	2-way	3-way	4-way	6-way
Cost [USD]	694	556	459	378
Power attenuation [dBm]	13.8	12.6	12	12.6

In both algorithms, the assignment of antennas to splitter ports is subjected to the constraint that the antenna with the highest power requirement must be connected to the splitter with the lowest accumulative power attenuation. This—while not guaranteeing the lowest possible power requirements for a floor—allows us to calculate a good (albeit not necessarily optimal) solution to the intra-floor DAS problem.

Considering the cost and attenuation for the splitter types (see Table 11.2), the 6-way splitting algorithm outperforms the 2-way splitting algorithm in both, minimizing the cost ($Cost^i_{CapEx}$) as well as insuring a low attenuation ($Att.floor_i$).

11.3.1.2 Inter-Floor Optimization

The inter-floor optimization problem can be seen as a variation on the intra-floor optimization, except we are connecting entire floors instead of, as before, individual antennas. Again we have a n power requirements (i.e. the values calculated for floor F_i, for all n floors) which must be satisfied and as before the connection is subject to the accumulative power attenuation incurred between the base station and the floor. The primary aim is to minimize the equipment and cabling costs with the difference that in this variation of the problem the cabling can change.

There is no central *telephone room* for all splitters required for the topology. Instead, they are distributed over the building (at most one splitter per floor): if the floor is a branching point in the inter-floor topology tree then there is a splitter on that floor with one of its outputs connected to the main splitter of the intra-floor topology of that floor. If, however, a floor is a leaf node then the cable directly connects to the floor's main splitter. For comparison, in the intra-floor problem the equivalent of this setup would be that each antenna can also be a splitter. But in this case, the distance that needs to be covered by the connecting cables is no longer fixed and, due to this, the decision on how to interconnect the individual floors will affect the amount of cable needed for the full implementation (cf. Fig. 11.5, above). For simplicity, we assume that the distance between floors is the same, so that any two adjacent floors can be connected using a constant length $\texttt{dist}_{floor}$ of cable.

We introduce a third boolean $\tau_{j,i}$, encoding the actual connections between individual floors: $\tau_{j,i}$ is 1 *iff* a splitter on floor j is directly connected to (feeding into) a splitter or the telephone room on floor i. $Cost_{CapEx}$—the cost for the equipment and the cabling for the inter-floor problem—can then be calculated as follows:

$$Cost_{cables} = \sum_{i,f \in \{1,\ldots,n\}} |i - j| \times \texttt{dist}_{floor} \times \texttt{c}_{cable} \times \tau_{j,i} \tag{11.3}$$

$$Cost_{splitters} = \sum_{\substack{t \in \{2,3,4,6\} \\ s \in S_t}} \texttt{c}_{splitter_t} \times \sigma^i_s + \tag{11.4}$$

With σ^i_s as above: a boolean that has the value 1 *iff* splitter s is located on floor F_i

$$Cost_{CapEx} = Cost_{cables} + Cost_{splitters} \tag{11.5}$$

Here, unlike before, we know the size of the tree: above we knew the set of *leaf nodes* and had to determine the number (and type) of splitters we wanted to use, thereby adding *branching nodes* to our tree. In the inter-floor problem, leafs can also be branched and thus the number of leafs is the final n. As convention, we fix the root node of our tree (in Fig. 11.3 this was node 3) to be node 0 and labelled individual floors by their numbers (cf. Fig. 11.5 for an example for $n = 4$).

Since inter-floor optimization consumes high cabling costs subject to variable power constraints, the heuristic methods are not suitable and a more refined model is required. In line with our approach to try to harness swarm intelligence methodology to solve this problem, our next major step is to convert the floors connection graph or spanning tree into a flat vector representation with which SI models can work.

11.3.2 PSO–DAS Design Model

We discussed that the intra-floor DAS is solved by a straightforward heuristic method that attempts the maximum utilization of the largest possible splitters. For a building with n floors and given the results of applying the intra-floor approach to the individual floors, we want to determine a spanning tree connecting all nodes to node 0 such that the individual power requirements min_i for floor F_i are met and the costs for the hardware and the cables are minimized.

Our approach encodes the spanning tree as Prüfer code (cf. Sect. 11.2.3) and applies a tailored Particle Swarm Optimization to search through the resulting solution space. Each position in this $n-2$-dimensional discrete Prüfer space corresponds to a spanning tree connecting all floors. We initialized all *particles* of the PSO with random locations (i.e. a random Prüfer code of length $n-2$) and a random velocity, the latter was bounded by (v_{min}, v_{max}) which were chosen in relation to the size of the solution space (i.e. in relation to n). The problem of particles leaving the relevant search space was addressed by a simple *bouncing rule*, i.e. a reversal of the sign of the element in the velocity vector that violated the boundaries (resulting in the particle bouncing back into the solution space and continuing as before but moving away from the *wall* into which it bounced).

In each iteration, all particles move from their current location in the direction and at the speed defined by their velocity vector. The resulting new position is decoded into a spanning tree which is first evaluated against the (hard) power requirements and, if these are met, the overall hardware cost is calculated.

Algorithm 3 Bottom-up power requirement for the inter-floor tree

```
function W = POWER-COST(INTER-FLOOR TREE)
    for each i^th tree level starting from bottom do:
        for each j^th node u_{i,j} at this level do:
            Calculate the required floor power W(u_{i,j})
            Add W_{i,j} ← W_{ij} + attenuation(cable, splitter)
            if W(u_{i,j}) > W(parent(u_{i,j})) then
                W(parent(u_{i,j})) = W(u_{i,j})
```

As mentioned, the inter-floor solutions are calculated using a heuristic (cf. Algorithm 3) starting from the first leaf of the bottom level and propagating backwards to the root, adding the total path power loss along the way. It continues for other leaves of the same parent and assigns maxima over all paths' costs that pass through this parent, thereby reducing the tree branch to the composite node, and continues that way recursively until the complete reduction to the single node is achieved.

Once the overall power requirement of the complete DAS solution is calculated, it is then rejected if it exceeds the power provided by the base station or accepted as valid otherwise. The total deployment cost $Cost$, that implicitly embodies the fitness function of the optimization problem, is then calculated by:

$$Cost = Cost_{CapEx} + \sum_{i \in \{1,\ldots,n\}} Cost^{i}_{CapEx} \tag{11.6}$$

Algorithm 4 was used to implement the PSO. The parameter α is set to 0.5 plus a random number between 0 and 0.5. This avoids the trap of local minimum by switching between the two modes of PSO, exploration mode ($0 < \alpha \leq 0.5$) and exploitation mode ($0.5 < \alpha \leq 1$). The parameters β and γ are set to 2 while r_p and r_b are random numbers drawn from the uniformly distributed range of (0, 1).

Algorithm 4 PSO-DAS

1: **function** (P, F) = PSO- DAS(PARTICLEN, FLOORN)
2:   **for** each particle i **do**:
3:     Init random position $x_i \in \{1, .., n\}$ and random velocity $v_i \in \{v_{min}, .., v_{max}\}$,
4:     Init best position and cost: $p_i \leftarrow x_i$; $f_i \leftarrow \infty$ (local) and $P \leftarrow x_i$; $F \leftarrow \infty$ (global)
5:   **while** not (termination criterion) **do**
6:     **for** each particle i **do**:
7:       **if** x_i passes power constraints **then**
8:         Evaluate cost function $F(x_i)$
9:         **if** $F(x_i) < f_i$ **then** $p_i \leftarrow x_i$, $f_i \leftarrow F(x_i)$
10:           **if** $f_i < F$ **then** $F \leftarrow f_i$, $P \leftarrow p_i$
11:       **for** each dimension $d = 1, .., n-2$ **do**
12:         $v_{i,d} = \alpha v_{i,d} + \beta r_p (p_{i,d} - x_{i,d}) + \gamma r_g (g_{i,d} - x_{i,d})$
13:       Update particle's position $x_i \leftarrow x_i + v_i$

The termination criterion (used in line 5) is defined as 'k consecutive iterations without improving on the global best solution'. We found $k = 10$ to be a good value (cf. Sect. 11.4.4 for the results of our investigations on k).

It is also important to mention that at both the intra-floor and inter-floor levels the presented PSO–DAS model does not leave any unused splitter outputs. Constraints in this problem are decided by the antenna power settings and the choice of splitters.

Summary

A hybrid model (combining intra- and inter-floor cost optimization for the design of I-DAS) was introduced. Addressing these two optimization problems separately reduces the complexity of the I-DAS problem. This is one of the reasons why our approach scales well for problems with large numbers of floors. In addition, using the PSO heuristic enables us to generate good solutions without having to worry about the combinatorial explosion of the solutions space.

Emphasis is placed on the elaboration of the model equations with a realistic examples of physical structures created in actual I-DAS implementation. A simple example of an actual spanning tree is also presented.

11.4 Results

Our approach was subjected to numerical experiments to evaluate its effectiveness of the PSO–DAS model and to investigate scalability. Our PSO was running with a number of swarm sizes of up to 500 particles.

11.4.1 Power Budget and Deployment Cost Relationship

To design of more sustainable infrastructures (i.e. to reduce power consumption), we imposed power budget restrictions on the solutions (i.e. we artificially restricted the maximum power available at the base station). This favours solutions with lower power requirements (at the price of a higher initial hardware cost).

Figure 11.7 illustrates the evolution of minimum-cost solutions found by PSO–DAS for buildings with a rising number of floors in the following power budget versions: 40, 45, 50 and 60 dBm. As can be seen, minimizing the power consumption and the cabling cost are two opposing objectives, and a trade-off between the two is needed to achieve a well-balanced solution. Limiting the power budget forces the model to return worse-cost solutions and vice versa. The limitations of this are illustrated by the fact that relaxing the power budget past a certain limit does no longer return less expensive solutions.

The balance between minimizing the power budget or the deployment cost can be controlled depending on the priorities of the building management. As we argue in the introduction, adopting a resource-aware stance is imperative, however, in a typical scenario the power budget is assumed upfront and set as a constraint for cost minimization. In our subsequent analysis, we have increased power budgets for taller buildings in line with the empirical observation of making the PSO search for good cost solutions challenging. We hope that—given this efficient tool—I-DAS design decision can now be made with an eye on the running costs of the installation; translating sustainable I-DAS deployment into a long-term financial benefit has the potential to make this part of the ICT infrastructure more power efficient.

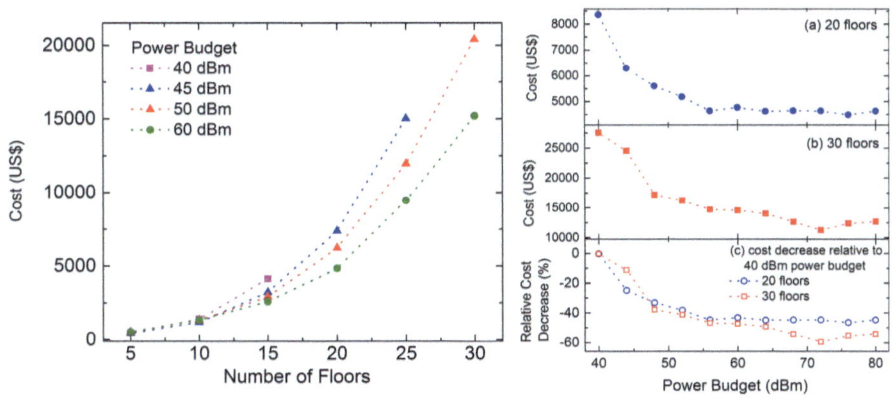

Fig. 11.7 The effect of different power constraints on cost with the last plotted point being the terminating data point (left) and the relationship between power budget and inter-floor deployment cost for the same size 20-floors building (right)

11.4.2 All-PSO Versus Hybrid Approaches

In general, an I-DAS deployment on a single floor is unlikely to exceed 20 antennas (usually between 8 and 20 in most practical realizations) and for these small problems our heuristic worked well, motivating the hybrid approach proposed and tested. However, in order to provide a well-rounded evaluation, we compared this to a non-hybrid approach where both, the intra- and the inter-floor optimization are performed with our PSO.

Figure 11.8 shows the outcome of this comparison: the hybrid approach outperformed the *all-PSO* approach in all cases, delivering lower deployment cost and requiring less computational power to obtain good solutions. In a hindsight, the *all-PSO* approach hinders the overall performance without the benefit to cost optimization outcome. This is happening because the all-PSO algorithm explores a much larger solution space resulting in a longer computing time. At the same time, no clear benefit of using the PSO-only approach was found.

11.4.3 Scalability and Quality of Solutions

Evaluating the approach against the global optimum (i.e. comparing *best found* versus *best possible* solution) was hampered by the $O(n^{n-2})$ complexity of the problem; however, for buildings with up to $n = 8$ floors, the best possible solution was determined through an exhaustive search and in all cases the PSO–DAS approach found the same optimal solution. While these are admittedly small problem instances, the resulting difference in computational cost already indicates the benefit of our approach: for the $n = 8$ problem, the exhaustive search took 15 min

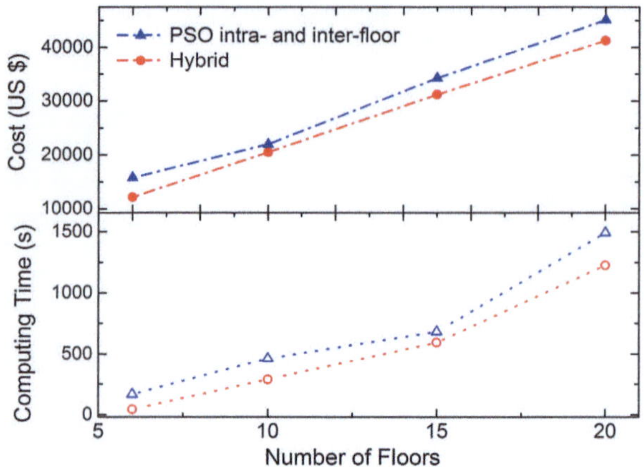

Fig. 11.8 Comparison between the hybrid and the all-PSO approaches in the context of deployment cost (upper panel) and computing time (lower panel)

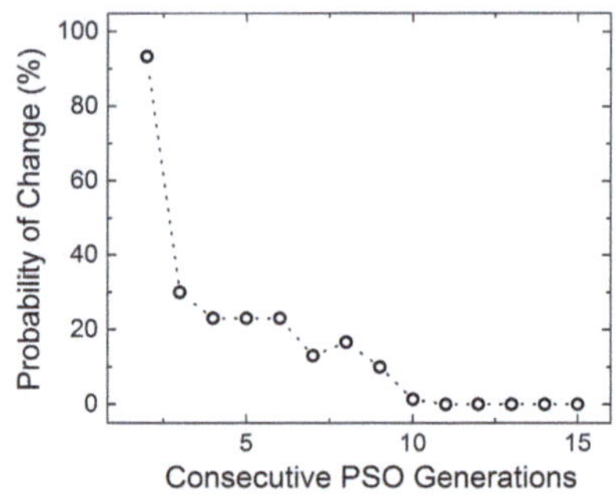

Fig. 11.9 Evaluation of the termination criterion used for the PSO (Algorithm 4, p. 185)

while our approach returned the optimal solution after 15 s. We project an exhaustive search for $n = 13$ to take around 75 years, and settle for the claim that our approach vastly outperformed the exhaustive search with regard to time while producing optimal solutions for $n \leq 8$ and very good or near-optimal solutions for a larger n.

11.4.4 PSO–DAS Termination Criterion

One of our initial investigations was designed to gain insight into what constitutes a suitable termination criterion for Algorithm 4. To this end 30 runs were conducted, applying the PSO–DAS model to problems with n ranging from $n = 5, \ldots, 100$. We recorded the probability of improving upon the current best-found solution in dependence to the number of previous consecutive iterations without improvement. The results are plotted in Fig. 11.9. We found that after ten consecutive generations without improvement the probability of improving further approaches zero.

11.5 Conclusions

ICT is becoming a basic necessity in our daily lives. Unfortunately, this is reflected in its resource consumption: the power requirement of ICT is in access of 10% of the combined global electricity consumption. A rapidly growing number of wireless access points and networks is facilitating the transition from wired to wireless communication. We provide a modelling for Indoor-Distributed Antenna Systems, which are widelyused to provide high-quality blanket coverage with low interference inside buildings. Due to the complexity of these networks, using traditional exhaustive approaches to determine an optimal network topology quickly becomes unfeasible. Distributed, nature-inspired approaches have been applied to a wide variety of computationally intractable problems and have been shown to reliably produce good (albeit not optimal) solutions. Such *heuristics* are scalable and robust, motivating their use for classes of problems such as the I-DAS design problem.

In this chapter, we have clearly demonstrated the effectiveness of combining a heuristic intra-floor equipment selection optimization with a Particle Swarm Optimization (PSO) implementation to optimize the inter-floor connectivity topology and to solve the Indoor-Distributed Antennas System design problems. In addition, we explained why PSO is not required for the intra-floor optimization due to a very small number of antennas practically installed on each floor. Within the limits of exhaustive search where determining the best solution is computationally feasible, we have shown the approach to find the best possible solutions, we have motivated the claim that the approach scales to larger problems. With regard to the use of PSO, we have also addressed the issue of encoding the solution space into n-dimensional Prüfer Space, motivated our choice for this encoding and provided efficient (linear time) algorithms to encode and decode topologies into number sequences.

The chapter provides details of our modelling as well as the algorithms required to reproduce our work. Design choices are motivated and common practices for the design of I-DAS, where we are aware of them, are described. We hope that with the ability to improve the design of I-DAS power cabling topology for hardware cost while ensuring the satisfaction of power budget constraints, network planners will consider optimizing their infrastructure not only for the CapEx (cost for hardware and deployment) but also for the OpEx (the cost of operating the hardware), as this directly translates into reducing the continuous power demand of their building.

In the near future, we would like to extend our approach to automate the design of active (or hybrid) I-DAS, in which optical fibres are used to connect a base station to a set of remote radio units (RRUs) within a tall building (usually over 20 floors). An active component found in the RRU is used to convert the optical signal to radio signal. From there, the RRU can be treated as a base station like our case to further distribute the radio signal to antennas through a set of passive splitters and coaxial cable to cover a dedicated number of floors within the tall building.

Finally, it has not escaped our attention that the approach, while detailed to address I-DAS problems, has the potential to be applied to similar problems in the domain of sustainable computing and infrastructure deployment and design.

Summary

The inter-floor connectivity has been solved with the Particle Swarm optimization model, using Prüfer code to map spanning trees to a solution space. Experimental tests using realistic scenarios (up to 100 floors and 1000 antennas) showed that PSO–DAS delivers good DAS solutions within seconds to minutes. Its supra-linear complexity (on a semi-log scale) in the number of floors can be further reduced by relaxing rather strict stoping conditions for the PSO algorithm. Our work directly supports the reduction of CapEx and, hopefully, will incentivise the DAS planner to consider imposing power budget constraints on the system to reduce the running cost (i.e. electricity consumption).

Acknowledgements This work was supported by UAE ICT Fund, and in part, by Khalifa Semiconductor Research Center and the Semiconductor Research Corporation under the Abu Dhabi SRC Center of Excellence on Energy-Efficient Electronic Systems (ACE^4S), Contract 2013 HJ2440. We acknowledge A. Ouali for his helpful remarks and F. Saffre for early stimulating discussions about the role of naturally inspired computing.

References

1. M. Naphade, G. Banavar, C. Harrison, J. Paraszczak, R. Morris, Smarter cities and their innovation challenges. Computer **44**(6), 32–39 (2011)
2. C. Woodside, It isn't easy being green. Nature Clim. Change **1**(1), 13–15 (2011)
3. M.P. Mills, The cloud begins with coal. Report, Digital Power Group (2013)
4. Editorial smart thinking. Nature **458**, 125–126 (2009)
5. World energy needs and nuclear power (2012), http://www.world-nuclear.org/info/inf16.html. Accessed 10 Dec 2012
6. J. Beyea, The smart electricity grid and scientific research. Science **328**(5981), 979–980 (2010)
7. F. Saffre, R. Gedge, Demand-side management for the smart grid, in *Proceedings of IEEE/IFIP Network Operations and Management Symposium Workshops (NOMS)* (2010), pp. 300–303
8. M.A. Marsan, L. Chiaraviglio, D. Ciullo, M. Meo, Optimal energy savings in cellular access networks, in *2009 IEEE International Conference on Communications Workshops, ICC Workshops 2009* (2009), pp. 1–5
9. S. Zhou, J. Gong, Z. Yang, Z. Niu, P. Yang, Green mobile access network with dynamic base station energy saving. Proc of ACM MobiCom **9**(262), 10–12 (2009)
10. S. Yinbiao, K. Lee, P. Lanctot, F. Jianbin, H. Hao, B. Chow, J-P. Desbenoit, G. Stephan, L. Hui, X. Guodong, S. Chen, D. Faulk, T. Kaiser, H. Satoh, O. Jinsong, W. Linkun, W. Shou, Z. Yan, S. Junping, Y. Haibin, Z. Peng L. Dong, W. Qin, Internet of things: wireless sensor networks. White paper, International Electrotechnical Commission (2014)
11. M. Erol-Kantarci, H.T. Mouftah, Wireless sensor networks for cost-efficient residential energy management in the smart grid. IEEE Trans. Smart Grid **2**(2), 314–325 (2011)
12. R. Yu, T. Watteyne, Reliable, low power wireless sensor networks for the internet of things: making wireless sensors as accessible as web servers. White paper, Linear Technology (2013)
13. Markets & Markets. Internet of things (IoT) & machine-to-machine (M2M) communication market by technologies & platforms (RFID, sensor nodes, gateways, cloud management, NFC, CEP, SCADA, ZigBee), M2M connections, IOT components - worldwide market forecasts (2014–2019). Top Market Reports TC 2351, Markets & Markets (2014)
14. F. Saffre, H. Hildmann, J.-L. Deneubourg, Can individual heterogeneity influence self-organised patterns in the termite nest construction model? Swarm Intell. **12**, 101 (2017)

15. N. Gershenfeld, S. Samouhos, B. Nordman, Intelligent infrastructure for energy efficiency. Science **327**(5969), 1086–1088 (2010)
16. F. Saffre, H. Hildmann, Don't step on the distribution's tail (investigating the impact of random fluctuations on efficient resource utilization), in *Proceedings of the 2014 Federated Conference on Computer Science and Information Systems*, ed. by M. Paprzycki, M. Ganzha, L. Maciaszek. Annals of Computer Science and Information Systems, vol. 2 (IEEE, 2014), pp. 1525–1527
17. L. Chen, D. Yuan, Mathematical modeling for optimal design of in-building distributed antenna systems. Comput. Netw. **57**(17), 3428–3445 (2013)
18. K. Hiltunen, B. Olin, M. Lundevall, Using dedicated in-building systems to improve HSDPA indoor coverage and capacity, in *61st IEEE Conference on Vehicular Technology* (2005), pp. 2379–2383
19. D. Adjiashvili, S. Bosio, Y. Li, D. Yuan, Exact and approximation algorithms for optimal equipment selection in deploying in-building distributed antenna systems. IEEE Trans. Mob. Comput. **14**(4), 702–713 (2015)
20. J.J. Bartholdi, D.D. Eisenstein, A production line that balances itself. Oper. Res. **44**(1), 21–34 (1996)
21. E. Bonabeau, M. Dorigo, G. Theraulaz, Inspiration for optimization from social insect behaviour. Nature **406**, 39–42 (2000)
22. S. Navlakha, Z. Bar-Joseph, Algorithms in nature: the convergence of systems biology and computational thinking. Mol. Syst. Biol. **7** (2011)
23. S. Camazine, J.-L. Deneubourg, N.R. Franks, J. Sneyd, G. Theraulaz, E. Bonabeau, *Self-Organization in Biological Systems* (Princeton University Press, Princeton, 2001)
24. A. Mugler, A.G. Bailey, K. Takahashi, P.R. ten Wolde, Membrane clustering and the role of rebinding in biochemical signaling. Biophys. J. **102**(5), 1069–1078 (2012)
25. J. Halloy, G. Sempo, G. Caprari, C. Rivault, M. Asadpour, F. Tâche, I. Saïd, V. Durier, S. Canonge, J.M. Amé, C. Detrain, N. Correll, A. Martinoli, F. Mondada, R. Siegwart, J.L. Deneubourg, Social integration of robots into groups of cockroaches to control self-organized choices. Science **318**(5853), 1155–1158 (2007)
26. M. Macaš, B. Gabrys, D. Ruta, L. Thotská, Particle swarm optimization of multiple classifier systems, in *9th International Work-Conference on Artificial Neural Networks* (2007), pp. 333–340
27. L-p. Zhou, B. Li, F. Wang, Particle swarm optimization model of distributed network planning. JNW **8**(10), 2263–2268 (2013)
28. X.H. You, D.M. Wang, B. Sheng, X.Q. Gao, X.S. Zhao, M. Chen, Cooperative distributed antenna systems for mobile communications [coordinated and distributed mimo]. IEEE Wirel. Commun. **17**(3), 35–43 (2010)
29. T.L. Marzetta, Noncooperative cellular wireless with unlimited numbers of base station antennas. IEEE Trans. Wirel. Commun. **9**(11), 3590–3600 (2010)
30. Q. Sun, S. Jin, J. Wang, Y. Zhang, X. Gao, K.K. Wong, On scheduling for massive distributed MIMO downlink, in *2013 IEEE Global Communications Conference (GLOBECOM)* (2013), pp. 4151–4156
31. H. Ren, N. Liu, C. Pan, C. He, Energy efficiency optimization for MIMO distributed antenna systems. IEEE Trans. Veh. Technol. **PP**(99), 1–1 (2016)
32. H. Beijner, The importance of in-building solutions in third-generation networks (Ericson, 2004). Ericson Review 2
33. R. Atawia, M. Ashour, T. El Shabrawy, H. Hammad, Indoor distributed antenna system planning with optimized antenna power using genetic algorithm, in *2013 IEEE 78th Vehicular Technology Conference (VTC Fall)* (2013), pp. 1–6
34. D.Y. Atia, Indoor distributed antenna systems deployment optimization with particle swarm optimization. M.Sc. thesis, Khalifa University of Science, Technology and Research (2015)
35. T. Paulden, D.K. Smith, Developing new locality results for the Prüfer code using a remarkable linear-time decoding algorithm. Electron. J. Comb. **14** (2007)
36. C.W. Borchardt, über eine Interpolationsformel für eine Art symmetrischer Funktionen und über deren Anwendung. In *Math. Abh. Akad. Wiss. zu Berlin* (Berlin, 1860), pp. 1–20

37. A. Cayley, On the theory of the analytical forms called trees. Philos. Mag. **13**, 172–6 (1857)
38. A. Cayley, *Cambridge Library Collection - Mathematics*, vol. 13 (Cambridge University Press, Cambridge, 2009), p. 2628
39. A. Berdahl, C.J. Torney, C.C. Ioannou, J.J. Faria, I.D. Couzin, Emergent sensing of complex environments by mobile animal groups. Science **339**(6119), 574–576 (2013)
40. S. Lim, D. Rus, Stochastic distributed multi-agent planning and applications to traffic. ICRA 2873–2879 (IEEE, 2012)
41. F. Saffre, R. Furey, B. Krafft, J.-L. Deneubourg, Collective decision-making in social spiders: dragline-mediated amplification process acts as a recruitment mechanism. J. Theor. Biol. **198**, 507–517 (1999)
42. C. Anderson, J.J. Boomsma, J.J. Bartholdi, Task partitioning in insect societies bucket brigades. Insectes Sociaux **49**, 171–180 (2002)
43. H. Hildmann, S. Nicolas, F. Saffre, A bio-inspired resource-saving approach to dynamic client-server association. IEEE Intell. Syst. **27**(6), 17–25 (2012)
44. R. Schoonderwoerd, J.L. Bruten, O.E. Holland, L.J.M. Rothkrantz, Ant-based load balancing in telecommunications networks. Adapt. Behav. **5**(2), 169–207 (1996)
45. E. Bonabeau, M. Dorigo, G. Theraulaz, *Swarm Intelligence: From Natural to Artificial Systems*. Santa Fe Institute Studies on the Sciences of Complexity (Oxford University Press, USA, 1999)
46. J. Werfel, K. Petersen, R. Nagpal, Designing collective behavior in a termite-inspired robot construction team. Science **343**(6172), 754–758 (2014)
47. F. Ducatelle, G.A. Di Caro, L.M. Gambardella, Principles and applications of swarm intelligence for adaptive routing in telecommunications networks. Swarm Intell. **4**(3), 173–198 (2010)
48. B. Gabrys, D. Ruta, Genetic algorithms in classifier fusion. Appl. Soft Comput. **6**(4), 337–347 (2006)
49. D. Renfrew, X.H. Yu, Traffic signal control with swarm intelligence, in *2009 Fifth International Conference on Natural Computation*, vol. 3 (2009), pp. 79–83
50. P. Kuila, P.K. Jana, Energy efficient clustering and routing algorithms for wireless sensor networks: particle swarm optimization approach. Eng. Appl. Artif. Intell. **33**, 127–140 (2014)
51. R.-J. Ma, N.-Y. Yu, J.-Y. Hu, Application of particle swarm optimization algorithm in the heating system planning problem. Sci. World J. (2013)
52. H. Prüfer, Neuer Beweis eines Satzes über Permutationen. Archiv der Mathematik und Physik **27**, 742–744 (1918)
53. B.A. Julstrom, Quick decoding and encoding of prfer strings: exercises in data structures (2005)
54. J. Gottlieb, B.A. Julstrom, G.R. Raidl, F. Rothlauf, Prüfer numbers: a poor representation of spanning trees for evolutionary search, in *Proceedings of the Genetic and Evolutionary Computation Conference (GECCO 2001)* (Morgan Kaufmann Publishers, San Francisco, California, 2001), pp. 343–350
55. H. Hildmann, D. Ruta, D.Y. Atia, A.F. Isakovic, Using branching-property preserving Prüfer code to encode solutions for particle swarm optimization, in *2017 Federated Conference on Computer Science and Information Systems (FedCSIS)* (Prague, Czech Republic, 2017)
56. D.Y. Atia, D. Ruta, K. Poon, A. Ouali, A.F. Isakovic, Cost effective, scalable design of indoor distributed antenna systems based on particle swarm optimization and prufer strings, in *IEEE Proceedings of 2016 IEEE Congress on Evolutionary Computation* (Vancouver, Canada, 2016)

Chapter 12
Low-Power, Dynamic-Data-Rate Protocol for IoT Communication

Shahzad Muzaffar and Ibrahim (Abe) M. Elfadel

12.1 Introduction

Networks made of 1-wire devices [10–12] use a single-wire protocol that does not require any clock and data recovery (CDR) circuit. This protocol is powered using the charge of a capacitor and its power requirements are quite minimal. These networks are called Microlans and are used in a variety of sensor applications [13–18]. The main disadvantage of these 1-wire protocols is their very low data rate—limited to 16 Kb/s. It is expected that IoT sensors such as imagers triggered by environmental events (motion, chemical concentrations, temperature, etc.) will require much higher data rates for the transmission of information. Some applications [19] transfer data in standard protocols such as the serial RS232-UART with data rate in the range of few Kb/s. High data rate, single-wire communication can be achieved by encoding/decoding bits onto the cycles of a carrier waveform [20]. To insure transmission reliability, the carrier signal needs to have a high voltage amplitude (±15 V) thus requiring voltage boosting and other special circuitry. Line coding techniques, like non-return to zero, Manchester, alternate mark inversion, are bit-time dependent and require either a CDR for clock synchronization or a higher clock rate to sample the line to detect edges or to measure the time between edges [27]. Another protocol is USB [26] which achieves high data rates but is known to be very power-hungry due to its complexity and to the need of an external controller to execute all the transfers. In general, digital systems operating at low clock rates (few MHz) and receiving serial data using a one-wire channel operating at a relatively high data rate will need a CDR to receive and recover data without errors. The block diagram in Fig. 12.1 shows the position of the CDR in relation to the serializer–deserializer

S. Muzaffar · Ibrahim (Abe) M. Elfadel (✉)
Masdar Institute at Khalifa University, Abu Dhabi, UAE
e-mail: ielfadel@masdar.ac.ae

S. Muzaffar
e-mail: smuzaffar@masdar.ac.ae

Ibrahim (Abe) M. Elfadel and M. Ismail (eds.), *The IoT Physical Layer*,
https://doi.org/10.1007/978-3-319-93100-5_12

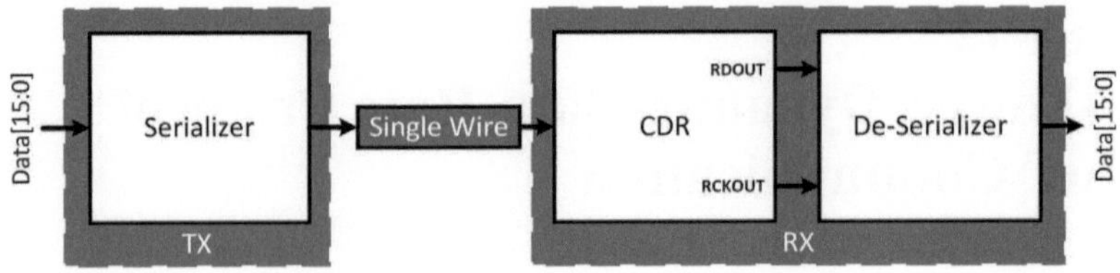

Fig. 12.1 Single-wire data transfer

(SERDES) of the communication channel. References [21] through [25] propose several variations on low-power CDR circuitry, but they all unfortunately fall short of the specifications of an area-efficient, ultra-low power, single-wire transmission protocol for rapid-burst IoT sensors.

The objective of this chapter is to fill up this gap and to present a novel signaling technique for data transfer over a single wire achieving high-data rates (in the MHz range), with low power consumption and small footprint. The protocol does not require a CDR, can operate with signals at low amplitude voltage (~1V), has simple encoding and decoding schemes, and can tolerate clocking differences between transmitter and receiver. We call this novel technique: Pulsed-Index Communication (PIC) [1–4] because its core idea is to transfer the indices of only the ON bits as a series of pulses. A very compact packet header gives information about the number of such indices, and the encoding operations to which the raw bits have been subjected. When the pulses are received, the receiver applies the appropriate decoding to infer the original data bits. PIC is dynamic in that it can accommodate several data rates in the range of 3.1–8.5 Mb/s with an average of 4.1 Mb/s using a 24-MHz clock. PIC exploits edge detection of incoming pulses to achieve remarkable robustness with respect to jitters, skews, and clock inaccuracies between the transmitter and receiver. PIC is also architecturally flexible in that it can be configured according to the signaling topology (Master–Slave, Ring, Star, etc.).

12.2 Pulsed-Index Communication

The proposed Pulsed-Index Communication (PIC) transfers data on one wire only without any additional wires except ground. The core of PIC is to transfer only the indices of the ON bits in the data stream while ignoring the OFF bits. The index is transferred as a series of pulses, hence the name of the protocol: Pulsed-Index Communication. High-data rates are achieved by encoding the raw bitstream so that the code word has as few ON bits as possible and these ON bits occupy the lowest possible index positions. The encoding process comprises three operations: segmentation, inversion, and reversal. The conceptual block diagram of the encoding and transmission process is shown in Fig. 12.2, whose blocks are explained in the following sections.

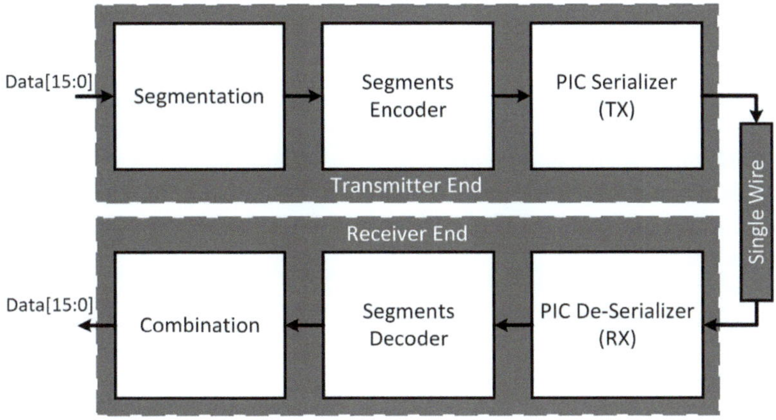

Fig. 12.2 Conceptual block diagram of PIC

12.2.1 Bit Selection and Pulse Count

As stated above, PIC transmits the indices of the ON bits only. This is illustrated in Fig. 12.3a which shows an example of an 8-bit data segment. Instead of transmitting all the ON and OFF bits, PIC transmits 2, 3, and 7 which are the index numbers of the ON bits. Each index will be sent in the form of i consecutive pulses where i is the index number (i.e., $i = 2$, 3, and 7). Each index's stream of pulses is followed by an inter-symbol delay of four clock cycles, as explained in Sect. 12.2.4.

If we have 16-bit data and all the bits are 1, then it would generate a large number of pulses to transmit. All the indices from 1 to 16 need to be sent as shown in Fig. 12.3b. Pulses would increase as the number of data bits increases. This is unacceptable as it will reduce the data rate significantly. Data segmentation can help address this problem.

Fig. 12.3 **a** Selection of ON bits **b** Problem of large number of pulses

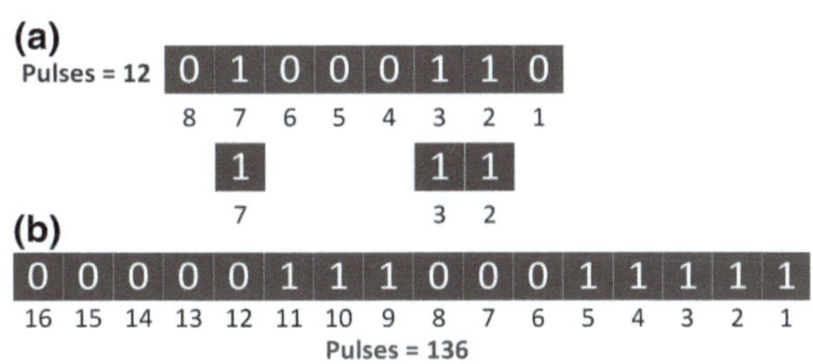

12.2.2 Segmentation

The number of transmitted pulses decreases quickly if the indices of the ON bits are small. This is achieved by introducing data segmentation in which the data are broken into small segments of equal size. Considering all the segments as distinct data words significantly lower the index values. The segmentation process and the resulting indices are shown in Fig. 12.4 for a 16-bit worst-case data stream in which all the bits are ON.

12.2.3 Encoding

Reducing the number of ON bits reduces the number of indices that need to be transmitted. This will result in a decrease of the total pulse count. Further reduction can be obtained if the ON bits, within each segment, are mapped to locations with small index numbers. This can be achieved using bitwise inversion and segment-wise reversal. Bitwise inversion is a conditional one's complement operation on the segment bits, the condition being that more than half of the length of the segment is occupied by ON bits. Reversal is a conditional right-left flipping operation, the condition being that the decimal value of the segment bits after flipping is less than its decimal value before flipping. Figure 12.5 illustrates the application of inversion and reversal on a 8-bit data word. In particular, it shows that inversion and reversal alone cannot minimize the number of pulses. On the other hand, combining inversion and reversal results in the smallest number of pulses, namely 6, for that particular

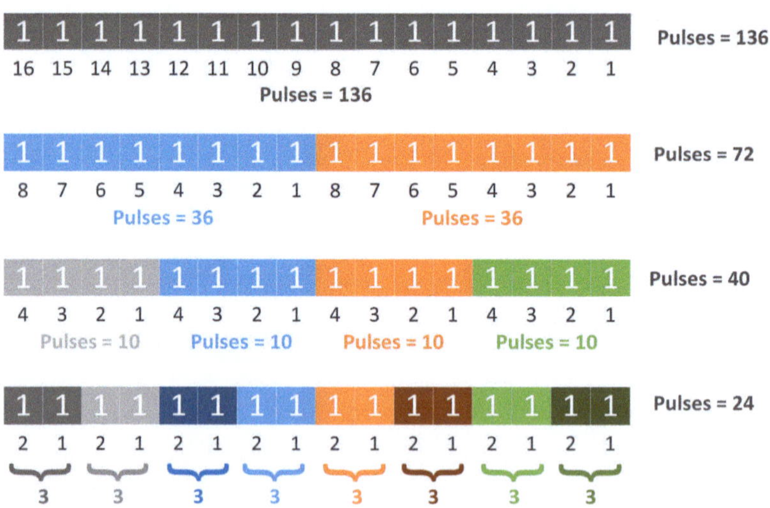

Fig. 12.4 Segmentation

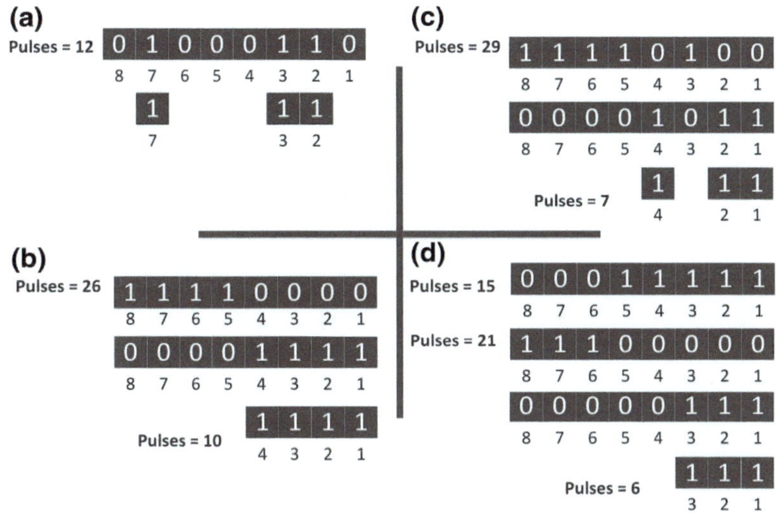

Fig. 12.5 Encoding: **a** No reversal and inversion **b** Reversal **c** Inversion **d** Inversion and reversal

examples. One's complement and flipping on short data words are not only very easy to implement in hardware but also very easy to invert (they are both idempotent operations). The latter property is very important for energy-efficient decoding. Section 12.5 is devoted to the description of an energy-efficient synthesis of these encoding/decoding procedures in both FPGA (Xilinx Virtex VII) and ASIC (GLOBALFOUNDRIES 65nm).

12.2.4 Serialization

The serializer collects the encoded data from the encoder and transmits it serially according to the format shown in Fig. 12.6. Each segment is transmitted along with a header, including encoding Flags and the number of indices (NOI). The body of the message includes the index pulses and a synch pulse. Flags is a 2-bit code representing one of the four encoding cases (Fig. 12.7). NOI represents the number of indices being transmitted. Each of these parts is sent in the form of a pulse stream (one pulse is equal to 1 clock cycle), and each stream is followed by a boundary tag made of 4 clock cycles. Transmission is initiated with a start pulse followed by all the message pulses. It ends with two stop pulses. There is boundary tag of 4 clock cycles after the start and stop pulses.

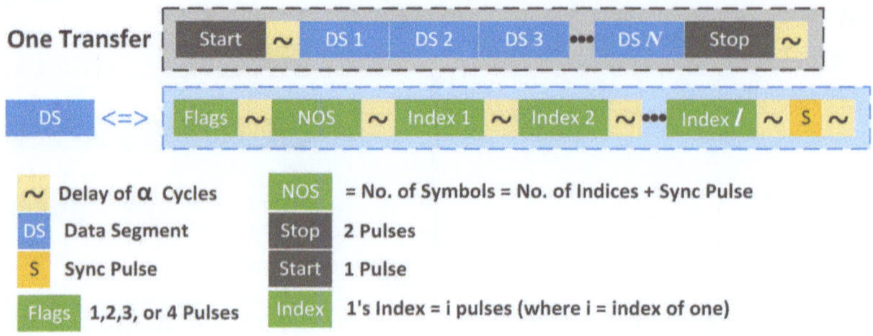

Fig. 12.6 PIC packet format

Fig. 12.7 Encoding flags

Pulses	Inverted	Reversed
1	NO	YES
2	YES	NO
3	YES	YES
4	NO	NO

12.2.5 De-Serialization, Decoding, and Combination

The receiver counts the pulses of each stream using rising edge detection which eliminates the need of CDR. The count of NOI pulses reveals the number of indices of the ON bits being received while the flag count is used to infer the segment encoding which will be used in the segment decoding process. The decoded segments are combined to assemble the full length of transmitted data.

12.3 PIC Protocol Analysis and Optimization

To model, analyze, and optimize the encoding/decoding procedure of segmentation, inversion and reversal, we first denote by b_i^s the ith bit in the sth segment. We let l be the number of bits per segment and N be the number of segments. Then, we have the following set of five equations, the last of which gives the data rate R of the PIC protocol for a clock period of T and total pulse count of C.

12.3.1 Equations

Let P_s is the number of pulses required for the ON bits, n_s is the number of ON-bits, C_s is the total cycles count, f_s is the flags' pulse count, and α is the inter-symbol delay, all in reference to segment s.

$$P_s = \sum_{i=0}^{l-1} (i + 1) b_i^s \tag{12.1}$$

$$n_s = \sum_{i=0}^{l-1} b_i^s \tag{12.2}$$

$$C_s = 2 + f_s + 3\alpha + (1 + \alpha) n_s + P_s \tag{12.3}$$

Summing the C_s over all the segments in the packet, we get

$$C = 2N + 3 + (3N + 2)\alpha + \sum_{S=1}^{N} f_s + (1 + \alpha) \sum_{S=1}^{N} n_s + \sum_{S=1}^{N} P_s \tag{12.4}$$

The data rate is then given by

$$R = \frac{lN}{TC} \tag{12.5}$$

The expression of C in (12.4) results from a summation over all segments in the data stream. The remainder of this section is devoted to showing how these equations have been used to analyze and optimize the PIC protocol for maximum data rate transmission, which is the major objective of this work.

12.3.2 Optimum Segment Size

The segment size is chosen to maximize data rate. For a small segment, delays inserted between pulse streams to separate symbols reduce the data rate. Similarly for large segments, ON bits with high indices require a large number of pulses to be transmitted, which in turn reduces the data rate. It is therefore intuitive that there is a segment size for which the data rate is maximum. This is illustrated in Fig. 12.8 which has the data rate plotted as function of the segment size for the three cases where the three levels of encoding are applied to the raw data. For this particular example, the plot shows that the data rate is maximized when the number of bits per segment is 8.

12.3.3 Data Rate

As mentioned in the previous sections, PIC is dynamic in that the actual data rate of the protocol is dictated by the pulse count which is very much data dependent (See Eqs. (12.4) and (12.5)). This paragraph analyzes the statistical distribution of

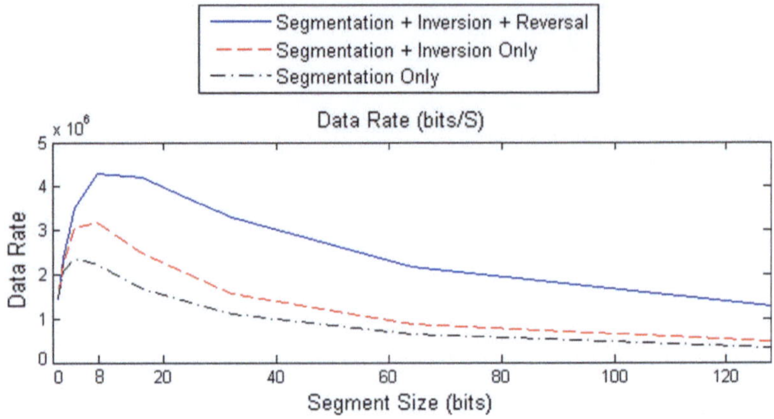

Fig. 12.8 Optimum segment size analysis

PIC data rates using exhaustive sampling of 16-bit data words ($2^{16} - 1$ PRBS), each segmented into two 8-bit segments. For data rate calculations, a 24 MHz clock is used. The objective of the analysis is to show that the full encoding process (Segmentation + Inversion + Reversal) is generally needed not only to maximize the (average) data rate but also to tighten the distribution of dynamic rates around the average data rate. For each data word, one computes both the raw pulse count (no encoding) and the pulse count that has resulted from full encoding. The histograms of pulse counts are shown in Fig. 12.9. Note that after full encoding, there is a significant reduction in the average number of pulse counts and a tightening of the pulse count distributions around the average. The histograms of data rates are shown in Fig. 12.10. Again, comparing the red and black histograms, there is an increase in the average data rate and a tightening of the data rate distribution around the average.

12.4 PIC Transmission System

The block diagram of PIC transmission system is shown in Fig. 12.11. Both ends of the system comprise the same transceiver along with its logical topology control (LTC) unit. The LTC can be used to implement a variety of transmission topologies (one-to-one, master–slave, ring, etc.) without any change in the underlying PIC encoding/decoding. This is because the PIC transceiver operates on raw bits and does not distinguish between data types (e.g., slave ID versus data packet).

The PIC transceiver flow is shown in Figs. 12.12 and 12.13. The transmitter and receiver are connected to the single-wire through tri-state buffers. The transmitter and receiver operate according to the protocols described below. Waveforms for transmission and reception are shown in Fig. 12.14.

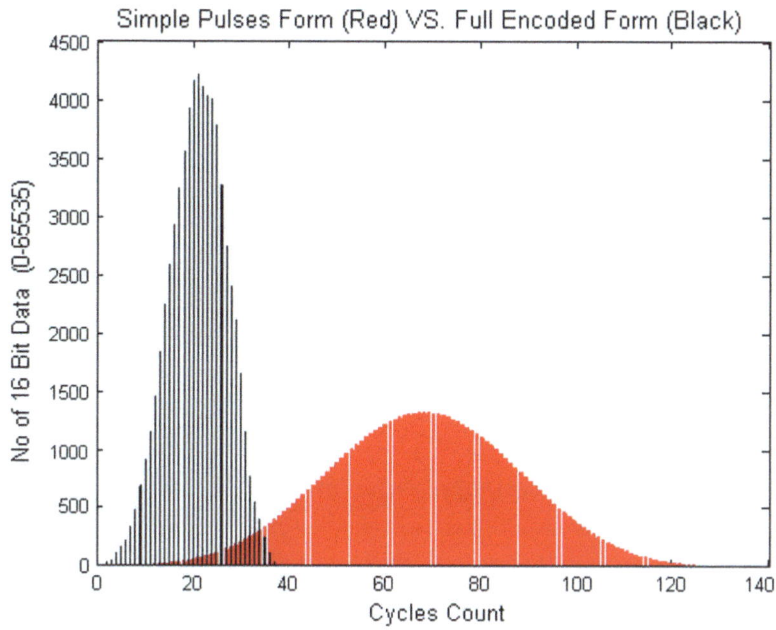

Fig. 12.9 Cycles/pulse count analysis

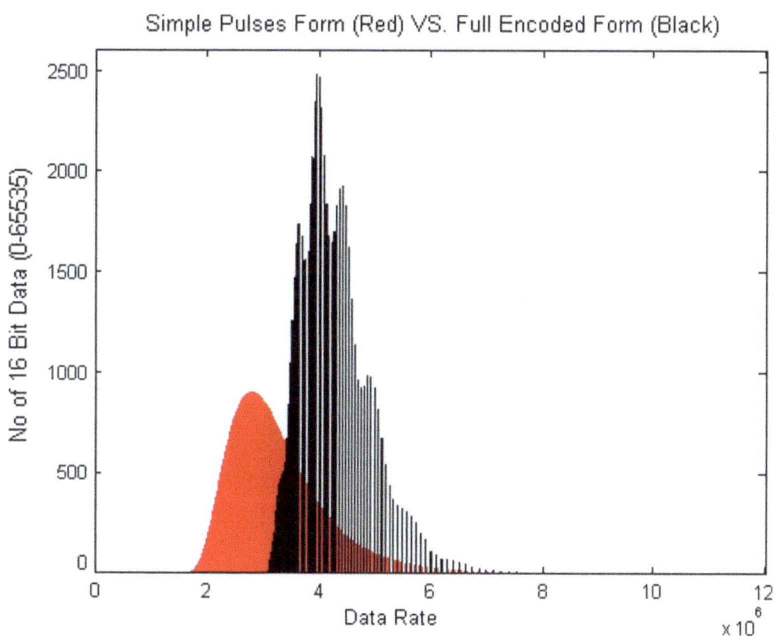

Fig. 12.10 Data rate analysis

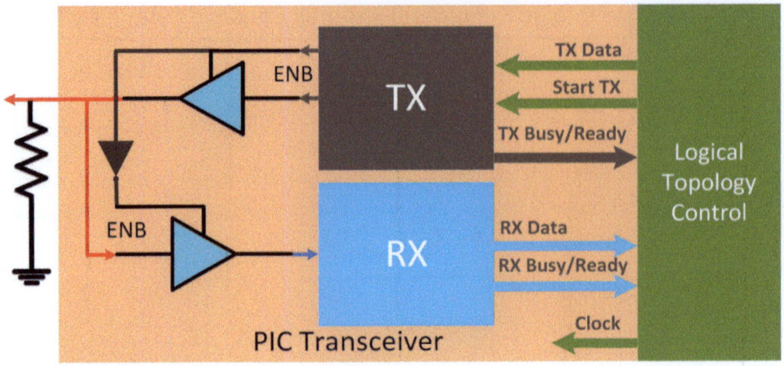

Fig. 12.11 One side of PIC transmission system

12.4.1 Transmitter Protocol

The PIC transmitter protocol is shown in Fig. 12.12. The receiver is the default mode when the transmitter is not active. To start transmission, it acquires access to the single-wire channel through tri-state buffers. Therefore, the system enters the TX mode on LTC request. Segmentation of data is performed, followed by the encoding of each segment. Inversion of segment bits is applied if the number of ON bits is larger than the number of OFF bits in the segment. The segment bits are reversed if the reversal results in a smaller binary number than the original segment. The encoded segments are then transmitted in the form of pulse streams separated by inter-symbol delays, according to the packet format shown in Fig. 12.6. TX leaves the line in high-impedance state after transmission is completed. The external pull-down technique is used to keep the line low in idle states.

12.4.2 Receiver Protocol

The PIC receiver protocol is shown in Fig. 12.13. A counter is used to count the receiver clock cycles. It gets reset on each of the falling edges of the input data pulses. At each of the falling edges, the count is compared with the delay threshold coefficient, α_{Th}. If the receiver clock count is greater than α_{Th}, it is considered as an inter-symbol delay. Otherwise, it is a data pulse (Fig. 12.13(Right)). The data pulses are counted using a separate counter and stored appropriately at the detection of inter-symbol delay. When all the pulses are received, the segment is decoded. At the end of a reception, all the decoded segments are rejoined to form the complete data word (Fig. 12.13(Left)).

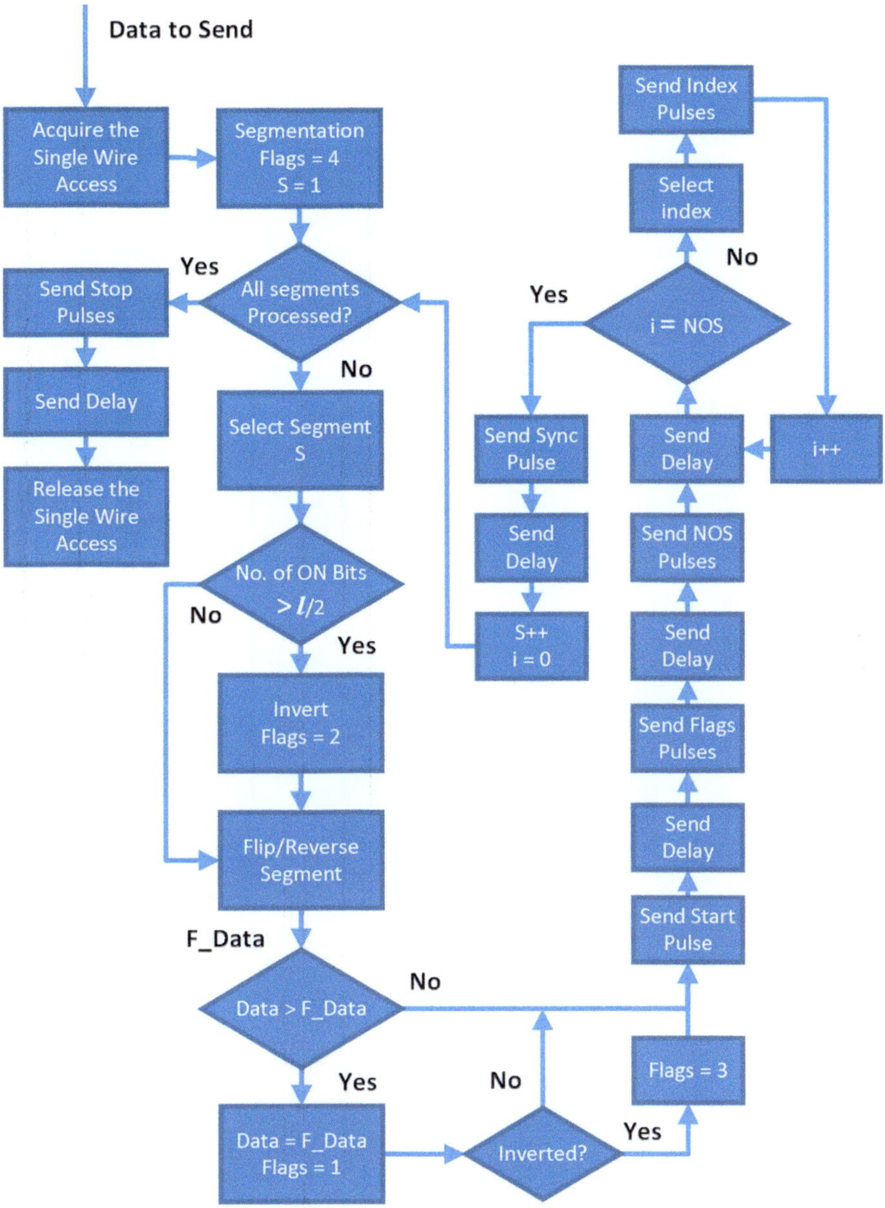

Fig. 12.12 PIC transceiver protocol diagram: the transmitter

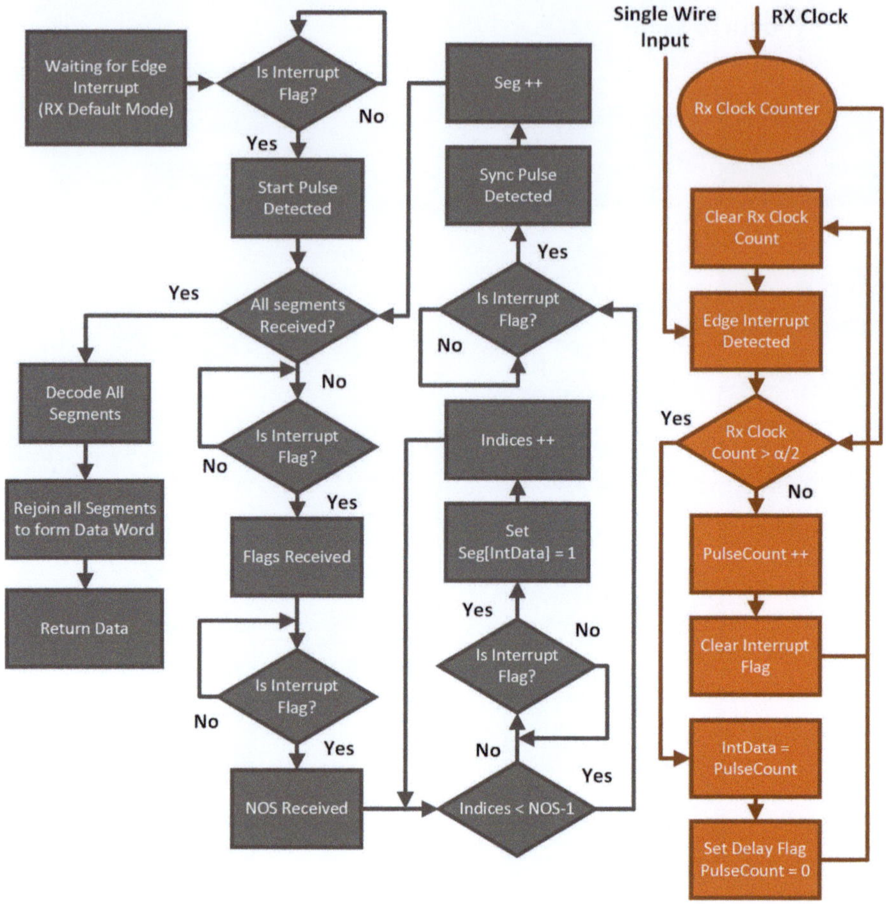

Fig. 12.13 PIC transceiver protocol diagram: the receiver (Left) and receiver edge detection (Right)

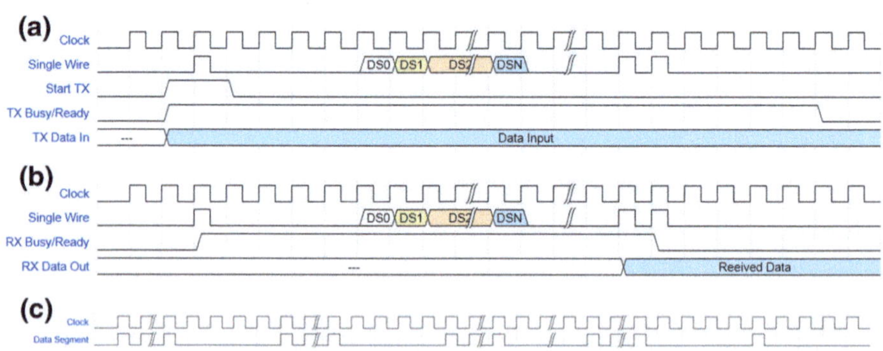

Fig. 12.14 **a** Transmission **b** Reception **c** Segment format

12.5 PIC Experimental Results

To verify PIC's functionality and performance, a prototype setup based on a Virtex-7 development board is utilized. All the blocks mentioned in Figs. 12.2 and 12.11 are implemented in Verilog hardware description language and tested on FPGA. Extensive simulations and real-time hardware checks are performed in order to verify the results. Two different implementations and their verifications are carried out and compared. In the first implementation, the data are sent in a simple serial bitstream of 1's and 0's without any encoding. In the second implementation, PIC is utilized to encode and serialize the data. 16-bit data words, two segments per word at a clock rate of 25 MHz are used. The transmitter sends the 16-bit data starting at 0 with an increment of 1 at each transmission. The receiver resends the same data back. The returned and original data words are compared to verify the complete round-trip chain. The number of perfect matches is logged.

Along with the FPGA prototype, the PIC system is also synthesized and verified using a synopsis logic-synthesis flow and a GLOBALFOUNDROES 65 nm process in order to have realistic area and power estimates. Table 12.1 summarizes and compares the data rates of few single-wire transmission techniques. It is clear that PIC outperforms [10, 19, 20] without any CDR or power conversion circuitry and with tolerance toward jitters, skew and clock inaccuracies.

Table 12.2 compares PIC with the normal serial transmission (NST) in terms of area and power. Note that NST needs CDR to recover data successfully while PIC does not. This is the main reason for the significant decrease in area and power when PIC is used. For small footprint applications (wireless sensor nodes, wearable computing, body-area networks, etc.), PIC is definitely the better choice. As for power, CDR is the main source of power consumption, and even if we use the recently published low-power CDR's proposed in [21–25], PIC still outperforms NST with low-power CDR since

$$P_{NST-SRL} + P_{CDR} > P_{PIC} \tag{12.6}$$

As mentioned above, to get the most realistic comparison with published literature, the PIC system is synthesized using GLOBALFOUNDRIES 65 nm technology and found out that PIC consumes 26.6 uW with a gate count of 2356, offering dynamic data rates in the range of 3.1–8.5 Mb/s (4.1 Mb/s average) with a 24-MHz clock. Compared with NST serialization and CDR, PIC can reduce area by more than 80% and power by more than 70%.

Table 12.1 Data rate comparison of one wire techniques

	PIC	1-Wire [10]	[19]	[20]
Data rate	3.1–8.5 Mb/s (Avg 4.1 Mb/s)	16 Kb/s	38.4 Kb/s	17 Kb/s

Table 12.2 PIC comparison with simple-serial

	Power (uW)			Area (gate count)			Process
	SRL[a]	CDR[c]	Total[d] (PI[e])	SRL[a]	CDR[c]	Total[d] (PI[e])	
PIC	26.6	N/A	26.6	2356	N/A	2356	65 nm
		70	102.1 (74%)		15600	16927 (86%)	90 nm [21]
		62.5	94.6 (72%)		60000	61327 (96%)	90 nm [22]
NST[b]	32.1	90	122.1 (78%)	1327	N/A	N/A	90 nm [23]
		57.5	89.6 (70%)		19800	21127 (89%)	65 nm [24]
		60.6	92.7 (71%)		N/A	N/A	28 nm [25]

[a] Serializer
[b] Normal Serial Transfer
[c] Estimated calculation
[d] SRL + CDR
[e] % Increase as compared to PIC

12.6 Timing and Robustness Analysis

The objective of this section is to provide a full theoretical framework that underlies the timing and robustness of PIC protocols. This framework is used to derive rigorously the performance parameters that we discovered empirically in Sects. 12.2 and 12.3, including those related to PIC robustness with respect to clocking uncertainties. The most important result of the PIC theory presented in this chapter is the crucial role played by the delay (expressed in transmitter clock cycles) between data segments in a given packet. For a given encoding scheme, this delay determines both the PIC average data rate and the maximum clock uncertainty tolerance. Rigorous timing and robustness analysis is provided to quantify the PIC robustness margin in the presence of transmitter-to-receiver clock variations as well as clock skew and jitter within each clock. The PIC robustness margin is verified experimentally using a full FPGA implementation that includes point-to-point transmission between two TI MSP430 microcontrollers, acting as two IoT sensor nodes over a single-wire connection.

Robustness is directly related to the capability of behaving appropriately in the presence of different sources of errors. In this section, we first survey the sources of errors in PIC channels. Then we describe the intrinsic timing parameters of the PIC protocols and introduce some of the constraints they must satisfy for error-free operation. The last subsection is devoted to deriving a closed-form upper bound on clock discrepancy between transmitter and receiver for error-free operation in the presence of clock variations.

12.6.1 Sources of Errors

Clock inaccuracy between transmitter and receiver is one of the main sources of errors and creates significant trouble in digital communication systems. The problem becomes even more serious when single channel communication is used. The clocks of both ends need to be synchronized to limit the errors. A variety of techniques, including efficient clock-and-data recovery (CDR) circuits, are available for such synchronization. PIC is an ultra-low power, single channel protocol, without CDR, but with unique robustness properties with respect to clock variations. This situation is expected to be very wide-spread in the IoT environment where IoT devices with different clocking and performance requirements need to communicate. Each end of the PIC communication link is comprised of a transmitter and a receiver. All the subsequent formulation and calculations are carried out assuming the transmitter is running at a slow clock rate f_S while the receiver is running at a fast clock rate f_F. The link setup is shown in Fig. 12.15a and b. In the remainder of this chapter, we adopt the convention that all the parameters with S subscript are for the slow-end and with F subscript are for the fast-end.

The sources of timing errors in PIC are as follows:

1. Data pulse jitter: This is the main source of time difference between two adjacent data pulses. If such jitter is high, the pulses start overlapping which introduces a missing-pulse error at the receiver. The extent to which PIC can tolerate these jitters is explained using Fig. 12.16, where T_T is the transmitter clock time period. The faster the clock, the closer the data pulses are to each other, and therefore,

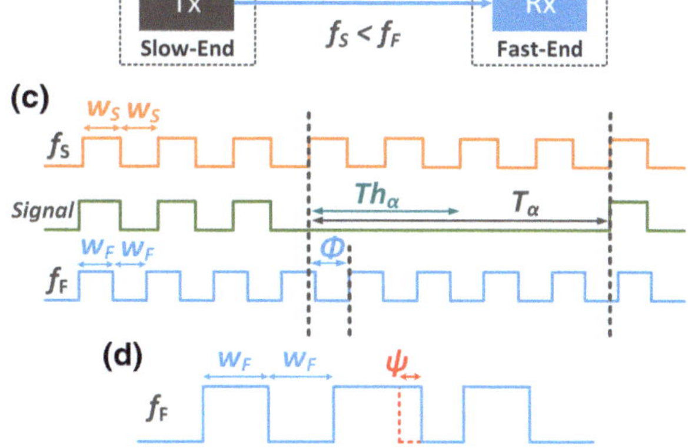

Fig. 12.15 **a** PIC communication link **b** PIC link with clock inaccuracy **c** PIC waveforms with clock inaccuracy **d** Receiver clock Jitter

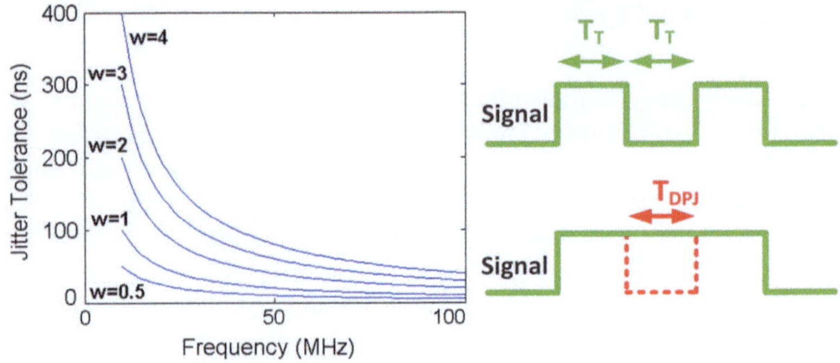

Fig. 12.16 Effect of frequency on Jitter tolerance

a small jitter may lead to successive pulse overlap. This explains the plot in Fig. 12.16 which shows that data pulse jitter tolerance is inversely proportional to transmitter clock rate. For a given clock rate, the data pulse jitter tolerance increases with the increase in pulse width coefficient w.

2. Phase shift Φ: This is the time difference between the edge of the receiver clock f_F and the transmitter clock edge marking the start of the inter-symbol delay T_α. The phase shift may affect the detection of inter-symbol delay especially in the presence of clock inaccuracies. Phase shift ranges from 0 to T_F, the period of the receiver clock, as shown in Fig. 12.15c. In terms of shift percentage φ, we have $\Phi = \varphi T_F, 0 \leq \varphi \leq 1$.
3. Receiver clock jitter Ψ: This may also affect the detection of inter-symbol delay. The jitter value is related to the receiver clock period T_F by $\Psi = \psi T_F, 0 \leq \psi \leq 0.5$, where ψ represents jitter percentage. Receiver clock jitter is shown in Fig. 12.15d.
4. Noise associated with the off-chip environment can have an effect on data pulses. The increase or decrease in pulse levels, due to external noise, makes it difficult for the receiver to detect pulses correctly. Depending on the noise level, an extra pulse may be detected or a pulse may be skipped. In both cases, one gets a decoding error. To analyze the performance of PIC in the presence of noise, the encoded pulse stream of data is exposed to white Gaussian noise. The noisy signal is filtered at the receiver end, then decoded according to PIC algorithm, and the number of errors encountered is counted. The results are plotted in Fig. 12.17 for different values of E_b/N_0 (the ratio of energy-per-bit to noise power spectral density) for both BPSK and PIC. PIC is less immune to noise as compared to BPSK, but its BER rapidly reduces to zero at E_b/N_0 of $\sim$10.5 dB. For the lower signal-to-noise ratios, a lightweight error correction scheme to improve BER may be used albeit at the expense of a small increase in power consumption. A full system-level analysis of this trade-off is the subject of on-going work. In this chapter, we assume that the BER is zero.

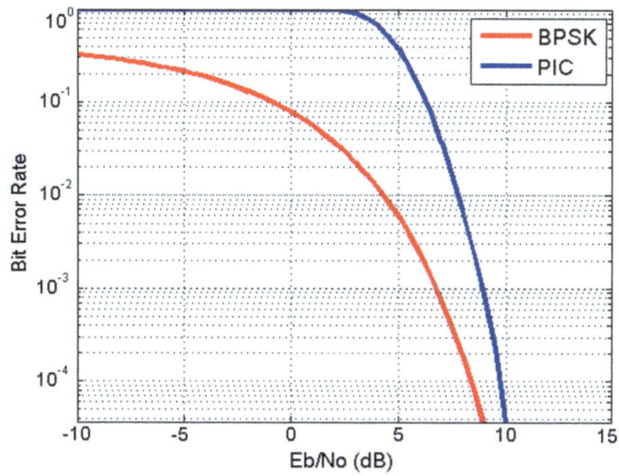

Fig. 12.17 BER analysis

12.6.2 Pulse Width Coefficient

The data pulse width coefficient, w, is the number of transmitter clock cycles during which the pulse remains high. An example is shown in Fig. 12.15c. The inter-symbol delay increases linearly with the pulse width as will be explained in the next subsection. On the other hand, the data rate reduces with inter-symbol delay. For a 50% duty cycle, the total pulse duration $\Pi = 2wT$, where T is the clock period of the transmitter, see Fig. 12.15c. A pulse width equal to half-clock cycle (i.e., $w = 0.5$) results in the highest data rate and lowest requirement on the inter-symbol delay coefficient (see next subsection). With the same pulse widths, for both the transmitter and receiver (i.e., $w_S = w_F$), the transmitted pulse time period would be equal to the time period at receiver (i.e., $T_S = T_F$). However, in the presence of clock inaccuracies, $w_S \neq w_F$, the transmitted data pulse with time period T_S will not map exactly to the time period T_F at the receiver. Instead, we have

$$T_S = \frac{w_S}{w_F} T_F \implies w_S = \frac{f_F}{f_S} w_F. \tag{12.7}$$

This relationship expresses the invariance of the *number* of transmitted pulses. This number depends solely on the product of the pulse width and the corresponding clock rate.

12.6.3 Inter-symbol Delay Coefficient

The inter-symbol delay coefficient, α, is the number of transmitter clock cycles used to separate the pulse streams of two adjacent symbols. This is the most important parameter in PIC design as it affects both its data rate and reliability. A badly selected

α will result in increased decoding errors and possibly complete failure. As will be seen in the following paragraphs, an efficient PIC protocol requires that α be an even number. The smallest possible α value is 4. T_{α_S}, the inter-symbol delay in terms of transmitter time period (T_S), is shown in Fig. 12.15c and is calculated as $T_{\alpha_S} = \alpha_S T_S$. To generate such delay, the transmitter pulls the line low and keeps it in that state for α number of clock cycles. For a successful reception, the inter-symbol delay at both ends must satisfy the *time* invariance principle:

$$\alpha_F T_F = \alpha_S T_S \tag{12.8}$$

In the presence of clock inaccuracies, phase shift, and clock jitter, the time invariance condition is expressed as

$$\alpha_F T_F = \alpha_S T_S + \varphi T_S + \psi T_S \implies \alpha_S = \frac{f_S}{f_F}\alpha_F - (\varphi + \psi) \tag{12.9}$$

As shown in Figs. 12.12 and 12.13, the correct PIC transmission and reception of data depends on several parameters that need to be selected judiciously. These parameters include the delay threshold and the inter-symbol delay.

Inter-Symbol Delay Threshold

A portion of inter-symbol delay, the delay threshold (Th_α), shown in Fig. 12.15c, is used at the receiver to discriminate between data pulses and inter-symbol delay. The optimal delay threshold is given by

$$Th_\alpha = \frac{T_\alpha}{2} = \frac{\alpha_F T_F}{2} \triangleq \alpha_{Th} T_F \tag{12.10}$$

where $\alpha_{Th} = \alpha_F/2$ is the delay threshold coefficient that ensures that the receiver clock cycle count does not decrease during the reception of inter-symbol delay to the extent that the receiver stops detecting it as inter-symbol delay. Also, the cycle count should not increase to the extent that the receiver starts detecting even data pulses as inter-symbol delay. The absence of such optimal threshold will result in decoding errors due to either pulse undercounting or inter-symbol interference.

Selection of Inter-Symbol Delay Coefficient

To distinguish the data pulses and the inter-symbol delay, the transmitter-generated delay should be longer than the duration of one data pulse. The inter-symbol delay coefficient should therefore satisfy

$$\alpha_S T_S > T_S + \varphi T_S + \psi T_F \implies \alpha_S > 1 + \varphi + \psi \frac{f_S}{f_F} \tag{12.11}$$

where local clock inaccuracies such as receiver phase shift and clock jitter are accounted for. Using the maximum possible values of 1 and 0.5 for φ and ψ, respectively, in (12.11) and assuming $f_F = f_S$, we get $\alpha_S > 2.5$.

In theory, the integer α_S can be chosen equal to 3, but from a hardware implementation view point, we set it equal to 4 as it is easier to implement multiplication and division of power 2 numbers using left or right shift operations. In the next subsection, we study the interplay between inter-symbol delay and robustness with respect to clock variations.

12.6.4 Clock Inaccuracy Tolerance

Given the clock frequency of the slow-end, f_S, and the parameters α_S, α_F, w_S, w_F, ψ, and φ of the PIC system, the goal now is to find the highest possible clock frequency for the fast-end, f_{Fmax}, above which decoding errors start to occur. To find f_{Fmax}, the condition for error-free operation should be fulfilled, namely the pulse duration should be less than the inter-symbol delay threshold

$$2w_S T_S \leq \frac{\alpha_S}{2} T_S \tag{12.12}$$

Using w_S from (12.7) and α_S from (12.9), we get

$$4w_F f_F^2 + (\varphi + \psi) f_S f_F - f_S^2 \alpha_F \leq 0 \tag{12.13}$$

which is satisfied if and only if

$$\beta \equiv \frac{f_F}{f_S} \leq \left[\frac{\sqrt{(\varphi + \psi)^2 + 16\, w_F\, \alpha_F} - (\varphi + \psi)}{8\, w_F} \right] \equiv \beta_{max} \tag{12.14}$$

This is the main theoretical result of this section. If $\beta \leq \beta_{max}$, PIC transmission will be error free. It is interesting to note that β_{max} is linear in the receiver clock jitter and phase shift but varies as the square root of receiver inter-symbol delay.

If we are to keep the inter-symbol delay coefficient α, the same for both ends, the rate of the fast clock should not exceed the limit imposed by $f_{Fmax} = \beta_{max} f_S$. In Fig. 12.18a, we plot f_{Fmax} for $f_S = 25\,\text{MHz}$. Of course, for error-free transmission, $f_F \leq f_{Fmax}$. In Fig. 12.18b, different regions of operation are identified. The safe region of operation is marked up to the limit calculated using f_{Fmax}. Beyond this, there is a region of uncertainty in which errors start occurring randomly. At a certain level of clock discranpency, total failure occurs due to the failure in detecting even a single inter-symbol delay. The recommend region of operation is of course the one delimited by f_{Fmax}.

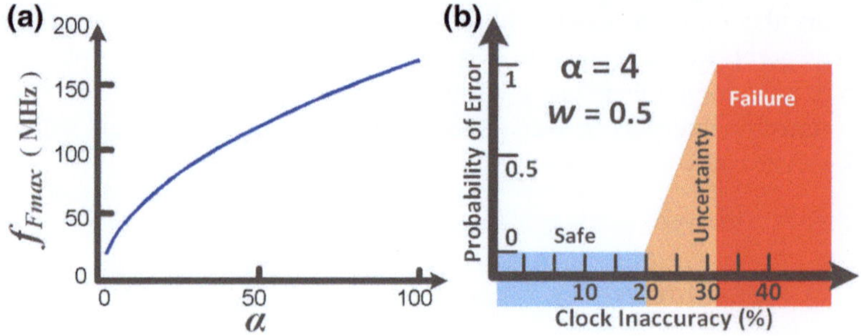

Fig. 12.18 **a** f_{Fmax} versus Inter-symbol delay **b** Regions of operation ($f_S = 25$ MHz)

12.6.5 Selection of Inter-symbol Delay Coefficient

Another interpretation of the inequality in (12.13) is as a design formula for selecting an appropriate value of receiver inter- symbol delay in the presence of clock discrepancies between transmitter and receiver and the presence of local clock inaccuracies at receiver. Solving for α_F, we get

$$\alpha_F \geq \lceil 4 w_F \beta^2 + \beta (\varphi + \psi) \rceil \equiv A_F \tag{12.15}$$

After selecting an inter-symbol delay coefficient for the slow-end α_S, the inter-symbol delay coefficient for the fast-end α_F is set equal to α_S if $f_F \leq f_{Fmax}$. Otherwise, α_F is a scaled version of α_S given by

$$\alpha_F = mod(A_F, 2) + A_F \tag{12.16}$$

This formula guarantees that α_F is the smallest even integer satisfying (12.13).

12.6.6 Summary on Inter-Symbol Delay

The larger α, the more resilient PIC is to timing variability. This, however, is achieved at the expense of significantly reduced data rates and increased power consumption. This trade-off is illustrated in Fig. 12.19. A careful adjustment of α is necessary to meet the requirements of a particular application. The α value used in all the experiments is 4, with $w = 0.5$. Values of α less than 4 will result in transmission failure *even* if transmitter and receiver are running on ideal clocks with the same frequency. Therefore, the minimum allowable value of α is 4.

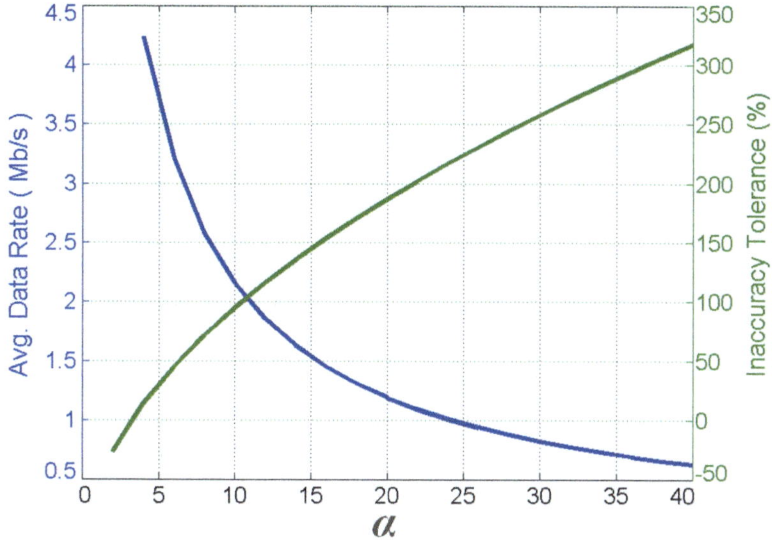

Fig. 12.19 Trade-off between robustness and data rate

12.7 Protocol Failure Modes and Error Correction

PIC transmits all the information in the form of a pulse train. A common case of erroneous transmission happens when there is a pulse miscount. In this case, PIC stops reception immediately and keeps it in a "halt" state until an explicit reset signal is sent, at which time transmission is resumed. With the use of a simple counter, the halt state provides a useful error-detection mechanism. The counter is activated via the receiver's busy signal which remains active when the error occurs. The counter gets reset at each falling edge and continues to count until the next falling edge. If the count reaches a threshold value, Th_{Error}, the state is considered an error and thus a reset signal is generated. An appropriate value of the error-detection threshold is

$$Th_{Error} = 2\,(\alpha + 2\,w\,l) \tag{12.17}$$

which is twice the number clock cycles needed for the inter-symbol delay followed by pulse train representing a full segment of OFF bits.

A variety of available techniques [6] can be employed to handle an error condition. One simple recommendation is to request the transmitter to resend the data. Another recommendation is for the configurable layer of the protocol to send an acknowledgment for each of the successful transfers. The only error that cannot be detected using the count method is the distortion of middle pulses in the index pulse train. In such cases, there will be no halt state and the receiver will infer a wrong index number. This transmission error can be handled using a simple parity check or other similar methods. However, such pulse distortion error is very unlikely as it occurs only in the presence of excessive external noise.

12.8 Robustness Analysis - Experimental Verification

An experimental setup comprised of two IoT nodes communicating using PIC is used to verify the limitations imposed by (12.14) on the maximum tolerable clock frequency for the fast end. Each node is composed of a PIC protocol module and a logical topology control (LTC) module based on a TI MSP430 microcontroller. The whole setup is implemented in Verilog on the Xilinx Virtex-7 FPGA platform. Two clocks, one for each node, are generated with the help of a Virtex-7 on-chip PLL. The slow-node clock is fixed at 25-MHz, but the rate of the fast-node clock is increased gradually from 25-MHz. Using $\alpha = 4$, $w = 0.5$, $\varphi = 1$, and $\psi = 0.01$ at both ends, we have $\beta_{max} = 1.2$. The LTC of the slow node directs the PIC transmitter to send the 16-bit data starting at 0 with an increment of 1 at each transmission. The fast end receives the data and replies back the same. The returned and original data words are compared to verify the complete round-trip chain. The experiment confirmed the results of (12.14) that the PIC transmission works flawlessly until the clock frequency of the fast node reaches $\approx$30-MHz, which is in agreement with the theoretical bound of (12.13).

12.9 PIC Power Management

In this section, a power-efficient PIC operation is discussed that includes the management of the power consumed in the physical layer (PHY). Furthermore, a mathematical model is developed and used to derive rigorously the performance parameters related to PIC PHY power management. The mathematical model is also used to derive a lower bound on the pulse width and therefore a lower bound on the power consumed in PIC PHY. The new power control policy is applied to a single-wire link with significant power saving achieved above and beyond those achieved due to absence of CDR circuits. These power savings are obtained without any impact on data rate and bit error rate (BER). The pulse control policy is implemented using 45 nm CMOS technology, verified on various communication links, involving single channels, and a rigorous power analysis is provided to quantify the proposed PIC power management.

12.9.1 *Sources of Power Consumption*

The PIC power consumption reported in Sect. 12.2, addresses only the PIC functional logic but ignores the power consumption in the PHY layer. For instance, the resistor connected to the single-wire may be an important source of power consumption in a full transceiver system. As shown in Fig. 12.20, the resistor contributes almost $\sim 7\,\mu W$ to the total PIC power dissipation (21% with 50% duty cycle of pulses). This is rather significant and opportunities for reducing the resistor power dissipation must be explored. This is the main objective of the following subsections.

12.9.2 *Proposed PIC PHY*

Approach

PIC is based on transmitting pulses on a single wire. The pull-down resistor attached to the single wire has power dissipation that is linear in the pulse duty cycle, λ. One approach to reduce PHY power is therefore to reduce the pulse duty cycle. This can be done within the pulse generation circuit with the constraint that the overhead in pulse width control should be much less than the projected power savings resulting from the narrower pulse. One approach to pulse generation is to AND the input pulse of duty cycle λ_{In} with its delayed version λ_D, where the delayed pulses are obtained using buffers as illustrated in Fig. 12.21a. More buffers result in a smaller λ at the output of the AND gate. The power overhead of such approach will of course increase with the number of inserted buffers and will therefore negate any power saving we may expect from a smaller λ. Another possible approach is to use a delay line [8, 9] instead of the inserted buffers, but again the power consumption of such delay line outstrips the $\sim 7\,\mu W$ power budget and so it is unable to achieve the narrow λ that is needed to reduce PHY power consumption.

To meet the power consumption requirements, an inverter coupled with a small capacitor is used to generate narrow pulses of λ duty cycle. This pulse generation and control circuit is shown in Fig. 12.21b. The single inverter is used to invert and delay the input pulse λ_{In} with the capacitor providing an additional load to increase the total delay. When λ_{In} and the delayed inverted pulse λ_{ID} pass through the AND gate, a narrow λ is generated. As shown in Fig. 12.22, the proposed PIC PHY includes the duty cycle generation circuit at the back of the tri-state buffer.

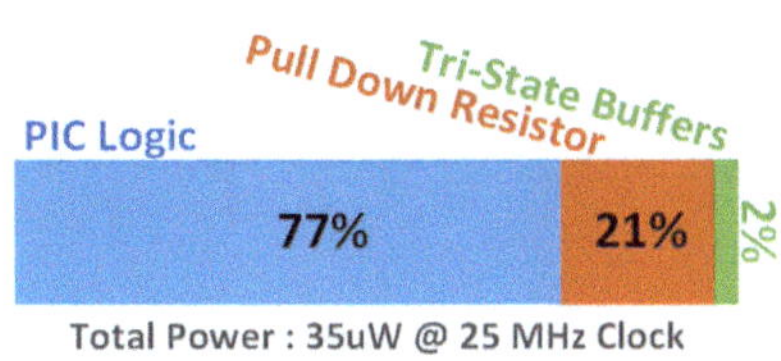

Fig. 12.20 PIC transceiver power consumption (50% duty cycle)

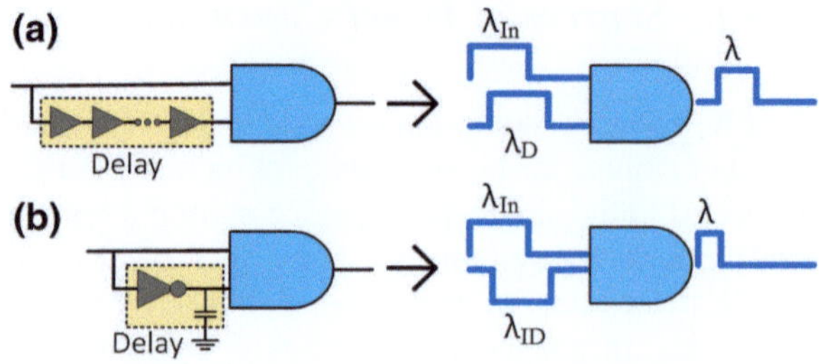

Fig. 12.21 Duty cycle generation **a** Buffers approach **b** Inverter and capacitor approach

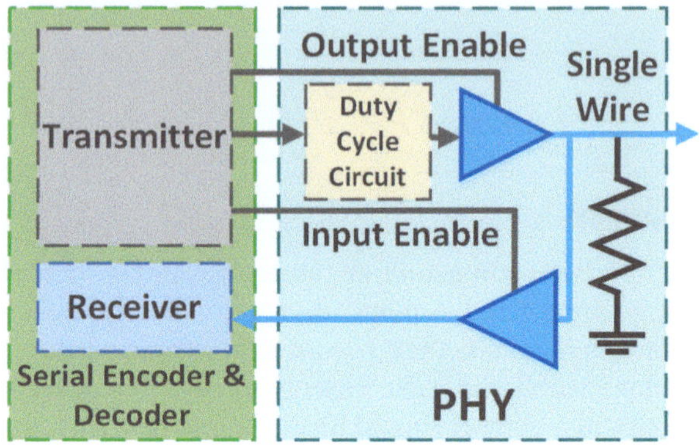

Fig. 12.22 Proposed PIC PHY layer

PHY Circuit Implementation

The proposed PIC PHY circuitry is implemented using a 45 nm process. The circuit diagram is shown in Fig. 12.23. The delay inverter is comprised of the $M1$ and $M2$ transistors loaded with a delay capacitor C_D. The AND operation is performed using transistors $M3$–$M8$, and the output is coupled with the pull-down resistor R_{PD} through a tri-state buffer. Table 12.3 catalogs the used parameter values. In the following subsections, we describe how to select the values of capacitor C_D and resistor R_{PD} to achieve a given duty cycle.

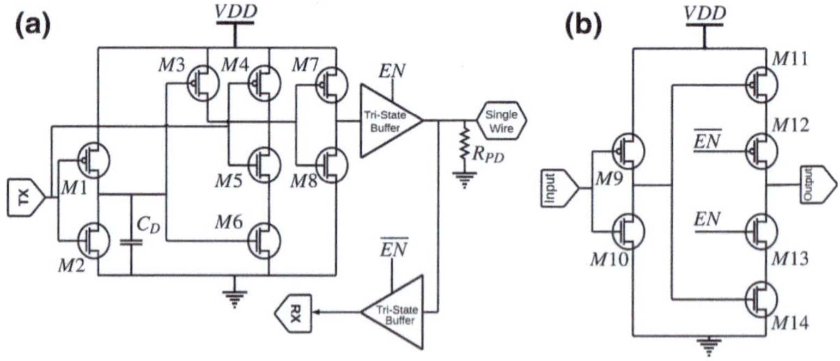

Fig. 12.23 **a** Proposed PIC PHY circuit **b** Tri-state buffer circuit

Table 12.3 Design parameters of duty-cycle generator

Parameters	Proposed	Unit
NMOS (W/L)	120/45	nm/nm
PMOS (W/L)	240/45	nm/nm
C_D (min)	4.6	f F
R_{PD} (min)	53.28	kΩ
λ_{In}	20	ns

12.9.3 Delay Capacitor

The pulse duty cycle λ is the time difference between the rising edge of the input pulse λ_{In} and the falling edge of the inverted and delayed pulse λ_{ID}. The time constant τ for the inverted output signal of the delay block (shown in Fig. 12.21b) is determined by the delay capacitor C_D and the ON resistance R_{on} of the transistor, $\tau = C_D R_{on}$. The smaller the time constant τ, the faster the discharging of C_D, and the narrower the λ. However, C_D cannot be reduced to zero as pulse detectability imposes both lower and upper bounds on C_D. The expressions of these bounds will be given in the next two paragraphs.

Lower Bound on Delay Capacitance

The lower bound on the selection of C_D is imposed by the time constant for the rising edge of the pulse at the output of inverter block of $M1$ and $M2$. During the rising edge, the $M1$ PMOS transistor is ON while the $M2$ NMOS is OFF. The time constant is then defined by the ON resistance of the PMOS transistor, $\tau = C_D R_{P-on}$. For a successful pulse detection at the receiver, the pulse voltage level should be $V_C \geq \frac{2}{3} V_{dd}$ where V_C is the voltage output across C_D:

$$V_C = V_{dd}\left(1 - e^{-\frac{t}{\tau}}\right) \tag{12.18}$$

Using $V_C = \frac{2}{3} V_{dd}$, $t = t_r$, $R_{on} = R_{P-on}$ and solving for C_D, one gets

$$C_D \geq -\frac{t_r}{ln\left(\frac{1}{3}\right) R_{P-on}} \implies C_D \geq 0.91\,\frac{t_r}{R_{P-on}} \tag{12.19}$$

where t_r is the rise time during which the capacitor is charged up to $\frac{2}{3} V_{dd}$. In 45 nm CMOS technology, $R_{P-on} = 13.32\,\text{K}\Omega$ which leads to the smallest possible value of $C_D = 4.6\,\text{fF}$. Violating this condition will result in a pulse detection failure at the receiver, that is, $\lambda = 0$ or 0% duty cycle.

Upper Bound on Delay Capacitance

The upper bound on C_D is imposed by the time constant for the pulse falling edge at the output of the inverter block of $M1$ and $M2$. During the falling edge, the $M1$

PMOS transistor is OFF while $M2$ NMOS is ON. The time constant is then defined by the ON resistance of NMOS transistor, $\tau = C_D R_{N-on}$. The higher τ is, the larger λ will be. Beyond a certain C_D value, λ saturates at 50%. For a duty cycle less than 50%, the λ_{ID} voltage level should be $\leq \frac{1}{3}V_{dd}$ and should be achieved while the input pulse λ_{In} is high. In other words, $t_f \leq \lambda_{In}$, where t_f is the output fall time during which the capacitor is discharged down $\frac{1}{3}V_{dd}$ and is determined by the capacitor discharge equation

$$\frac{1}{3}V_{dd} = V_{dd}\, e^{-\frac{t_f}{\tau}} \implies t_f = -ln\left(\frac{1}{3}\right) R_{N-on} C_D \leq \lambda_{In} \tag{12.20}$$

which leads to the upper bound

$$C_D \leq 0.91\,\frac{\lambda_{In}}{R_{N-on}} \tag{12.21}$$

In 45 nm CMOS technology, $R_{N-on} = 17.22\,\text{k}\Omega$ which leads to the largest possible value of $C_D = 1pF$ in the case of a 20 ns long input pulse. Beyond this upper bound on capacitance, λ saturates to 50%.

12.9.4 Sizing the Pull-Down Resistor

The pull-down resistor (R_{PD}) attached to the single-wire at the output of tri-state buffer keeps the line low during the high-impedance state of the buffer. R_{PD} directly impacts the output voltage at the output rising edge during which the transistors $M11$, $M12$, and $M13$ are ON while $M14$ is OFF. A voltage divider comprised of two PMOS on-resistances (i.e. $2R_{P-on}$) and R_{PD} is formed. For the successful detection of the pulse at the receiver, the output pulse voltage V_{Pulse} should satisfy $V_{Pulse} \geq \frac{2}{3}V_{dd}$ where

$$V_{Pulse} = V_{dd}\,\frac{R_{PD}}{R_{PD} + 2R_{P-on}} \tag{12.22}$$

which means $R_{PD} \geq 4R_{P-on}$. In 45 nm CMOS technology, the $R_{P-on} = 13.32\,\text{k}\Omega$ which leads to the smallest possible value of $R_{PD} = 53.28\,\text{k}\Omega$. Violating this condition will result in small pulse voltages at the output which may fail detection at the receiver. The larger R_{PD} is, the larger the output pulse swing.

12.9.5 Duty Cycle

The pulse duty cycle λ at the output is given by

$$\lambda = t_r + t_f \tag{12.23}$$

which yields

$$\lambda = 1.0986\, C_D\, (R_{P-on} + R_{N-on}) \tag{12.24}$$

The percentage duty cycle, $\lambda_{\%}$, is determined by

$$\lambda_{\%} = \frac{\lambda}{2\lambda_{In}} \times 100 \tag{12.25}$$

The minimum duty cycle λ_{min} is determined using the smallest possible delay capacitance (i.e., $C_D = C_{D-min} = 4.6$ fF) in (12.24) and is given as

$$\lambda_{min} = 1.0986\, C_{D-min}\, (R_{P-on} + R_{N-on}) = 154.34 \text{ ps}$$

where R_{P-on} and R_{N-on} are 13.32 and 17.22 kΩ, respectively, for a 45 nm process. In particular,

$$\lambda_{min-\%} = \frac{\lambda_{min}}{2\lambda_{In}} \times 100 = 0.39\ \%$$

12.10 Power Management - Verification

12.10.1 Power Analysis

The proposed policy to manage PIC power consumption, through controlling the pulses duty cycle, is verified rigorously using Spice-level simulations for a 45 nm CMOS process in the Cadence design environment. The power analysis is performed using the minimum duty cycle of 0.39% which is generated with $C_D = 4.6$ fF, $R_{PD} = 54$ kΩ, and $\lambda_{In} = 20$ ns (i.e., clock = 25 MHz). The resistor, with the proposed PHY, consumes only 0.1% (27.22 nW) of the total PHY power (265.4 nW). The tri-state buffers power consumption is also reduced to 0.9% (238 nW) only. The total power saving, as compared to the one shown in Fig. 12.20, is 20%. The improved power consumption ratios are shown in Fig. 12.24. The total PHY power consumption is linearly related to the duty cycle as shown in Fig. 12.25a. Clearly the pulse duty cycle greatly impacts the overall power consumption. For the experimental prototype, the duty cycle generator should not consume more than 7 μW ($\lambda \leq 5\%$ for the experiment performed). On the other hand, the power consumption reduces with the increase in pull-down resistance, as shown in Fig. 12.25b. If the PHY is intended to

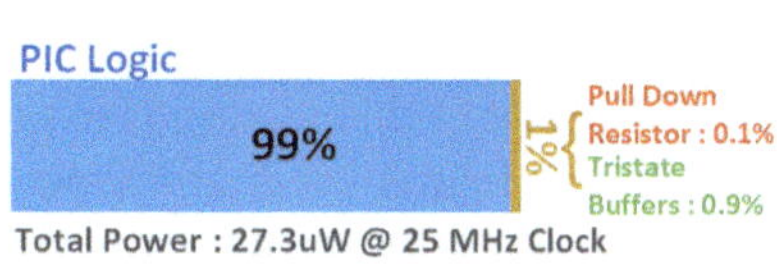

Fig. 12.24 Proposed PIC transceiver power consumption (0.4% duty cycle)

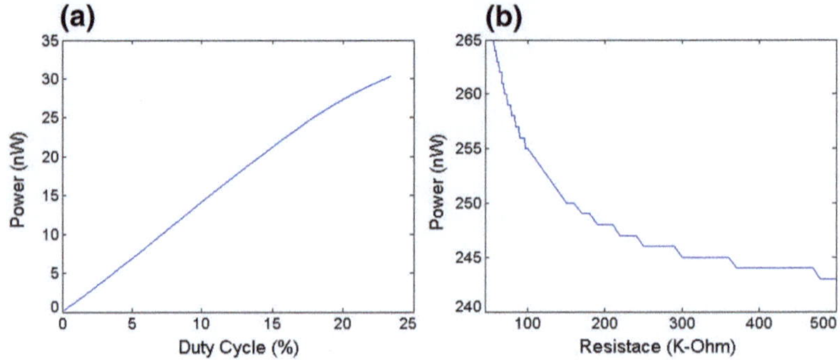

Fig. 12.25 PHY power consumption (**a**) versus Duty cycle ($R_{PD} = 54\ \mathrm{k\Omega}$) (**b**) versus Resistance ($C_D = 4.6\ \mathrm{fF}$)

connect to downstream loads presenting an input capacitance in parallel with the pull-down resistance, then increasing the resistance will increase the rise and fall slews of the pulse. Therefore, the upper bound on the resistance is application-dependent and should be determined based on the overall timing performance of the full link.

12.10.2 BER Analysis

Noise associated with the off-chip environment can have an effect on data pulses. The increase or decrease in pulse levels, due to the external noise, makes it difficult for the receiver to detect pulses correctly. Depending on the noise level, an extra pulse may be detected or a pulse may be skipped. In both cases, one gets a decoding error. To analyze the performance of PIC in the presence of noise, the encoded pulse stream of data is exposed to white Gaussian noise. The noisy signal is filtered at the receiver end, then decoded according to PIC algorithm, and the number of errors encountered is counted. The results are plotted in Fig. 12.26 for different values of E_b/N_0 (the ratio of energy-per-bit to noise power spectral density) for both BPSK and PIC. For PIC, the results are plotted with variations in the pulse duty cycle. PIC is less immune to noise as compared to BPSK, but its BER reduces rapidly to zero at E_b/N_0 of $\sim$10.5 dB. Variations in pulse duty cycle have no significant impact on PIC BER.

12.11 Automatic Protocol Configuration

The objective of this section is to provide an algorithm for automatically setting the PIC protocol parameters at the power-on phase. To achieve this, the section introduces a new algorithm that rigorously specifies the protocol configuration procedure

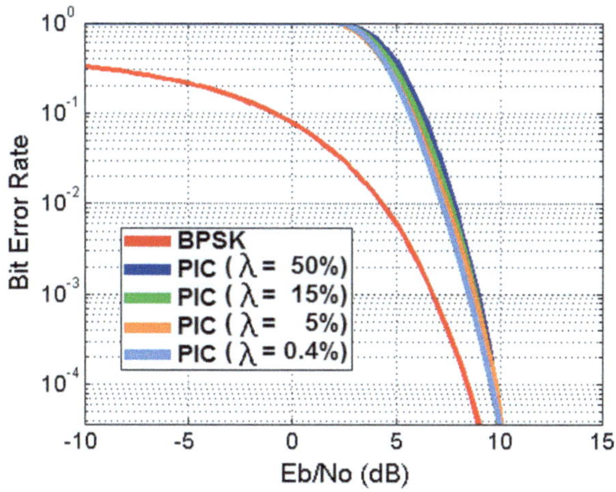

Fig. 12.26 BER analysis

and uses closed-form formulas to assign suitable protocol parameters to both ends of the transmission link based on clock-rate differences. Pulse count differences between the pulse trains at the transmitter and receiver are used to find a suitable inter-symbol delay coefficient so as to eliminate the need to know the exact clock rates at both ends of the link. Moreover, a power-efficient realization with very low hardware complexity of the inter-symbol delay coefficient calculations is discussed to enable efficient protocol configuration. The hardware realization is evaluated in both the ASIC and FPGA design flows. The FPGA implementation is experimentally verified and tested on a point-to-point communication link between two IoT devices connected via a single PIC channel in a master-slave mode.

12.11.1 Algorithm

Consider a single-channel link between two IoT devices that are configured in a master-slave topology, as in Fig. 12.27. Each device uses its own clock frequency. Also each device generates a reference clock of 100 kHz along with the local device clock. The reference clock frequency is defined as the default clock for the PIC transceivers and can be used anytime to establish a communication link between the master and slave devices. All the devices power up with the default inter-symbol delay coefficient $\alpha = 4$. The master device also maintains a table of α settings for all the slave devices, and by default, all the α entries are equal to 4. The master device selects a corresponding α for a slave device from the table and considers this as its own α when it needs to communicate with that particular slave. The objective of the algorithm is to update not only the master table entries but also the local α settings of all the connected slave devices. These updated values are then used to carry out a successful communication at the local clock rate considered as a PIC clock at each

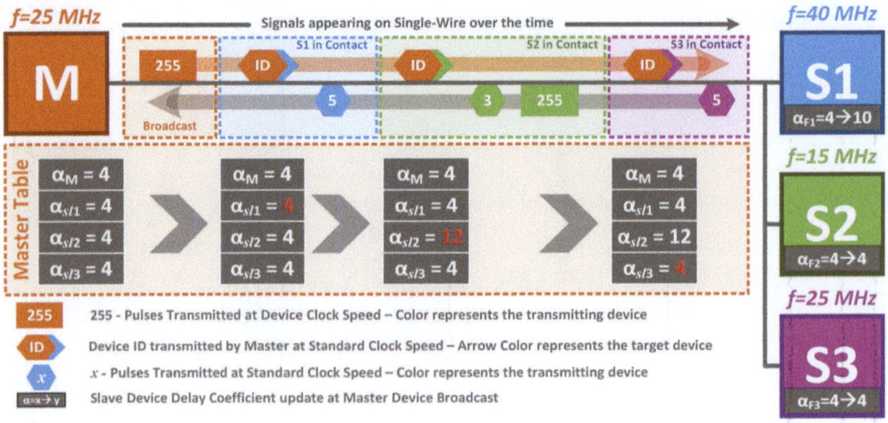

Fig. 12.27 An example of automatic parameter detection

end of the link. The PIC α value for the reference clock is $\alpha = 4$ irrespective of the local settings in the slave device and its table entries in the master device.

In the power-on phase, the master device transmits system clock pulses for one second to all the slave devices. At first rising edge of the transmitted pulse stream, slave devices start counting both the received pulses f_S and their own device clock pulses f_F. The counting ends with the pulses from master device. Using pulses count difference ($\Delta f = f_F - f_S$), the high-speed node of the link at f_F is recognized. Slave sets its local α_F to a new *suitable* value if it is the high-speed node, or keeps the default value of 4 otherwise[1]. Next, the master device starts the second part of parameter detection by asking each slave device about the status of parameter setting. This process is carried out using the reference clock of 100 kHz. The master device sends a slave ID followed by the reception of either 5 or 3 pulses from the target device. If the target slave was successful in setting a new α_F, 5 pulses would be received and the master will update the table entry for this device with $\alpha = 4$. If the target slave has failed in setting a new α_F, 3 pulses at 100 kHz would be received followed by the reception of system clock pulses of slave device for one second. The 3 or 5 pulses in response to slave ID are followed by a 6 clock cycle delay to distinguish these pulses correctly from the following system clock pulses. On reception of these pulses, the master device follows the same counting and α_F setting process as described for the slave and updates the table entry for this slave with a new suitable α_F. The process continues until all the slave devices are covered. The master sets one α entry in its table for each slave in the network. The power-on automatic detection process allows the master device to configure all slave devices before they start communication using

[1]The process to find a suitable inter-symbol delay coefficient value is presented in the next subsection.

their device clocks. The process steps are given in Algorithm 1 (Fig. 12.29) and an example of such configuration process for a single-channel link with three slave devices is illustrated in Fig. 12.27.

12.11.2 Inter-symbol Delay Coefficient Calculation

Given the frequency information of the slow node, the inter-symbol delay coefficient α is always calculated by the fast node. This is because the fast node can adjust its delay coefficient to reduce its transmission speed, but it is impossible for the slow node to go beyond its device clock rate. To find a suitable α, first we need to set the maximum supported clock ratio (β_{max}). The phase shift φ and the clock jitter ψ are the knobs in (12.14) to adjust the supported percentage of clock ratio (β_{max}) assuming the default values of $\alpha_F = 4$ and $w_F = 0.5$. The ideal case where both φ and ψ are zero, the ratio reaches to $\approx$41%. However, this is not the case in real-world applications where there are always a phase shift and clock jitter that reduces the clock ratio. Empirical results and the information provided in Sect. 12.2 show that the safer values of $\varphi = 1$ and $\psi = 0.01$ lead to a ratio of $\approx$18%. Once the frequency information of the slow node is shared with the fast node, the β_{max} is calculated using (12.14). The equations (12.15) and (12.16) are then used to find the suitable α_F as per the steps given in Algorithm 1 (Fig. 12.29). The plot in Fig. 12.28 shows the variation in α_F due to the increase in fast node frequency (f_F) when the slower node is at 25 MHz of clock (f_S).

While Algorithm 1 (Figs. 12.29 and 12.30) is suitable for finding α_F, it does have one important limitation. Indeed, to share the precise frequency information, a device needs to transmit pulses at least for one second. This information exchange tends to be time-consuming when there are many devices attached to a single-channel link.

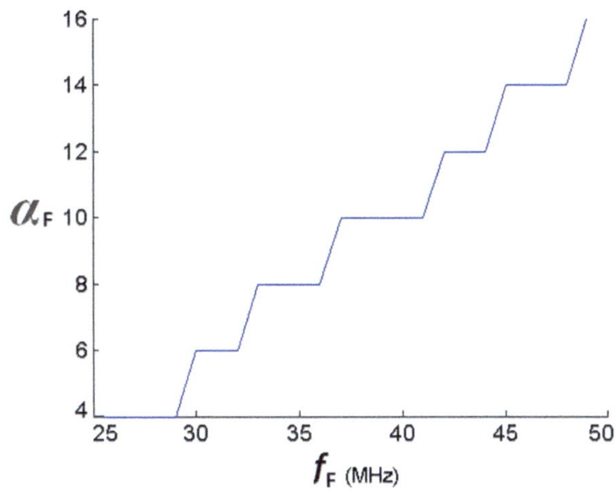

Fig. 12.28 α_F versus f_F when $f_S = 25$ MHz

Fig. 12.29 Algorithm 1: automatic parameter detection—master device

$f_{d,master}$: Master Device Clock
$f_{d,slave}$: Slave Device Clock
f_S : Received Pulse Count[1]
f_F : Device's Local Clock Pulse Count[1]
f_{100} : 100KHz reference Clock
$\alpha_{sl}[n]$: α settings of n Slave Devices in Master Device Table
$\alpha_{F,n}$: Local α Setting of nth Slave Device

Master Device Algorithm

Set $n = \#of\, slaves$, $i = 0$
Broadcast pulses for one second[2] at $f_{d,master}$
while loop$(i < n)$
  Send $Slave\_ID[i]$ at f_{100}
  $Pulses =$ **Receive** Pulses at f_{100}
  if $(Pulses = 5)$
    $\alpha_{sl}[i] = 4$
  else
    $f_S =$ Receive Pulses at $f_{d,slave}$ for one second[2]
    $f_F = f_{d,master}$
    if $(f_F \leq f_S \beta_{max})$
      $\alpha_{sl}[i] = 4$
    else
      $f_{FN} = f_F - (\beta_{max} - 1) f_S$
      $\beta = f_{FN}/f_S$
      $A_F = \lceil 4 w_F \beta^2 + \beta(\varphi + \psi) \rceil$
      $\alpha_{sl}[i] = mod(A_F, 2) + A_F$
  i++
end loop

---

[1]See Subsection 12.11.2 for more specific settings.
[2]A more efficient alternative is proposed in Subsection 12.11.2.

Therefore, instead of transmitting the pulses for one second, the transmission of just 255 pulses is adopted from one node to another. Hence, at receiver node, both the received pulse count N_S and the device clock pulse count N_F become known. Algorithm 1 (Figs. 12.29 and 12.30) applies without any change in the formulation except replacing the f_F, f_S, and f_{FN} with N_F, N_S, and N_{FN} respectively. This method of α_F calculation by considering solely the pulse counts is efficient and suitable for all IoT devices including low-end microcontrollers.

12.11.3 Low-overhead Hardware for α_F Calculation

The calculations to find a suitable α_F can be readily implemented in a microcontroller using the C language. On the other hand, for a VLSI or an FPGA implementation, the floating point operations involved in such calculations present a significant area and power overhead due to the additional hardware resources needed to handle floating

Fig. 12.30 Algorithm 1: automatic parameter detection—slave device

$f_{d,master}$: Master Device Clock
$f_{d,slave}$: Slave Device Clock
f_S : Received Pulse Count[1]
f_F : Device's Local Clock Pulse Count[1]
f_{100} : $100KHz$ reference Clock
$\alpha_{sl}[n]$: α settings of n Slave Devices in Master Device Table
$\alpha_{F,n}$: Local α Setting of nth Slave Device

Slave Device Algorithm

---

f_S = **Receive** pulses for one second[2] at $f_{d-master}$
$f_F = f_{d,slave}$
if $(f_F < f_S)$
    $\alpha_{F,n} = 4$
    Setting Failed
else
    if $(f_F \leq f_S\beta_{max})$
        $\alpha_{F,n} = 4$
    else
        $f_{FN} = f_F - (\beta_{max} - 1)\, f_S$
        $\beta = f_{FN}/f_S$
        $A_F = \lceil 4w_F\beta^2 + \beta\,(\varphi + \psi)\rceil$
        $\alpha_{F,n} = mod(A_F, 2) + A_F$
    Setting Passed
Wait for Slave ID at f_{100}
if (Setting Failed)
    Send 3 Pulses at f_{100}
    Send pulses for one second[2] at $f_{d,slave}$
else
    Send 5 Pulses at f_{100}

---

[1]See Subsection 12.11.2 for more specific settings.
[2]A more efficient alternative is proposed in Subsection 12.11.2.

point arithmetic. To mitigate this overhead in the context of the automatic detection algorithm, we now present a hardware decoding scheme that maps the pulses count difference ($\Delta N = N_F - N_S$) to a suitable α_F.

The HW decoder works as follows. For each α_F, there exists a range of ΔN as shown in Table 12.4. Algorithm 2 (Fig. 12.31) illustrates the process of finding these ranges. In the algorithm, N_S is set to 255 because α_F is always calculated by the fast node and it is guaranteed that 255 pulses will be transmitted by the slow node. Figure 12.32 shows the result of an example with $f_S = 25$ MHz and $25 \leq f_F \leq 125$ MHz. The number of α_F ranges sets an upper bound on the maximum clock ratio supported by the automatic baud rate feature. The sweep ranges for f_F can be increased or decreased as per the application requirement. A custom HW decoder is designed that accepts a 10-bit ΔN, finds the corresponding range as per Table 12.4, and sets an 8-bit α_F at the output.

Table 12.4 α_F ranges for decoder implementation

$\triangle N_{min}$	$\triangle N_{max}$	β clock ratio (%)	α_F
1	47	$\leq$ 18.4	4
48	74	$\leq$ 28.8	6
75	122	$\leq$ 47.6	8
123	164	$\leq$ 64.0	10
165	202	$\leq$ 79.2	12
203	238	$\leq$ 93.2	14
⋮	⋮	⋮	⋮
997	1009	$\leq$ 395.6	96
1010	1020	$\leq$ 400.0	98
⋮	⋮	⋮	⋮

Fig. 12.31 Algorithm 2: α_F ranges for decoder implementation

Find β_{max}
Select f_S, f_{F-max}, and $\triangle f$
Set $N_S = 255$, $T_S = (1/f_S) \times N_S$, $f_F = f_S$, and $i = 0$
while loop ($f_F \leq f_{F-max}$)
    $N_F = T_S \times f_F$
    if ($N_F \leq N_S \beta_{max}$)
        $\alpha_F[i] = 4$
    else
        Find N_{FN}, β, A_F
        $\alpha_F[i] = mod(A_F, 2) + A_F$
    $\triangle N[i] = N_F - N_S$
    $i{+}{+}$
    $f_F = f_F + \triangle f$
end loop
Find $\triangle N_{min}$ and $\triangle N_{max}$ from $\{\triangle N\}$ for each of the unique α_F values in $\{\alpha_F\}$

12.12 Experimental Verification

An experimental setup comprised of two IoT nodes communicating using PIC is used to verify the operation of the automatic parameter detection algorithm. Each node is composed of a PIC protocol module, an automatic parameter detector (APD), a PIC PHY layer, PHY switch, and a clock generator, as shown in Fig. 12.33. The clock generator provides two clocks to a node, a device clock and a reference clock (100 kHz). The PHY switch, controlled by APD, is used to allow either the PIC module or the APD to access the PHY layer for establishing the physical link. Both the master and slave nodes have a similar implementation except for the differences present in APD module. The master APD is composed of three main modules: a con-

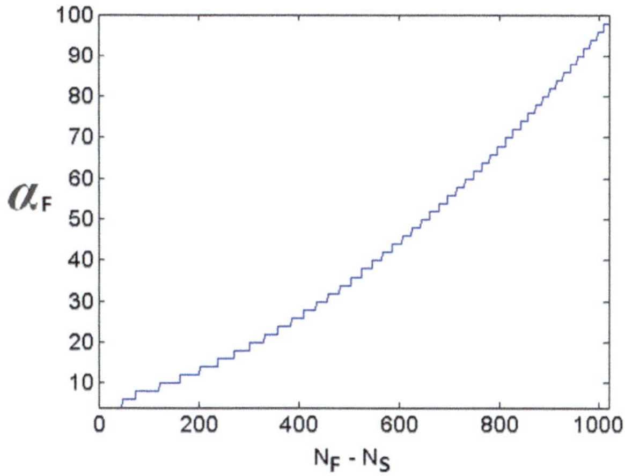

Fig. 12.32 α_F versus $\triangle N$ when $f_S = 25$ MHz and $25 \leq f_F \leq 125$ MHz

troller, a decoder, and an α_F table. The slave APD is composed of the same modules except that there is only one local α_F setting instead of a table. The controller module takes care of all the power-on configuration process, controls the PHY access, and connects the PIC to a suitable clock during communication. The decoder accepts $\triangle N$ as an input that is generated by the controller and outputs a suitable α_F for table entry at the master node or for local setting at the slave node. During communication after completing the power-on configuration process, the master node directs a particular slave device using the 100 kHz reference clock, fetches the corresponding α_F from the table, and using this α_F communicates further with the said slave device. This on-the-fly change in the inter-symbol delay coefficient allows the slave devices to use their device clock to adjust their data rates and thus, an adaptive baud rate network is established.

The full experimental setup is implemented in Verilog on the Xilinx Virtex-7 FPGA platform. Two separate clocks with different frequencies and phase shifts, one for each node, are generated with the help of a Virtex-7 on-chip PLL. The master clock is fixed at 25 MHz, while the clock rate for the slave node is allowed to deviate from 25 MHz. The master does power-on configuration and then verifies two-way communication using the automatically detected parameters. The experiments confirm that the PIC transmission works flawlessly. Along with the FPGA prototype, the automatic parameter detection system is also synthesized using a Synopsis logic-synthesis flow and a GLOBALFOUNDRIES 65 nm process. We found out that the system shown in Fig. 12.33 (excluding PIC) maintains the low-power operation of PIC consuming only 4.35 μW with a gate count of ≈1500 at a clock frequency of 25 MHz. The synthesis results are shown in Table 12.5.

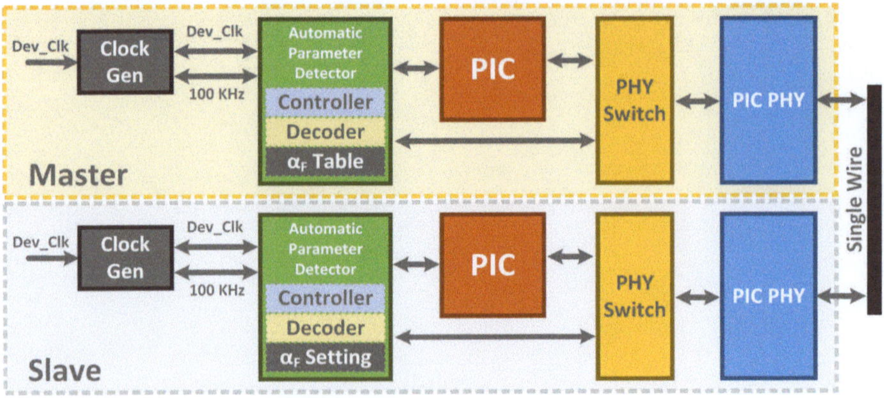

Fig. 12.33 Block diagram of the experimental setup

Table 12.5 Synthesis results

	Power (μW)	Area (*gate count*)
APD system (excluding PIC)	4.35	≈1500
PIC	26.6	2356
Total	30.95	≈3856

12.13 PIC for Wired and Wire-less Transmission

The Pulsed-Index Communication protocol lies at the MAC layer of the system and communicates with the help of PIC PHY layer as described in Sect. 12.4. The discussion so far has assumed the single wire as a communication medium. However, PIC can be used with a variety of single-channel signaling techniques that accept simple digital input, as shown in Fig. 12.34. The line-of-sight optical communications such as infrared (IR) and lasers are few of these commonly used single-channel techniques. Simply replacing the PIC PHY layer only with the IR or laser PHY will enable the system to transmit over single-channel using PIC protocol. Similarly, simple RF modules are easy to interface and can replace a wire without any extra circuitry. Other wired communication techniques such as power-line communication (PLC) can be enabled by interfacing suitable PLC PHY modules. Figure 12.34 explains how these techniques can easily be used with PIC MAC layer.

12.14 Conclusions

Pulsed-Index Communication (PIC) [1–4] is a novel, simple yet robust, method of signaling over single-wire channels that fills up the need for high-data rate, ultra-low power, protocols for IoT sensors. PIC offers up to a 4.1 Mb/s average data rate

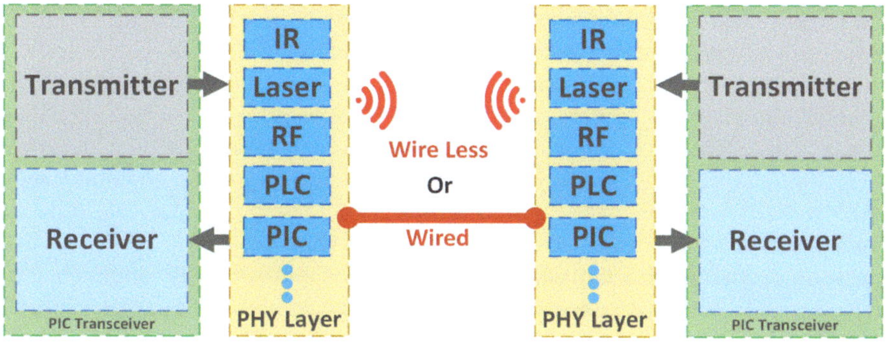

Fig. 12.34 PIC over different communication media

with a 24-MHz clock at 26 uW in 65 nm technology. PIC reduces silicon area and power consumption significantly by eliminating the need of a CDR. It is also robust toward skews, jitters, and clock inaccuracies as its decoding is based on counting the rising edges of the transmitted pulses. This is because jitters and skews impact edge locations but not their number. The higher data rate is achieved by encoding the data so that the ON bits are minimized and compacted toward the leading edge of the data segment.

This chapter has presented a detailed timing analysis for Pulsed-Index Communication protocols to meet robustness requirements. In the analysis, the inter-symbol delay parameter used to delimit the boundary between data segments in a packet plays a significant role. In addition to the gained insight into PIC, the analysis helps quantifying the trade-off between data rate and robustness. Based on the analysis, an inter-symbol delay coefficient α of 4 clock cycles and a pulse width w of half-clock cycle are recommended. These recommended design parameters have been experimentally verified using a Xilinx Virtex7 FPGA platform that illustrates both the simplicity, efficiency, and reliability of using PIC as a single-channel communication protocol for IoT devices.

The chapter has also presented a detailed power management policy for Pulsed-Index Communication protocols to improve further their ultra-low-power characteristics without impacting their data or bit error rates. In this policy, pulse duty cycle control is used to reduce the power consumption of the PIC PHY layer. In addition, a mathematical model has been used to select the physical design parameters related to PIC PHY power management. Based on the implementation and analysis, a minimum capacitance of 4.6 fF and a minimum resistance of 54 kΩ are recommended to generate the smallest possible duty cycle of 0.39%. These recommended design parameters have been experimentally verified using Spice-level simulations of a 45 nm process. An additional power saving of 22% has been achieved in the PIC PHY layer that is added to the already significant power savings in the PIC functional logic as reported in Sect. 12.2.

The power-on algorithmic process for automatically detecting the PIC protocol parameters allows the master device to configure all the slave devices connected to a single-channel network prior to the start of any device-to-device communication. The proposed automatic configuration methodology eliminates the need for compile time or manual assignment of protocol parameters. Moreover, the methodology removes the restriction on all the devices in the network to agree on a single communication speed and allows the devices with different capabilities to communicate reliably. Toward this end, the master device selects right parameter values on-the-fly that enable it to communicate with a particular slave. This makes the PIC single-channel network an adaptive baud rate network. The proposed architecture is experimentally verified and tested on a point-to-point communication link using a Xilinx Virtex7 FPGA platform that illustrates the simplicity, efficiency, and reliability of using automatic PIC parameter detection. In particular, the chapter shows that the efficient hardware realization of the algorithm maintains the low-power operation of PIC protocol while consuming only 4.35 uW (65 nm process).

Acknowledgements This work has been supported by the Semiconductor Research Corporation (SRC) under the Abu Dhabi SRC Center of Excellence on Energy-Efficient Electronic Systems (ACE^4S), Contract 2013 HJ2440, with funding from the Mubadala Development Company, Abu Dhabi, UAE.

References

1. S. Muzaffar, J. Yoo, A. Shabra, I.M. Elfadel, A pulsed-index technique for single-channel, low-power, dynamic signaling, in *Design, Automation and Test in Europe Conference and Exhibition (DATE'15)* (2015), pp. 1485–1490
2. S. Muzaffar, I.M. Elfadel, Timing and robustness analysis of pulsed-index protocols for single-channel IOT communications, in *IFIP/IEEE International Conference on Very Large Scale Integration (VLSI-SoC)* (2015), pp. 225–230
3. S. Muzaffar, I.M. Elfadel, Power management of pulsed-index communication protocols, in *IEEE International Conference on Computer Design (ICCD)* (2015), pp. 375–378
4. S. Muzaffar, N. Saeed, I.M. Elfadel, Automatic protocol configuration in single-channel low-power dynamic signaling for IOT devices, in *IFIP/IEEE International Conference on Very Large Scale Integration (VLSI-SoC)* (2016)
5. ETSI, *UICC - Contactless Front-end (CLF) Interface, Technical Specification, Version 7.3.0* (2008–09)
6. K.V.K.K. Prasad, *Principles of Digital Communication System and Computer Network* (Dreamtech Press, New Delhi, 2003)
7. S.-J. Song, N. Cho, H.-J. Yoo, A 0.2-mw 2-mb/s digital transceiver based on wideband signaling for human body communications. J. Solid State Circuits **42**(9), 2021–2033 (2007)
8. T. Mehrabi, K. Raahemifar, V. Geurkov, Design of a 4-bit programmable delay with TDC-based BIST for use in serial data links, in *International Symposium on Integrated Circuits (ISIC)* (2014), pp. 580–583
9. S. Jun-Ren, L. Te-Wen, H. Chung-Chih, Delay-line based fast-locking all-digital pulsewidth control circuit with programmable duty cycle, in *IEEE Asian Solid State Circuits Conference (A-SSCC)* (2012), pp. 305–308
10. MAXIM, *OneWireViewer User's Guide, Version 1.4*. AN3358 (2009)
11. MAXIM, *Reading and Writing 1-Wire Devices Through Serial Interfaces*. AN74 (2009)

12. MAXIM, *Overview of 1-Wire Technology and Its Use*. AN1796 (2008)
13. S. Wang, K.W.E. Cheng, K. Ding, Design of the temperature and humidity instrument based on 1-wire sensor for electric vehicle motors, in *International Conference on Power Electronics Systems and Applications* (2009), pp. 1–5
14. B. Huang, J. Lei, Y. Bo, The reading data error analysis of 1-wire bus digital temperature sensor DS18B20, in *International Conference on Modelling, Identification and Control* (2012), pp. 433–436
15. H. Xue, Research and development of a intelligent temperature-measuring system based on 1-wire bus, in *International Conference on Intelligent Computation Technology and Automation*, vol. 2 (2008), pp. 30–33
16. J. Dudak, G. Gaspar, G. Michalconok, Extension of 1-wire measuring system SenSys, in *International Symposium on MECHATRONIKA* (2012), pp. 1–4
17. ShouHuan Jun, Liang Xi, Application of one-wire bus on dynamic testing system of refrigeration devices, in *International Conference on Computer Science and Information Processing*, (2012), pp. 1114–1116
18. E. Diaconescu, C. Spirleanu, An identifying and authorizing application using 1-wire technology, in *International Symposium for Design and Technology in Electronic Packaging* (2010), pp. 243–248
19. C.A. dos Reis Filho, E.P. da Silva, E. de L. Azevedo, J.A.p. Seminario, L. Dibb, Monolithic data circuit-terminating unit (DCU) for a one-wire vehicle network, in *Solid-State Circuits Conference* (1998), pp. 228–231
20. C. Jia, D. Wu, I. Hawkins, A. Forsyth, One-wire communication system for cryogenic converter control, in *International Conference on Power Electronics, Machines and Drives* (2012), pp. 1–5
21. M. Loh, A. Emami-Neyestanak, All-digital CDR for high-density, high-speed I/O, in *IEEE Symposium on VLSI Circuits (VLSIC)* (2010), pp. 147–148
22. M. Loh, A. Emami-Neyestanak, A 3x9 Gb/s shared, all-digital CDR for high-speed, high-density I/O. IEEE J. Solid-State Circuits (JSSC) **47**(3), 641–651 (2012)
23. Q. Du, J. Zhuang, T. Kwasniewski, A 2.5 Gb/s, low power clock and data recovery circuit, in *Canadian Conference on Electrical and Computer Engineering* (2007), pp. 526–529
24. Y. Urano, W-J. Yun, T. Kuroda, H. Ishikuro, A 1.26mW/Gbps 8 locking cycles versatile all-digital CDR with TDC combined DLL, in *International Symposium on Circuits and Systems (ISCAS)* (2013), pp 1576–1579
25. L.-K. Soh, W.-T. Wong, A 2.5-12.5Gbps interpolator-based clock and data recovery circuit for FPGA, in *Asia Symposium on Quality Electronic Design (ASQED)* (2012), pp. 373–376
26. K.-B. Shin, K.-H. Seong, D.-H. Yeo, B. Kim, J.-Y. Sim, H.J. Park, Verilog synthesis of usb 2.0 full-speed device phy ip, in *International SoC Design Conference(ISOCC)* (2013), pp. 162–165
27. W.L. Couch, Baseband pulse and digital signaling, in *Digital and Analog Communication Systems*, 8th edn. (Prentice Hall, New Jersey, 2012), p. 784

Chapter 13
Efficient Algorithm for VT/VF Prediction for IoT SoCs

Temesghen Tekeste and Hani Saleh

13.1 Introduction

Sudden cardiac death (SCD) accounts for approximately 300,000 deaths in the USA per year and in most cases is the final result of ventricular arrhythmias including ventricular tachycardia (VT) or ventricular fibrillation (VF) [1]. Ventricular arrhythmia is an abnormal rhythm that targets the heart and is responsible for 75 to 85% of heart-related sudden deaths unless treated within seconds [1]. Most ventricular arrhythmias are caused by coronary heart diseases, hypertension, or cardiomyopathy, and if not accurately diagnosed or treated, immediate death occurs [2]. VT is a fast rhythm of more than 3 consecutive beats originating from the ventricles at a rate more than 100 beats per minute [3]. VF is another rhythm characterized by chaotic activation of ventricles, and it causes immediate cessation of blood circulation and degenerates further into pulse-less or flat ECG record indicating no cardiac electrical activity.

Machine learning techniques such as neural networks [4] and support vector machine (SVM) [5] have been suggested as a useful tool to improve the VT/VF detection efficiency. However, these tools are computationally demanding and complex to implement. Moreover, they may increase the overall requirements of the detection system if not utilized properly. For example, the selected ECG features should be relevant and show significant potential in the detection of ventricular arrhythmia. Otherwise, the efficiency of the classifier would decrease and thus degrade the overall performance.

In this chapter, a high-performance method for the prediction of the onset of ventricular arrhythmia by using six ECG features is presented. The set of features

T. Tekeste · H. Saleh (✉)
Khalifa University, Abu Dhabi, UAE
e-mail: hani.saleh@kustar.ac.ae

T. Tekeste
e-mail: temesghen.habte@kustar.ac.ae

Ibrahim (Abe) M. Elfadel and M. Ismail (eds.), *The IoT Physical Layer*,
https://doi.org/10.1007/978-3-319-93100-5_13

includes the PQ, QP, RT, TR, PS, and SP interval variability. The proposed method is demonstrated to be computationally simple-to-realize. The combination of these features is unique and is proved to be the most significant combination for the prediction of VT/VF episodes. Furthermore, all the signals used in validating the algorithm are annotated as NORMAL from the database. Hence, in this work the novelty is in finding which of these normal beats indicate the future occurrence of VT/VF.

13.2 Proposed Ventricular Arrhythmia System

This work proposes a fully integrated ECG signal processing system suitable for real-time applications. The proposed system consists of three main stages as shown in Fig. 13.1. The digitized ECG signal is applied at the input to the processing stage. The processing stage is responsible for removing the noise coupled with the ECG. This stage includes adaptive detection and delineation technique of the P, QRS, and T waves. The proposed technique takes into account the different ECG waveform morphologies and utilizes adaptive search windows and thresholds to accurately detect the fiducial points in each heartbeat.

In the second stage, six features are extracted from a five-second window of the ECG signal. Then, different unique combinations of these features are constructed and used as the input to the final stage. In the final stage, Linear Discriminant Analysis (LDA) is used to identify the signals that are susceptible to ventricular arrhythmia. There are many reasons for choosing LDA. First, the ECG features have shown strong potential in the prediction of ventricular arrhythmia with a p value < 0.001. Second, it was intended to investigate the performance of the system without introducing the strong biasing effect of a classifier. Lastly, LDA is a simple classification method that can be easily implemented in hardware.

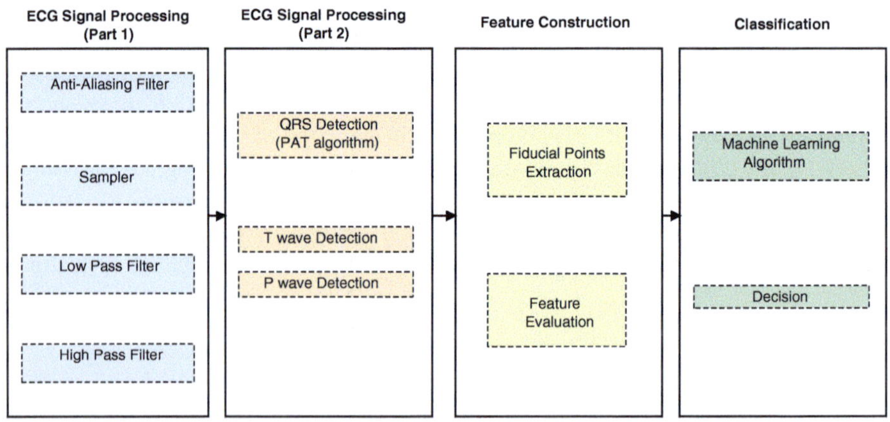

Fig. 13.1 Schematic representation of the proposed ventricular arrhythmia prediction system

13.2.1 Short-term ECG features

The initial step in defining parameters for classification is to extract all the fiducial points of ECG, which are the wave limits and peaks of the P, QRS, T wave components of the ECG. All of these points are used to determine intervals for a classifier system.

The performance of the learning method is strongly affected by the number and relevance of the input variables. Hence, all the extracted features are analyzed and assessed independently using several techniques to discriminate the best ones. A unique set of six morphological features were found to be the most indicative characteristics of ventricular arrhythmia, which are PQ, QP, RT, TR, PS, and SP intervals variability. The PQ interval represents the interval from the atrial depolarization to the ventricular depolarization and is measured from the beginning of the P wave to the onset of the QRS complex. The QP interval is measured from the onset of the QRS complex to the beginning of the P wave of the next cardiac cycle. The RT interval is the duration of ventricular systole in which the ventricles remain in a depolarized state [6]. The interval is measured from the R peak of the QRS complex to the start of the T wave. In contrast, the TR interval defines the ventricular diastole interval that is related to the duration the ventricles refill with blood after contraction [6]. The TR interval is measured from the start of the T wave of one cardiac cycle to the R peak of the next cycle. The time intervals between the beginning of the P wave and the offset of the QRS complex and between the offset of the QRS complex of one cycle and the beginning of the P wave of the next cycle define PS interval and SP interval, respectively.

13.2.2 Statistical Analysis

Table 13.1 shows a statistical analysis of the ECG features by comparing between two groups of ECG datasets, named as GROUP A and GROUP B. The features demonstrated a good potential in the rate of change where the mean value differs significantly between the two groups for $p < 0.001$. The mean value of the PQ

Table 13.1 Statistical analysis of the ECG features

Feature	$\mu \pm \sigma$ (s) GROUP - A	$\mu \pm \sigma$ (s) GROUP - B	p value
RR interval	0.9389±0.1161	1.775±0.0897	<0.001
PQ interval	0.0690±0.0456	0.044±0.0586	<0.001
QP interval	0.8691±0.1396	0.9625±0.1684	<0.001
RT interval	0.4356±0.0352	0.615±0.3695	<0.001
TR interval	0.5026±0.1039	0.3923±0.4001	<0.001
PS interval	0.1695±0.0435	0.1215±0.0885	<0.001
SP interval	0.7686±0.1373	0.885±0.1553	<0.001

interval is slightly greater in GROUP A than GROUP B. A similar observation can be made for the TR and PS intervals. On the other hand, the mean value of the heart rate variability (RR interval) is doubled in GROUP B compared to GROUP A, which is expected. The rest of the features including QP, RT, and SP intervals have significantly higher mean values in GROUP B compared to GROUP A.

13.2.3 Information Gain Attribute Evaluation

Feature selection (FS) by filtering has been used to rank the importance of the seven selected features. The best feature that gives highest accuracy is selected during the feature selection analysis. This feature has been combined with other individual features of high relevance, and thus, the best combinations have been produced.

FS by filter is independent of the machine learner and uses an attribute evaluator. Moreover, a ranker is used to rank all the features in the original dataset [7]. In this work, the Information Gain (IG) attribute evaluator was applied.

The foundation of the IG attribute evaluator is the entropy, which measures a system's unpredictability [8]. The entropy of Y, $H(Y)$, is given by [7].

$$H(Y) = -\sum_{y \in Y} p(y) \;\; log_2(p(y)) \tag{13.1}$$

where $p(y)$ is the marginal probability density function for the random variable Y. Sometimes, the observed values of Y in the training dataset are partitioned according to the values of a second feature X. In this case, the entropy of Y after the observation of X is given by [7].

$$H(Y|X) = -\sum_{x \in X} p(x) \;\; \sum_{y \in Y} p(y|x) \;\; log_2(p(y|x)) \tag{13.2}$$

where $p(y|x)$ is the conditional probability of y given x (further details are given in Sect. 13.3. The IG measure reflects information about Y provided by X and is given below. In this study, Y is the class (GROUP A or GROUP B) and X is the vector containing the seven ECG features.

$$IG = H(Y) - H(Y|X) \tag{13.3}$$

13.3 Classification using Naive Bayes

The Naive Bayes classifier is particularly useful for hardware implementation as its iterative parameter estimation is not complicated. It is assumed that the proposed features are independent from each other for the task of VT/VF prediction. In this work, all the correlations between the features extracted from the same waveform are negligible. Weka/MATLAB was used to rank the developed features, and then VT/VF prediction accuracy was computed for various combinations of the developed features. The proposed combination achieved the highest accuracy of 99.25%.

As the name indicates, the Naive Bayes classifier is a probabilistic model which is based on Bayes' theorem to compute the probability of a data belonging to each class given some observations from the training dataset [9]. For a set of feature vectors d and class c_i, the Bayes theorem is given as:

$$P(c_i \mid d) = \frac{P(d \mid c_i)\, P(c_i)}{P(d)} \tag{13.4}$$

The best class to assign a data to is the one that maximizes the conditional probability out of all the classes. This can be represented by Eqs. (13.5)–(13.7).

$$c = argmax\ \ P(c_i \mid d) \tag{13.5}$$

$$= argmax\ \ \frac{P(d \mid c_i)\, P(c_i)}{P(d)} \tag{13.6}$$

$$= argmax\ \ P(d \mid c_i)\, P(c_i) \tag{13.7}$$

For multiple independent feature vectors, the joint probability likelihood can be calculated as the product of the individual conditional probabilities, as given below:

$$c = argmax\ P(c_i) \prod_{x} P(x \mid c_i) \tag{13.8}$$

The values of each feature vector associated with each class is governed by *Gaussian Distribution*. Hence, the conditional probability likelihood of a given value v from a feature vector x given a class c_i, P($v \mid c_i$), is evaluated using equation, where μ and σ^2 are the mean and the variance of the values of x associated with class c_i and are given as:

$$P(x = v \mid c_i) = \frac{1}{\sigma\sqrt{2\pi}} e^{-(v-\mu)^2 / 2\sigma^2} \tag{13.9}$$

$$\mu = \sum_{i=1}^{N} (x_i) \tag{13.10}$$

$$\sigma^2 = \frac{1}{N} \sum_{i=1}^{N} (x_i - \mu)^2 \tag{13.11}$$

For any unclassified value, the Gaussian model computes the above equations from the training dataset of each class. For falsification purposes, two classes are considered: GROUP A and GROUP B.

13.3.1 Classification Procedure

The classification procedure is shown in Fig. 13.2. The signals from GROUP A and GROUP B were divided randomly into different sets; training, testing and validation sets to verify the performance; 64% of the data was used for training, and the remaining 36% was split equally into testing and validation sets. The training and testing procedure was repeated more than once to make sure that the results were independent of the randomization process.

Different combinations of the selected training features were fed into the classification model as input, and then, the models were evaluated on the corresponding combination test data as shown in Fig. 13.3. In addition, each combination was

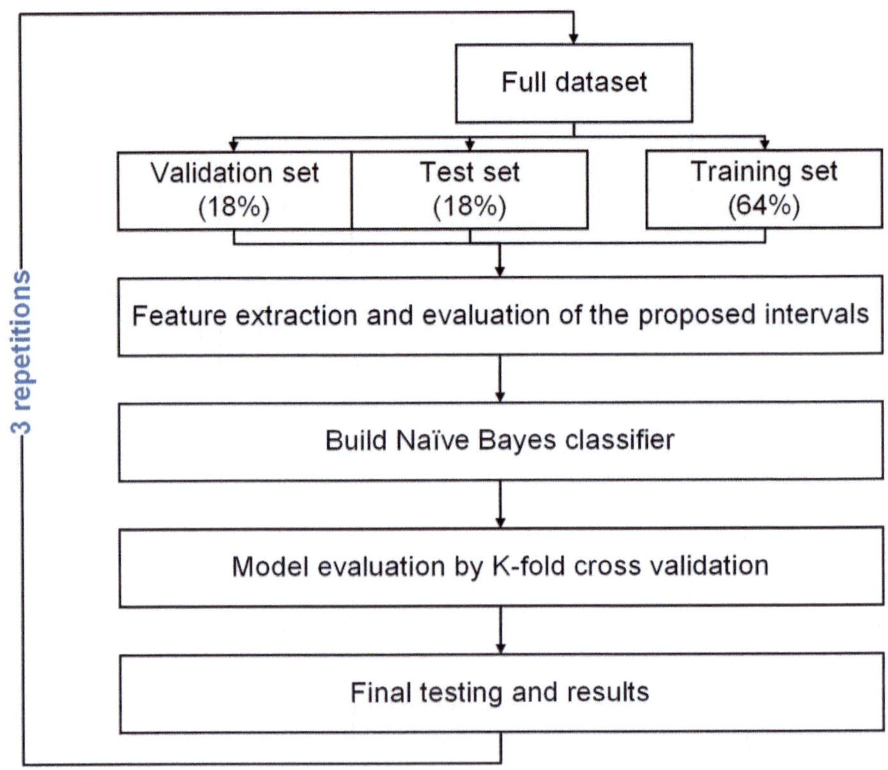

Fig. 13.2 Learning flowchart of Naive Bayes classifier

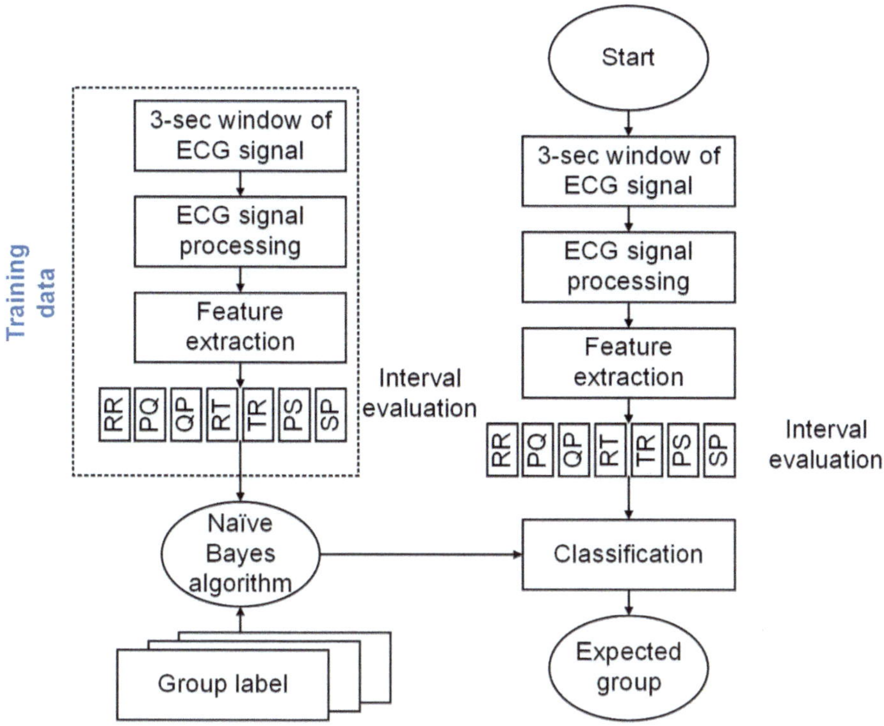

Fig. 13.3 Sequence diagram of training classifier

validated using multiple k-fold cross-validations on the study dataset. The k-fold cross-validation procedure was repeated more than once. The average of the performance was used for evaluation.

13.3.2 Performance of the Classification Stage: Part 1

The performance of the classifier was analyzed using each combination independently. Different k-fold cross-validations were investigated and repeated more than once for each procedure. The out-of-sample average performance of the fivefold, sevenfold, and tenfold cross-validations are shown in Tables 13.2, 13.3, and 13.4, respectively. An accuracy (ACC) of 99.98%, SE of 98.9%, and P of 99% were obtained by the tenfold cross-validation. Furthermore, the AUC values for most of the combinations by any k-fold cross-validations were significant as well.

Table 13.2 Fivefold cross-validation results of the feature combinations (window size = 3 s)

		Validation set (%)			
Combination number	Combination intervals	ACC	SE	P	AUC
#1	TR, SP	95.6	96	94.4	79.8
#2	TR, SP, QP	94.73	94.7	94.5	77.8
#3	TR, SP, QP, RT	96.65	96.7	96.2	82.9
#4	TR, SP, QP, RT, PS	95.48	95.5	95.7	92.2
#5	TR, SP, QP, RT, PS, PQ	95.76	95.8	95.9	92.4
#6	TR, SP, QP, RT, PS, PQ, RR	99.98	99.9	100	99

Table 13.3 Sevenfold cross-validation results of the feature combinations (window size = 3 s)

		Validation set (%)			
Combination number	Combination intervals	ACC	SE	P	AUC
#1	TR, SP	95.54	96	94	79.7
#2	TR, SP, QP	94.8	94.8	94.5	77.6
#3	TR, SP, QP, RT	96.6	96.7	96.3	82.5
#4	TR, SP, QP, RT, PS	95.36	95.4	95.6	92.1
#5	TR, SP, QP, RT, PS, PQ	95.3	95.4	95.7	92.7
#6	TR, SP, QP, RT, PS, PQ, RR	99.98	99.9	100	99

Table 13.4 Tenfold cross-validation results of the feature combinations (window size = 3 s)

		Validation set (%)			
Combination number	Combination intervals	ACC	SE	P	AUC
#1	TR, SP	95.52	96	94.4	79.9
#2	TR, SP, QP	94.7	94.7	94	77.8
#3	TR, SP, QP, RT	96.7	96.7	96.2	83
#4	TR, SP, QP, RT, PS	95.44	95.4	95.7	92.2
#5	TR, SP, QP, RT, PS, PQ	95.42	95.4	96	93
#6	TR, SP, QP, RT, PS, PQ, RR	99.98	98.9	99	99.97

13.3.3 Comparison to Published Work: Part 1

The proposed system is compared against other systems reported in [5, 10–13] in terms of accuracy (ACC), sensitivity (SE), precision (P), area under characteristics curve (AUC), calculation time (CalT), length of processing window in seconds, and the database used. The results of this comparison are shown in Table 13.5. The tenfold cross-validation yielded the best values for the ACC, SE, P, and AUC as compared

Table 13.5 Comparison between the proposed system and prior art

Reference	Method	ACC (%)	SE (%)	P (%)	AUC	CalT (s)	Window Size (s)	Database
[10], Chazal et al.	ECG morphology, heartbeats interval, RR-interval	97.4	77.7	81.9	NA	NA	NA	MIT-BIH
[11], Amann et al.	Time-delay	96.2	79.0	97.8	94.0%	1.7	8	MIT-BIH, AHA, CUDB
[12], Alonso et al.	FS and SVM	96.8	97.1	61.6	98.7%	NA	8	MIT-BIH, CUDB, VFDB
[5], Q. Li et al.	SVM	98.1	98.4	98	NA	NA	5	AHA, CUDB, VFDB
[13], Sayadi et al.	Bayesian framework	99.1	98.77	97.47	NA	NA	NA	MIT-BIH
This work [14]	**Unique set of ECG morphologies and Naive Bayes**	**99.98**	**98.9**	**99**	**99.97 %**	**1**	**3**	**NSRDB, AHA, MIT-BIH**

to prior art. In addition, most of the reported systems fall in the category of detection of the ventricular arrhythmia while the proposed system is used for prediction as the processing of the signal was done before the VT/VF onset. Furthermore, our system has the fastest performance (CalT = 1 s) with the earliest prediction interval (up to 3 h). Note that the comparison was made on the sixth combination.

13.4 Conclusions

A new prediction method of ventricular arrhythmia that combines a unique set of ECG features with LDA was presented. The developed ECG features were extracted from two consecutive heartbeats whose combination has never been used in any previous detection or prediction method. Based on statistical analysis, each feature individually and in combinations showed a good potential in the prediction of ventricular arrhythmia with p value <0.001.

The proposed method achieved an outstanding capability of predicting the arrhythmia up to 3 h before the onset. An accuracy (ACC) of 99.1%, sensitivity (SE) of 98.95%, precision (P) of 98.39%, and area under the receiver operating characteristics curve (AUC) of 99.97% were obtained on the out-of-sample validation data by tenfold cross-validation with 3 s window size. The high performance and simple computations of the proposed method make our system unique and suitable for inclusion in System on Chips (SoCs) targeting wearable mobile medical devices.

Acknowledgements This work has been supported by the Semiconductor Research Corporation (SRC) under the Abu Dhabi SRC Center of Excellence on Energy-Efficient Electronic Systems (ACE^4S), Contract 2013 HJ2440, with funding from the Mubadala Development Company, Abu Dhabi, UAE.

References

1. J.W. Schleifer, K. Srivathsan, Ventricular Arrhythmias: state of the art. Cardiol. Clin. **31**(4), 595–605 (2013)
2. D.P. Zipes, H.J. Wellens, Sudden cardiac death. Circulation **98**(21), 2334–2351 (1998)
3. C.J. Garratt, C. Statter, D. Statters, *Mechanisms and Management of Cardiac Arrhythmias* (BMJ Books, London, 2001)
4. J. Pardey, Detection of ventricular fibrillation by sequential hypothesis testing of binary sequences. IEEE Comput. Cardiol. 573–576 (2007)
5. Q. Li, C. Rajagopalan, G. Clifford, Ventricular fibrillation and tachycardia classification using machine learning method (2013)
6. M.S. Finkler, Lab 10: Cardiovascular physiology, Indian University, pp. 1–5
7. H. Wang, T.M. Khoshgoftaar, K. Gao, A comparative study of filter-based feature ranking techniques, in *2010 IEEE International Conference on Information Reuse and Integration (IRI)* (IEEE, 2010), pp. 43–48
8. G. Forman, An extensive empirical study of feature selection metrics for text classification. J. Mach. Learn. Res. **3**, 1289–1305 (2003)

9. C. Pillers, Dobler, Mathematical statistics: basic ideas and selected topics. Am. Stat. **56**(4), 332–332 (2002)
10. P. De Chazal, M. O'Dwyer, R.B. Reilly, Automatic classification of heartbeats using ecg morphology and heartbeat interval features. IEEE Trans. Biomed. Eng. **51**(7), 1196–1206 (2004)
11. A. Amann, R. Tratnig, K. Unterkofler, Detecting ventricular fibrillation by time-delay methods. IEEE Trans. Biomed. Eng. **54**(1), 174–177 (2007)
12. F. Alonso-Atienza, E. Morgado, L. Fernandez-Martinez, A. García-Alberola, J. Rojo-Alvarez, Detection of life-threatening arrhythmias using feature selection and support vector machines (2014)
13. O. Sayadi, M.B. Shamsollahi, G.D. Clifford, Robust detection of premature ventricular contractions using a wave-based bayesian framework. IEEE Trans. Biomed. Eng. **57**(2), 353–362 (2010)
14. N. Bayasi, T. Tekeste, H. Saleh, B. Mohammad, A. Khandoker, M. Ismail, Low-power ecg-based processor for predicting ventricular arrhythmia. IEEE Trans. Very Larg. Scale Integr. (VLSI) Syst. **PP**(99), 1–13 (2015)

Chapter 14
MSER-in-Chip: An Efficient Vision Tool for IoT Devices

Andrzej Sluzek, Hani Saleh, Baker Mohammad, Mahmoud Al-Qutayri and Mohammed Ismail

14.1 Introduction

MSER features (originally proposed in [1] and computationally improved in [2]) continue to attract attention of machine vision researchers and practitioners. MSERs are approximately invariant to affine transformations (i.e., perspective distortions) which makes them particularly suitable for the analysis of images of 3D real worlds. In comparison with other affine-invariant features, their main advantages are as follows: (1) moderate computational complexity and the algorithmic structure suitable for hardware implementations, e.g., [3, 4], and (2) a good identification of significant image parts usually combined with high repeatability under typical image distortions (as reported in [5]).

Additionally, MSERs are generally believed to be less numerous and larger than other affine-invariant keypoints. We have confirmed (using standard setups of MSER detectors over many thousands of random images) that MSER keypoints are actually 48% less numerous than Harris-Affine keypoints, and 32% less numerous than Hessian-Affine (*hesaff*) keypoints, see [6].

A. Sluzek (✉) · H. Saleh · B. Mohammad · M. Al-Qutayri · M. Ismail
Khalifa University, Abu Dhabi, UAE
e-mail: andrzej.sluzek@kustar.ac.ae

H. Saleh
e-mail: hani.saleh@kustar.ac.ae

B. Mohammad
e-mail: baker.mohammad@kustar.ac.ae

M. Al-Qutayri
e-mail: mqutayri@kustar.ac.ae

M. Ismail
e-mail: ismail@kustar.ac.ae

A. Sluzek
Warsaw University of Life Sciences-SGGW, Warsaw, Poland

Ibrahim (Abe) M. Elfadel and M. Ismail (eds.), *The IoT Physical Layer*,
https://doi.org/10.1007/978-3-319-93100-5_14

Nevertheless, some disadvantages of MSER features have been identified as well. In particular, MSERs have limited performances on blurred and/or textured images. Both cases are actually related to the image scale, because image blurring (which can distort shapes of extracted MSER regions) is equivalent to image downscaling. Similarly, shapes of fine texture details can vary irregularly under image rescaling. Protruding fragments of non-convex MSERs are particularly vulnerable to scale variations (as discussed in [7]).

In this chapter, we focus on the advantageous characteristics of MSER features which make them an important tool for prospective applications of IoT devices equipped with vision capabilities. We also propose how to overcome certain limitations of MSERs.

In Sect. 14.2, basic definition and fundamental characteristics of MSER features are briefly summarized. Based on that, hardware-friendly algorithms for MSER detection and the actual design of SoC implementation of MSER detector are presented in Sect. 14.3.

Section 14.4 focuses on newly developed SIMSER (*scale-insensitive MSER*) features, which have exhibited much better performances for textured and scaled-down (blurred) images than MSERs. Again, hardware-friendly algorithms for SIMSER detection are of primary interest.

The final Sect. 14.5 surveys a number of current and prospective applications of MSER (and SIMSER) features, especially in the context of growing popularity, diversity, and complexity of IoT devices.

14.2 MSER Features: Basic Definitions and Properties

In machine vision, one of the most fundamental properties of systems designed for real-world applications is their adaptability to perspective distortions which inevitably deform acquired data. This is particularly critical in case of vision-equipped IoT devices, which would be typically used for identification, retrieval, detection, tracking, recognition, etc., of natural objects, which are usually subject to such distortions in their visual representations.

Invariant local features (e.g., [8]) are one of the most important tools for such applications. Therefore, invariance to affine transformations (which sufficiently accurately models perspective distortions in typical tasks of machine vision) is a very useful characteristic of those features.

MSER (*maximally stable extremal region*) features are considered one of the most successful examples of affine-invariant local features. They were originally proposed in [1] for stereovision, but subsequently MSERs have been applied to many other problems of machine vision, including image alignment (e.g., [9]), object tracking and detection (e.g., [10, 11]), robotic vision (e.g., [12]), or visual data recognition and retrieval (e.g., [13, 14]).

Given a gray-level image, MSER features are found within a family of binary images obtained by applying incrementally growing thresholds to the original image (which generates gradually shrinking white regions) or to the inverted image (which effectively generates gradually shrinking black regions). *Maximally stable extremal regions* (MSERs) are those white or black regions which are least sensitive (in terms of their area) to threshold increments. Formally, a binary region $Q(t)$ (where t indicates the current threshold level) is considered a MSER feature if the growth rate function $q(t)$ defined by the derivative of the region area over the threshold values:

$$q(t) = \frac{\frac{d}{dt}\|Q(t)\|}{\|Q(t)\|}, \tag{14.1}$$

reaches a local minimum for that region (where $\|\cdot\|$ indicates the region area).

In case of color images, MSER features can be separately detected in each channel or in the color space with appropriately defined order of vector data (e.g., [15]).

An exemplary image and its white and black MSERs are shown in Fig. 14.1. It can be noticed that the shapes of MSERs quite accurately correspond to what a human vision would perceive as the most distinctive fragments of the original image.

Shapes of MSER features outline visually prominent image patches. However, other popular local features (see the survey in [8]) are usually represented by either circles or ellipses (which approximate sizes and shapes of visually distinctive image fragments). Additionally, circular/elliptic representations of local features are necessary to build feature *descriptors* (i.e., numerical or symbolic specification of the feature contents needed for the evaluation of visual similarities between individual features and between images).

Thus, MSERs are also converted into ellipses. For each MSER feature, the ellipse center is defined by the center of mass of the MSER patch, while the ellipse itself is *the best-fit* ellipse of the patch. Several methods exist to find the best-fit elliptic approximations, but in machine vision the most popular algorithm is based on the second-order central moments of the approximated shapes (e.g., [16]). Figure 14.2 shows an exemplary image with elliptic approximations of MSERs found in this image.

Fig. 14.1 Exemplary image **a** and its black **b** and white **c** MSER features

Fig. 14.2 Elliptic approximations of MSER features found in an exemplary image. In some cases, the shapes of the corresponding MSER regions can be identified from the ellipse outlines, but for many ellipses it is rather difficult to guess the original shapes of MSERs

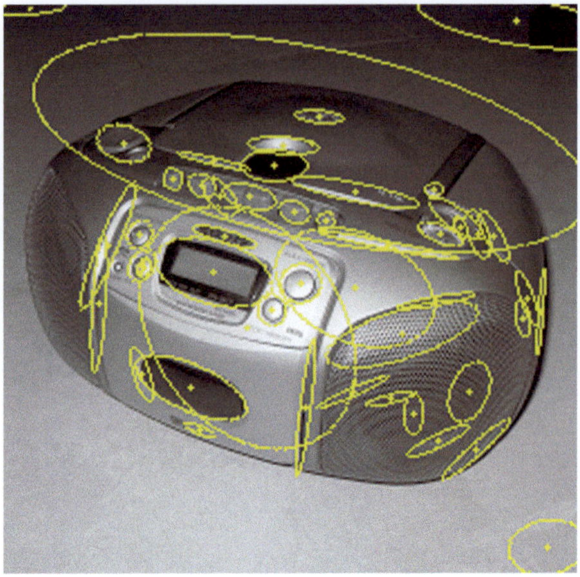

14.3 Algorithms and Architectures

Compared to other affine-invariant local features, MSERs have low complexity (which makes them so attractive for IoT applications) and the underlying computational steps have regular structures, suitable for hardware implementations.

Since digital images are quantized with a finite (usually 255) number of intensities, the growth rate function defined by Eq. 14.2 can be substituted by one of its discrete approximations:

$$q(t_j) = \frac{\|Q(t_j) - Q(t_{j-1})\|}{\|Q(t_j)\|} \quad \text{or} \quad q(t_j) = \frac{\|Q(t_{j+1}) - Q(t_{j-1})\|}{\|Q(t_j)\|}, \tag{14.2}$$

where the difference $t_j - t_{j-1}$ defines the threshold increment Δt. For standard images of 255 levels of intensity, Δt typically ranges between 2 and 7 (the lower the value of Δt, the more accurate but slower the algorithm is).

For each threshold, the images have to be segmented into white/black regions and the growth of those regions should be tracked over the subsequent thresholds in order to find the local minima of the discretized growth rate function (defined by Eq. 14.2). Intuitively, tracking can be performed straightforwardly because with the incremented threshold levels the corresponding binary regions are nested. Then, the regions which are least different (in terms of their area) from their counterparts at previous and next threshold levels are identified as MSER features.

In two alternative MSER detectors (i.e., [1, 2]), the detection algorithm is run twice, on the original image and on its inversion, so that both white and black MSERs are detected by the identical subroutines. However, the algorithmic structures of both detectors are different. In [1], the image is thresholded at each level, and the tree of nested regions (*white* for original images and *black* for inverted ones) is built using a *union-find* algorithm, e.g., [17]. Computational complexity of this MSER detector is near-linear at $O(n \times log(log(n)))$, where n is the number of image pixels (resolution). The detector proposed in [2] has the linear complexity $O(n)$. It resembles the *flood-fill* algorithm adapted to gray-level images. The second algorithm is generally faster for high-resolution images, but the differences are not very significant so that both variants are feasible candidates for hardware-based implementations.

In this chapter, hardware designs for both MSER detectors are addressed, but we focus on the first one (it will be referred to as the *classical detector*) for which the implementation is completed, including the manufactured chip prototypes.

14.3.1 Hardware-Friendly Algorithms for MSER Detector

The need to run the MSER detection algorithm twice (on the original image to identify bright MSERs and on the inverted image to detect dark MSERs) is inconvenient and actually redundant. Therefore, a single-run variant of the classical MSER detector has been proposed in [18] and subsequently used in [4, 19].

In this single-run algorithm, binary images are simultaneously segmented into both black and white regions using either four- or eight-pixel connectivity (in the designed hardware, four-pixel connectivity is used). Assuming that a binary image (i.e., the image of interest thresholded at a selected level t) has $M \times N$ resolution, the single-run MSER detector utilizes the following data structures in the *modified union-find* algorithm:

- *Seed* matrix of regions *SM* (of the same size as the image, i.e., $M \times N$). The initial content of *Seed* is $SM_{i,j} = M \times (i-1) + j$, i.e., each pixel is a seed for itself. After the image is processed, all pixels belonging to the same region share the same seed number k.
- *Region Size* matrix *RS* (also of $M \times N$ size) which specifies the size of region to which each pixel belongs. Initially, the matrix contains only ONES, i.e., each pixel is considered a separate region.
- *Map-of-regions* structure, which lists all regions (represented by their seeds) in the binary image, and their sizes and binary colors (black or white).

A small exemplary binary image and the contents (initial and final) of its *SM* and *RS* matrices are shown in Fig. 14.3, while its *map-of-regions* array is given in Table 14.1.

When the image is thresholded using a sequence of thresholds ranging from 0 to *MaxValue* (typically 255) with Δ increment, the detected regions are always nested, i.e., a black (or white) region found at t threshold is always fully contained within the corresponding black (or white) region at $t + \Delta$ (or $t - \Delta$) threshold. By using this property, minima of the growth rate function specified by Eq. 14.2 can be easily obtained by processing *maps-of-regions* over three consecutive threshold values.

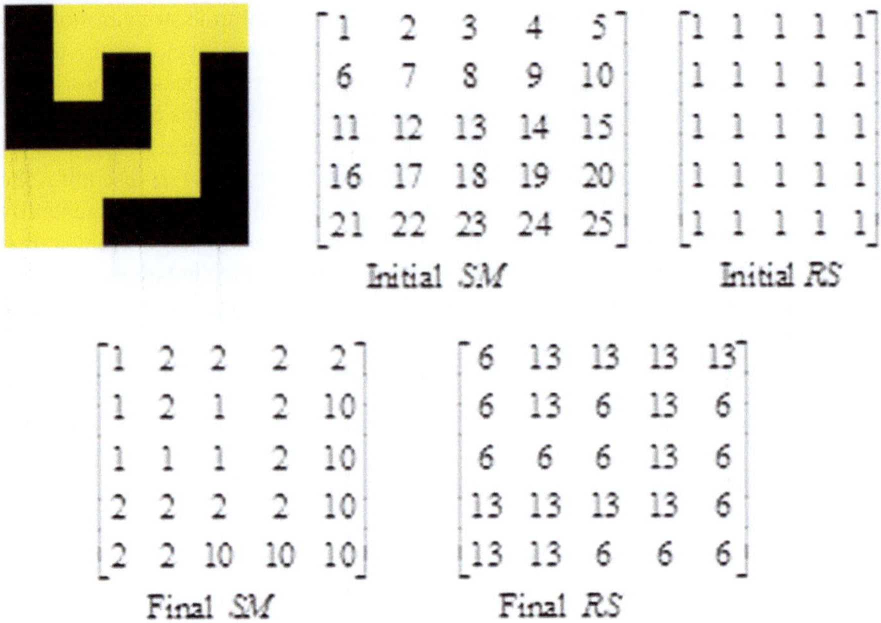

Fig. 14.3 A binary image and the contents (initial and final) of its *SM* and *RS* matrices

Table 14.1 Map-of-regions for Fig. 14.3 example

Region no.	Seed	Size	Color
1	1	6	Black
2	2	13	White
3	10	6	Black

Regions corresponding to those minima are considered MSERs if they satisfy some additional conditions which specify the acceptable minimum and maximum areas of MSER regions, and other requirements of practical significance, see [1, 2].

The detected MSER regions are eventually approximated by the best-fit ellipses of the equation:

$$A(x - x_0)^2 + 2B(x - x_0)(y - y_0) + C(y - y_0)^2 = 1, \tag{14.3}$$

where (x_0, y_0) is the region center of mass (calculated from the moments of order 0 an 1), while A, B, and C are derived from the central moments of order 2.

Since moments (of any order) for all regions of a segmented image can be precomputed in a single scan of the image, the computational costs are negligible (mainly conversion of region moments into ellipse parameters).

Altogether, the single-run MSER detection algorithm can be specified by the following pseudo-code (details of the modified union-find algorithm can be found in [18]):

==

Pseudo-code for single-run MSER detection algorithm

```
Input: Im(M,N), MaxArea, MinArea, delta; Storage: I_bin(M,N)
Storage: SM(M,N) %Seed matrix
Storage: RS(M,N) %Region size matrix
Storage: SeedList Storage: Map_of_regions Storage:
         number_of_Seeds
Storage: list_of_MSERs
% list of elliptic approximations of detected MSERs
for t = 0:delta:255
    I_b (M,N) = (Image(M,N) <= t);
    [SM,RS,Seeds] = Modified_UnionFind(I_bin);
    if Seeds contains region seed(s) which are not in SeedList
        Append new seed(s) to SeedList
    endif;
    Update Map_of_regions by new Q(t+),Q(t) and Q(t-)
    if t >= 3*delta \newline
        for j=1:number_of_Seeds
            if MSER\index{MSER} detection conditions satisfied
                find moments of the region, calculate ellipse
                 and add the ellipse to the list_of_MSERs
            endif;
        endfor;
    endif;
endfor;
```

==

14.3.2 System-on-Chip Implementation

The algorithm outlined in Sect. 14.3.1 has been eventually converted into a *system-on-chip* (SoC) hardware design. The applied model was ASIC (Application-Specific Integrated Circuit) design flow.

The single-run MSER detector design was synthesized, placed, and routed using 65nm CMOS technology and standard cell library. The high-level architecture of this SoC is summarized in Fig. 14.4, while more details of the incorporated *finite-state machines* (FSMs) are highlighted in Fig. 14.5.

The final design details and the actual chip are shown in Fig. 14.6.

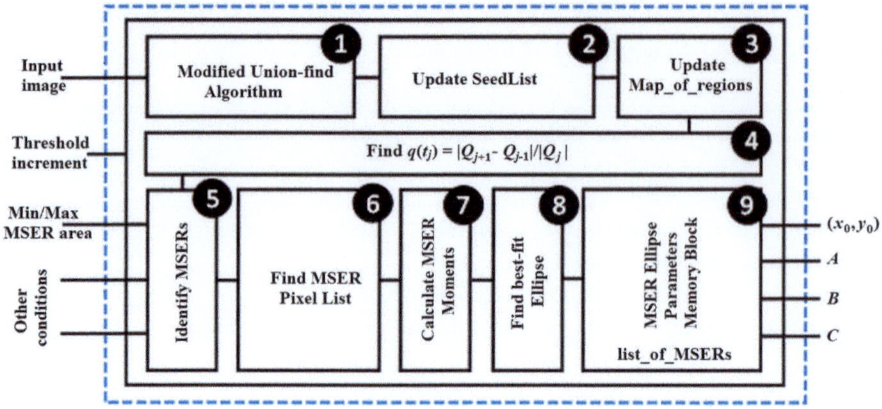

Fig. 14.4 High-level architecture of SoC implementation from [18]

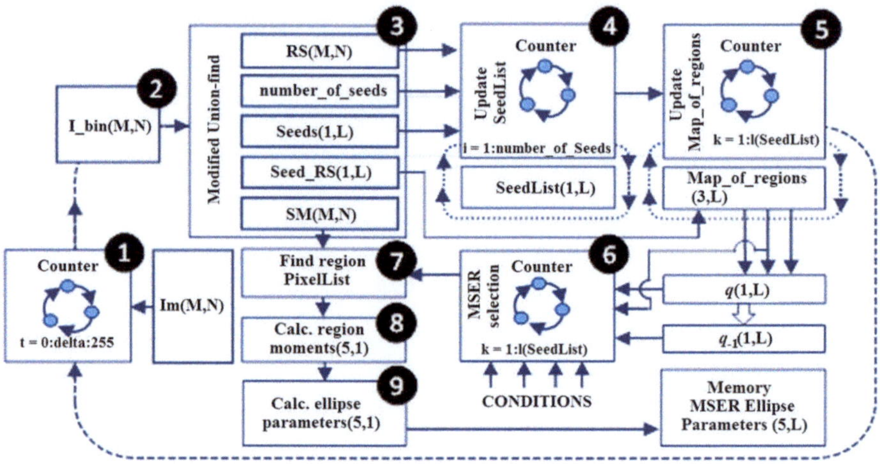

Fig. 14.5 FSM-level model of SoC implementation from [18]

The chip has been designed for the image resolution 256 × 256, but implementations with higher resolutions have been successfully simulated as well (up to high-definition TV standards). Another significant advantage of the design is its processing rate evaluated by the *frames-per-second* (FPS) metric. For 256 × 256 resolution, the average (as it depends on the complexity of the processed frames) performance is around 183 FPS which outperforms, by a wide margin, alternative solutions in the literature (e.g., [3]).

At the current resolution, the chip has a very small area (approx. 57, 100 μ m^2) and low power consumption (approx. 2.25 mW). The required memory (storage space) is approximately equal to

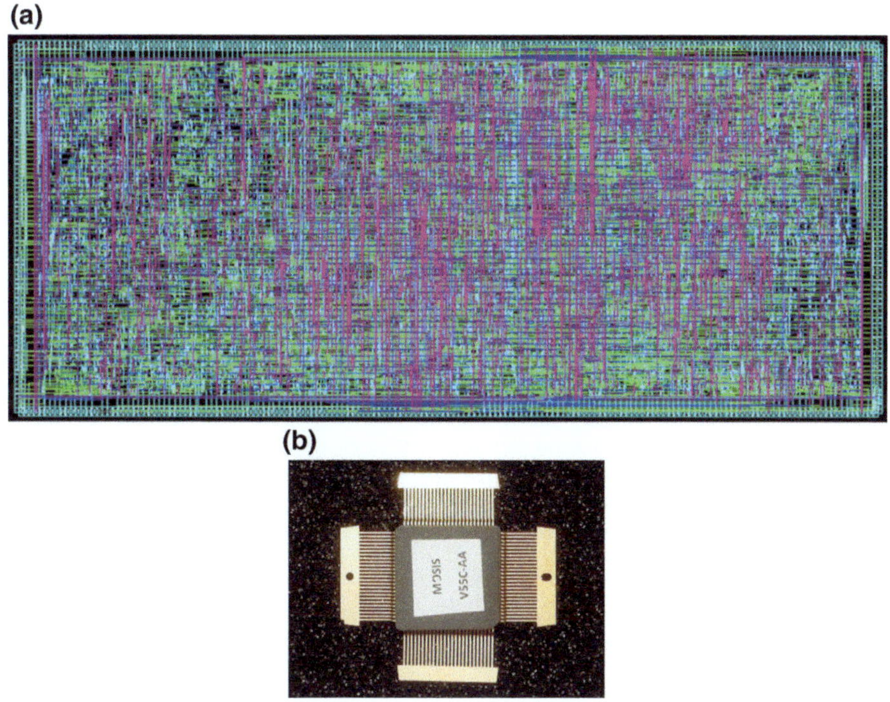

Fig. 14.6 Chip design details (synthesized, placed, and routed) of the MSER detector **a**, and the actual chip **b**

$$(1.125 + 0.25 \lceil log_2(n) \rceil) \times n, \tag{14.4}$$

where n is the image resolution (in pixels) and $\lceil \rceil$ denotes the *ceiling* operator.

Altogether, the obtained performances make our design suitable for a diversity of IoT devices with limited power resources, even in environments with very rapid changes of scene contents.

It has been mentioned in Sect. 14.1 that the alternative MSER detection algorithm proposed in [2] has slightly lower complexity (and because of that became more popular, e.g., it is used in the MATLAB implementation of MSER detector). The hardware architecture for this detector has been developed as well (details are available in [20].

The circuit-level details of this solution have not been fully investigated yet, so that it is difficult to predict its performances compared to the presented SoC. Nevertheless, the method requires two passes (the original and the inverted images) to detect both bright and dark MSERs which may affect its frame rate.

14.4 Scale-Insensitive MSER Features

Although MSER features have many advantages (confirmed in numerous applications), some disadvantages have been identified as well. In particular, MSERs have limited performances on blurred and/or textured images which might affect their usefulness in visually difficult conditions and in certain environments.

One of the recent attempts to rectify those disadvantages has been presented in [21]. The proposed *scale-insensitive maximally stable extremal regions* (SIMSER features) extend the mechanism of MSER detection into a 2D space $Threshold \times Scale$. The scale dimension is added because both issues are actually related to the image scale. Blur (which can distort shapes of extracted MSERs) is equivalent to image downscaling, e.g., [22]. Similarly, shapes of fine texture details can vary irregularly under image rescaling.

In general, the proposed modification identifies regions which are maximally stable both under the threshold changes and under the scale variations (i.e., blur). Therefore, instead of Eqs. 14.1 and 14.2, more sophisticated mechanisms for minima detection should be used.

First, the processed images are convolved with smoothing filters representing diversified levels of blurs (scales) so that a family of images at different scales is built. Given an image presented over a range of scales $s \in S$ (i.e., a family of images) and binarized using a range of thresholds $t \in T$, a binary regions $Q(s, t)$ (where s defines the current scale and t indicates the current threshold) is considered SIMSER, if two growth rate functions $q_1(s, t)$ and $q_2(s, t)$ jointly reach the local minimum at that region:

$$q_1(s,t) = \frac{\frac{\partial}{\partial t} \|Q(s,t)\|}{\|Q(s,t)\|} \quad \text{and} \quad q_2(s,t) = \frac{\frac{\partial}{\partial s} \|Q(s,t)\|}{\|Q(s,t)\|}. \tag{14.5}$$

In practice, the discrete equivalents of Eq. 14.5 are used:

$$q_1(t_j, s_k) = \frac{\left\| Q(t_{j+1}, s_k) - Q(t_{j-1}, s_k) \right\|}{\left\| Q(t_j, s_k) \right\|}, \tag{14.6}$$

$$q_2(t_j, s_k) = \frac{\left\| Q(t_j, s_{k+1}) - Q(t_j, s_{k-1}) \right\|}{\left\| Q(t_j, s_k) \right\|}. \tag{14.7}$$

It has been shown in [21] that SIMSER features better identify visually prominent fragments of images (an example given in Fig. 14.7) and generally perform better in various matching, retrieval, and tracking tasks.

What makes SIMSERs even more attractive is their computational similarity to MSERs, including hardware implementation. The SoC solution presented in Sect. 14.3.2 can be directly used for processing images at gradually reduced scales (such images can be generated instantaneously by simple downscaling) and for

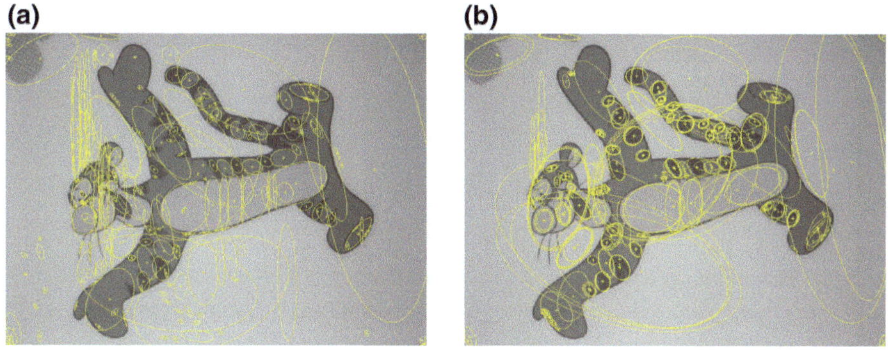

Fig. 14.7 Best-fit ellipses for MSERs **a** and SIMSERs **b** detected in an exemplary image

detecting minima in the threshold dimension ($q_1(t_j, s_k)$ growth rate function in Eq. 14.6).

However, the problematic issue is whether the growth rate function $q_2(t_j, s_k)$ (see Eq. 14.7) which does not exist in the original MSER algorithm can be evaluated at the same computational complexity. Unfortunately, the binary regions generally do not nest over the sequence of scales and their topology may unpredictably change under image smoothing. Nevertheless, a simple algorithm has been proposed to track correspondences between regions in the neighboring scales and to simultaneously compute the growth rate function $q_2(t_j, s_k)$. A commented pseudo-code which clearly confirms $O(n)$ complexity of this algorithm is as follows:

==

Pseudo-code for computing $q_2(t_j, s_k)$ growth rate function

```
Input: Im1(M,N), Im0(M,N)
% two binary images of M*N size (at two neighbouring scales)
Input: Rs1(M,N), Rs0(M,N)
% two labeled images of M*N size
Input: SoR1(K1), SoR0(K0)
% list of region sizes (K1 regions in Im1 and K0 regions
in Im0)
Storage: Ints(K1,K0) <- zeros
% size of intersections between regions from Im1 and Im0
Output storage: q2(K1) <- large value
% q2 function for Im1 regions initialized with very
large values
Output storage: previous(K1) <- zeros
% numbers of the matching regions from the previous scale

for i = 1:M
    for j = 1:N
```

```
        if Im1(i,j)==Im0(i,j)
            Ints(Rs1(i,j),Rs0(i,j))++;
        endif
    endfor
endfor for i = 1:M
 for j = 1:N
   temp=SoR1(Rs1(i,j))+SoR0(Rs0(i,j))-2*Ints(Rs1(i,j),
   Rs0(i,j));
   temp=temp/SoR1(Rs1(i,j)); \newline
   if temp $<$ q2(Rs1(i,j)) && Ints(Rs1(i,j),Rs0(i,j)) > 0
      q2(Rs1(i,j)) = temp; previous(Rs1(i,j)) = Rs0(i,j);
   endif
 endfor
endfor

==========================================================
```

Further details of the intended hardware implementation of SIMSER feature detector are discussed in [23].

14.5 Intended Applications

As highlighted in Sect. 14.2, MSER features have been found instrumental in highly diversified applications of machine vision, but from the IoT perspective the most important areas are apparently object detection/recognition and object tracking. They incorporate a wide range of tasks where various IoT devices (equipped with digital cameras) might have to visually detect or identify known (or previously encountered) physical objects, to track their motions (often in complex and unpredictable environments), and to report activities of those objects.

Currently, both detection and tracking tasks are often considered jointly because as pointed out in [24]: *Automatic object detection has become quite successful to the degree that tracking of a single object may be achieved by performing detection in every frame.*

Achieved characteristics of the designed SoC MSER detector fully satisfy (in terms of size, power consumption, and frame rate of image processing) requirements of small, low-cost devices performing autonomously. Since feature detection is typically the most computation-intensive part of feature-based image retrieval and matching (which, in the intended applications, effectively mean *detection* and *tracking*), the remaining computational costs can be handled by the resources (usually limited) available on board of such devices.

The system described in [25] can be mentioned as an exemplary *proof-of-concept* study, even though the developed hardware was not physically used there (replaced by the equivalent software module). In this study, a low-cost method for visual detection and tracking of a mobile ground object (randomly changing its location) by a drone with an onboard camera and inexpensive computational resources has been presented.

All elements of the detection/tracking process have been optimized to reduce computational costs and energy consumption. In particular, MSER detection would be performed (in the intended final version) by the developed hardware. Secondly, the visual vocabulary used for detection and tracking is built and used in a simplified

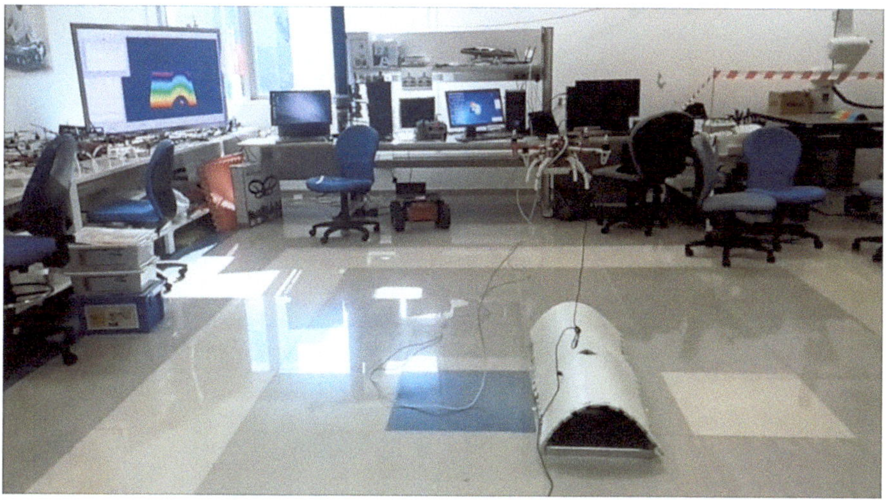

Fig. 14.8 Exemplary workspace for the *proof-of-concept* study from [25]

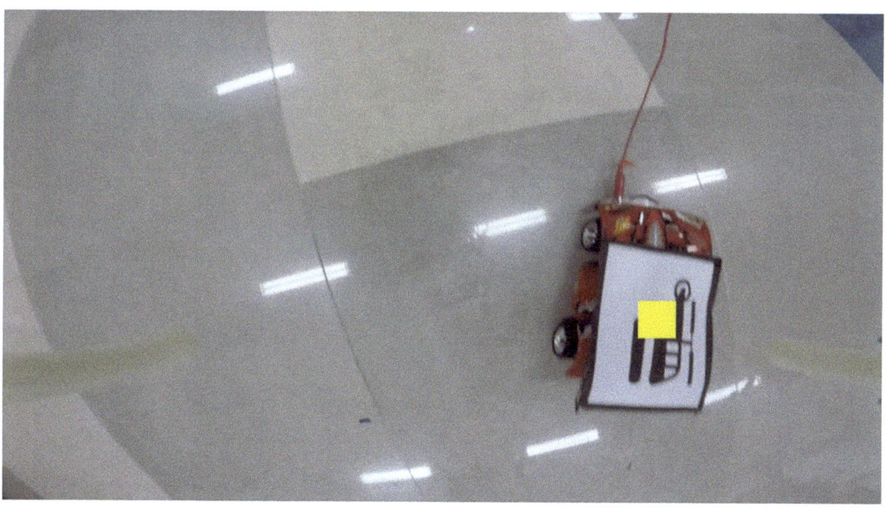

Fig. 14.9 Exemplary detection results of the *proof-of-concept* study from [25]

way and, finally, the target localization module employs a statistical approximation of RANSAC algorithm, [26], with much lower complexity.

Figures 14.8 and 14.9 show illustrative examples of the system setup and its performances.

14.6 Conclusions

The content of this chapter should be considered from a wider perspective. Although we focus on a hardware implementation of a specific machine vision algorithm (which actually is one of several popular algorithms used for invariant feature detection in camera-captured images or video frames), the actual message conveyed in the chapter is more general.

IoT devices are becoming more sophisticated, both in terms of the amount of captured/registered data and in terms of complexity of algorithms used to process those data. With the exponential growth of IoT devices (it is predicted that 24 billion of such devices will be deployed by 2020), even the most advanced data transmission channels would be saturated rather quickly. Thus, there is an inevitable need for algorithms which can be embedded in IoT devices (even very small and cheap) to perform tasks which have been until recently reserved for much more powerful computational systems. Then, only the aggregated results (which are, in terms of the size, a tiny fraction of the originally acquired data) would be transmitted over the communication channels.

The presented example shows that even highly complicated tasks of machine vision (which are often combined with AI methods) can be successfully encapsulated into tiny chips with negligibly small power consumption. Thus, in our opinion, a small step leading to development of IoIT (*Internet of Intelligent Things*) has been made.

Acknowledgements This work has been supported by the Semiconductor Research Corporation (SRC) under the Abu Dhabi SRC Center of Excellence on Energy-Efficient Electronic Systems (ACE^4S), Contract 2013 HJ2440, with funding from the Mubadala Development Company, Abu Dhabi, UAE.

References

1. J. Matas, O. Chum, M. Urban, T. Pajdla, Robust wide baseline stereo from maximally stable extremal regions, in *13th British Machine Vision Conference BMVC'02* (2002), pp. 384–393
2. D. Nistér, H. Stewénius, Linear time maximally stable extremal regions, in *10th European Conference on Computer Vision ECCV'08*, vol.2 (2008), pp. 183–196
3. F. Kristensen, W. MacLean, Real-time extraction of maximally stable extremal regions on an FPGA, in *2007 IEEE International Symposium on Circuits and Systems ISCAS'07* (2007), pp. 165–168
4. E. Salahat, H. Saleh, A. Sluzek, M. Al-Qutayri, B. Mohammed, M. Ismail, A maximally stable extremal regions system-on-chip for real-time visual surveillance, in *41st IEEE Industrial Electronics Society Conference IECON'15* (2015), pp. 2812–2815

5. K. Mikolajczyk, T. Tuytelaars, C. Schmid, A. Zisserman, J. Matas, F. Schaffalitzky, T. Kadir, L. Van Gool, A comparison of affine region detectors. Int. J. Comput. Vis. **65**(1), 43–72 (2005)
6. A. Śluzek, Multi-distinctive MSER features and their Descriptors: a low-complexity tool for image matching, in *2015 International Conference on Advanced Concepts for Intelligent Vision Systems, Lecture Notes in Computer Science*, vol. 9396 (2015)
7. R. Kimmel, C. Zhang, A.M. Bronstein, M.M. Bronstein, Are MSER features really interesting? IEEE Trans. PAMI **33**(11), 2316–2320 (2011)
8. T. Tuytelaars, K. Mikolajczyk, Local invariant feature detector: a survey. Found. Trends Comput. Graphics Vis. **3**(3), 177–280 (2007)
9. R. Szeliski, Image alignment and stitching: a tutorial. Found. Trends Comput. Graphics Vis. **2**(1), 1–104 (2006)
10. M. Donoser, C. Arth, H. Bischof, Detecting, tracking and recognizing license plates, in *8th Asian Conference on Computer Vision ACCV'07*, vol. 2 (2007), pp. 447–456
11. V. Lepetit, P. Fua, Monocular model-based 3D tracking of rigid objects. Found. Trends Comput. Graphics Vis. **1**(1), 1–99 (2005)
12. D. Meger, P.-E. Forssén, K. Lai, S. Helmer, S. McCann, T. Southey, M. Baumann, J.J. Little, D.G. Love, Curious George: an attentive semantic robot. Robot. Auton. Syst. **56**(1), 43–72 (2005)
13. J. Sivic, A. Zisserman, Video data mining using configurations of viewpoint invariant regions, in *IEEE Conference on Computer Vision and Pattern Recognition CVPR'04* (2004), pp. I.488–I.495
14. W. Zhao, Y.-G. Jiang, C.-W. Ngo, Keyframe retrieval by keypoints: can point-to-point matching help? in *5th International Conference CIVR'06* (2006), pp. 72–81
15. M. Donoser, H. Bischof, M. Wiltsche, Color blob segmentation by MSER analysis, in *IEEE International Conference on Image Processing ICIP'06* (2006)
16. A. Śluzek, *Zastosowanie metod momentowych do identyfikacji obiektow w cyfrowych systemach wizyjnych (in Polish)* (WPW, Warszawa, 1990)
17. A. Meijster, M.H. Wilkinson, A comparison of algorithms for connected set openings and closings. IEEE Trans. PAMI **24**(4), 484–492 (2002)
18. E. Salahat, Real-time extraction of maximally stable extremal regions on SoC for visual surveillance, MSc thesis, Khalifa University, ECE Department (2014)
19. E. Salahat, H. Saleh, A. Sluzek, M. Al-Qutayri, B. Mohammed, M. Ismail, Architecture and method for real-time parallel detection and extraction of maximally stable extremal regions (MSERs). United States Patent no. 9,311,555 (2016)
20. S. Alyammahi, E. Salahat, H. Saleh, A. Sluzek, A hardware accelerator for real-time extraction of the linear-time MSER algorithm, in *41st IEEE Industrial Electronics Society Conference IECON'15* (2015), pp. 65–69
21. A. Śluzek, Improving performances of MSER features in matching and retrieval tasks, in *14th European Conference on Computer Vision ECCV'16, Lecture Notes in Computer Science*, vol. 9915 (2016)
22. T. Lindeberg, Scale-space theory: a basic tool for analyzing structures at different scales. J. Appl. Stat. **21**(2), 224–270 (1994)
23. A. Sluzek, H. Saleh, Algorithmic foundations for hardware implementation of scale-insensitive MSER features, in *59th IEEE Interantional Midwest Symposium on Circuits and Systems (MWSCAS'16)* (2016)
24. A. Smeulders, D. Chu, R. Cucchiara, S. Calderara, A. Dehghan, M. Shah, Visual tracking: an experimental survey. IEEE Trans. PAMI **36**(7), 1442–1468 (2014)
25. A. Śluzek, A. Alali, A. Alzaabi, A. Aljasmi, Visual target tracking using a low-cost methodology based on visual words, in *2015 International Conference on Computer Vision and Graphics ICCVG, Lecture Notes in Computer Science*, vol. 9971 (2016)
26. M.A. Fischler, R.C. Bolles, Random sample consensus: a paradigm for model fitting with applications to image analysis and automated cartography, in *4th European Conference on Computer Vision ECCV'06* (1996), pp. 683–695

Part IV
Power Management

"In this country everyone must pay for everything he gets. If you wish me to use my magic power to send you home again you must do something for me first. Help me and I will help you." L. Frank Baum, The Wonderful Wizard of Oz

Chapter 15
A Low-Power, High-Resolution ZCS Control for Inductor-Based Converters

Mohammad Alhawari, Baker Mohammad, Hani Saleh and Mohammed Ismail

15.1 Introduction

Advancements in CMOS technology and circuit design have led into the era of *micro*/*nano* watt-based systems [1]. These systems support IoT applications such as health care where energy harvesting sources could assist or replace batteries. A thermoelectric generator (TEG) which converts a temperature gradient into an electrical energy has been an excellent candidate as an energy source [2]. A TEG is based on the generation of an electron–hole pair in a semiconductor due to a temperature gradient as a result of the Seebeck effect as shown in Eq. 15.1 [2].

$$V_{TEG} = S\Delta T \quad (15.1)$$

where ΔT is the difference in temperature across the TEG, and S is the Seebeck coefficient of the material used. Human body heat offers a continuous source of energy since the human body temperature is regulated. A typical TEG device generates 20–50 mV when placed on a human skin at room temperature [3]. At this low voltage range, an interface circuit should be designed in order to boost the voltage efficiently. Figure 15.1a shows a typical architecture of an energy harvesting system which consists of a maximum power extraction stage that ensures impedance matching with the harvester. Further, a regulation stage is used to regulate the

M. Alhawari · B. Mohammad (✉) · H. Saleh · M. Ismail
Khalifa University, Abu Dhabi, UAE
e-mail: baker.mohammad@kustar.ac.ae

M. Alhawari
e-mail: mhawary85@gmail.com

H. Saleh
e-mail: hani.saleh@kustar.ac.ae

M. Ismail
e-mail: ismail@kustar.ac.ae

Ibrahim (Abe) M. Elfadel and M. Ismail (eds.), *The IoT Physical Layer*,
https://doi.org/10.1007/978-3-319-93100-5_15

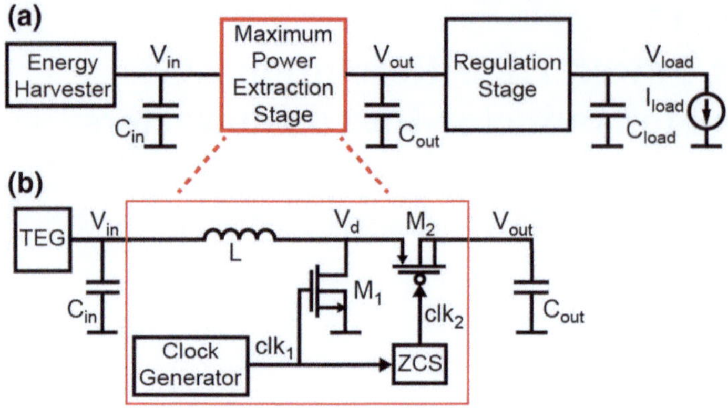

Fig. 15.1 **a** A typical architecture of an energy harvesting system **b** A power conversion stage, utilizing an inductor-based DC-DC converter

output voltage for the load. Figure 15.1b illustrates an inductor-based DC-DC converter which is typically utilized to achieve high efficiency at a high conversion ratio [4–8]. It consists of two power switches, an inductor, a clock generator and a zero current switching (ZCS) circuit which ensures a proper operation of the converter and achieves high efficiency.

The ZCS control is an essential circuit for the synchronous inductor-based DC-DC converter. The ZCS controls the PMOS switch on the high side of the converter by turning it off when the inductor current reaches zero. If the PMOS switch is turned off before or after the zero current point, then the efficiency and the output voltage will be affected which results to performance degradation. Further, measuring the inductor current directly is not an easy task especially in low-power application.

In this chapter, an improved control of the ZCS circuit is presented. Using a simple FSM, the proposed ZCS control uses a coarse and fine delay steps to cover a wide delay range and a high resolution to locate the zero point of the inductor current using few control bits. Measured results in 65 nm CMOS confirm 81% peak efficiency of the overall system.

15.2 Background

An inductor-based DC-DC converter is widely used in energy harvesting applications due to its high efficiency [7, 8]. The efficiency of the converter depends on the inductor quality factor, the technology used, losses in the converter and the control circuit. In addition, a synchronous inductor-based DC-DC converter is used to reduce the losses associated with the diode in the asynchronous converter [4]. However, the synchronization (using the ZCS control) between the switches in the synchronous converter is essential to maintain a high efficiency. The ZCS is used to turn off the PMOS when the inductor current reaches zero. Different techniques can be implemented to detect the zero current point such as sensing the voltage polarity across the PMOS switch using a comparator-controlled MOSFET (active diode) [7, 9]. Further, in [10], the zero current is detected by forming an inductor current loop using a switch connected in parallel with the inductor. This switch is activated at the end of the discharging phase, so that the current polarity is sampled and the zero current point is evaluated. Another technique is to monitor the V_d node between the inductor and the PMOS [11]. In [11], if the PMOS is turned off before or after the zero current point, then V_d becomes positive or negative, respectively. As such, the current polarity can be evaluated by adjusting the PMOS turn-on time to locate the zero current point.

There are some challenges when using the V_d method reported in [11]. When the converter is designed in an open loop (not regulated converter), then the zero current point will vary with the load current. As such, the ZCS circuit should cover a wide delay range such that the zero current point can be always detected. To cover a wide delay range, many delay elements should be used which translates to a large area and complex design. Further, the resolution of the delay plays an important role in maintaining the efficiency. As such, the main issue is whether the ZCS circuit should use a coarse delay to cover wide delay range, which compromises the efficiency, or to use fine delay elements to maintain the efficiency while increasing the chip area as many delay elements will be used.

To explain more, Fig. 15.2 shows a SPICE simulation of an inductor-based DC-DC converter efficiency with the ZCS circuit at load of 10 kΩ. The x-axis represents the time difference between the zero current point and the turn-off time of the PMOS switch. At this particular case, the pulse width clk_2 of the PMOS switch at which the zero current is achieved is equal to 520 ns. The negative and the positive time difference T_{diff} represent early and late conditions, respectively. As depicted in the figure, the efficiency drops if the time difference gets larger for both cases. In the late case, a negative inductor current flows from the output capacitor which degrades the efficiency. In contrast, in the early case, the inductor current passes through the parasitic diode of the PMOS which causes a loss in the efficiency due to the internal resistance and the voltage drop of the diode. Note that in the early case, the converter becomes asynchronous at T_{diff} less than −520 ns.

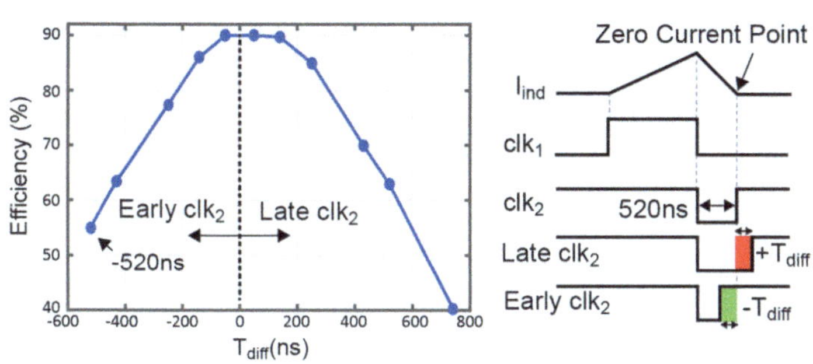

Fig. 15.2 SPICE simulation of the DC-DC converter efficiency versus the time difference between clk_2 and the zero current point

15.3 TEG Energy Harvesting System Architecture

15.3.1 Inductor-Based DC-DC Converter Design

One approach to design the synchronous inductor-based DC-DC converter is by first choosing the inductance value. A higher inductance gives lower losses for the same source impedance [1]. As such, there is a trade-off between the form factor of the inductor and the efficiency. In addition, the inductance value affects the impedance matching with the harvester. Hence, the value of the inductor in our system is chosen to be 47 µH (Part Number: SRR6603, 6.60 × 4.40 × 3 mm) which has a high-quality factor and satisfies the small form factor requirement of our system.

The voltage gain of the converter is an important factor in TEG harvesting since the input voltage range is small (20–50 mV). Different losses in the inductor-based DC-DC converter (inductor quality factor, conduction and switching losses, etc.) limit the voltage gain where the highest gain is at light load. For the synchronous converter, the voltage gain is shown in Eq. (15.2).

$$\frac{V_{out}}{V_{in}} = \frac{T_{nmos,on}}{T_{pmos,on}} + 1 \tag{15.2}$$

where $T_{nmos,on}$ and $T_{pmos,on}$ are the on-time of the NMOS and PMOS, respectively.

Typically, a low duty cycle (D) of clk_1 for the bottom switch (NMOS) is preferred for low-power applications. A duty cycle can be made fixed in our system since the inductor-based DC-DC converter does not regulate the output voltage. A duty cycle of 0.5 is chosen for the following reasons. First, the implementation of a 0.5 duty cycle clock generator is fairly easy where a ring oscillator can be used. In contrast, lower duty cycle circuit requires extra circuits which contribute to the cost of the overall power consumption [7]. Second, for a fixed inductance and TEG series resistance (R_{TEG}), decreasing the charging phase (lower duty cycle) of the inductor, $T_{nmos,on}$, will require lowering the discharging phase $T_{pmos,on}$ assuming a fixed ratio in Eq. (15.2) for a fixed voltage gain. As such, the ZCS circuit needs to be faster for lower $T_{pmos,on}$ so that it turns off the PMOS faster which makes the design of the ZCS circuit elusive. After choosing the duty cycle of clk_1, the frequency of clk_1 can be calculated using Eq. (15.3) such that the maximum power transfer is achieved.

$$f_{clk_1} = \frac{R_{TEG} x D^2}{2L} \tag{15.3}$$

The clock frequency equals ∼18 kHz at a 0.5 duty cycle (D) of clk_1 for the bottom switch, TEG series resistance, R_{TEG}, of 7 Ω and L of 47 µH. Note that since the frequency and the duty cycle of clk_1 are fixed, for a non-regulated output voltage of the converter, the inductor current remains zero for a finite interval of time where no power is drawn from the input source. As such, the average value of the inductor current is always smaller than the critical value of continuous conduction mode - discontinuous conduction mode current (CCM-DCM) ($I_{ind,critical}=\Delta I_{ind}/2$), which places the converter always in the DCM mode.

15.3.2 Proposed ZCS Control Circuit

Figure 15.3a shows the proposed system that includes the TEG model, an inductor-based DC-DC converter and the proposed ZCS control. In addition, Fig. 15.3b shows the key waveforms of the main control signals. The behavior of the circuit is as follows. Initially, the FSM activates the coarse delay mode to change the width of the clk_2 pulse so that the zero current point is located. First, the clk_1 activates the NMOS switch to charge the inductor. Then, at the falling edge of clk_1, clk_2 becomes low to turn on the PMOS. The pulse width of clk_2 depends on the previous value of the counter. Once clk_2 is high, then V_d node will be evaluated using a D flip-flop to indicate early or late condition. As such, the $Detect$ signal will be changed and the counter value will be updated accordingly. Once the zero current is located, the FSM stores the coarse delay value in a register and changes the mode to the fine delay mode in order to optimize the zero current location. The same counter is used in the fine delay mode and its output value is updated based on the location of the zero current point. Finally, the signal $Flip$ is changed from a logic high to low to indicate the fine delay mode. Even if the zero current point is moved due to a load change, the FSM has the flexibility to track the new zero current point. Note that the delay to generate $clk_{2,delayed}$ is defined as the time required for V_d to settle (early or late) which should be set right for proper evaluation. Extra bits from a serial register are added to the design to account for different variations. Finally, a ring oscillator is utilized to produce the clock signal clk_1.

The 3-bit counter is shared between the coarse and fine modes so that 56 fine delay steps (8 fine steps in each coarse step which is equal to 7) are achieved with only 3-bit design. As such, our system has an effective number of bits equal to 5.8 bits ($ln56 = 5.8$). Note that two external power supplies are used. The digital part operates at 0.6 V to reduce the power consumption. Further, the power switches are driven by 1 V (using two level shifters and buffers) such that their losses are minimized. The power consumption from both 0.6 and 1 V is included in the overall efficiency measurements.

Figure 15.4 shows a SPICE simulation of the system dynamics when ZCS is operated. As shown in Fig. 15.4, the coarse delay searches for the zero current point and toggles around it. Then, the FSM switches to the fine mode for the final settling. The coarse and fine steps are 250 and 50 ns, respectively. Note that the settling time of the proposed ZCS control is faster than a fixed delay system for the same effective number of bits. To explain, a system that has fixed delays with the same resolution of our proposed system (56 delay steps) needs 5.8 bits ($2^{5.8}$). Even if the zero current location is in the middle range, the fixed delay system needs 28 cycles (56/2) to reach the zero current point. In contrast, the proposed system needs a $maximum$

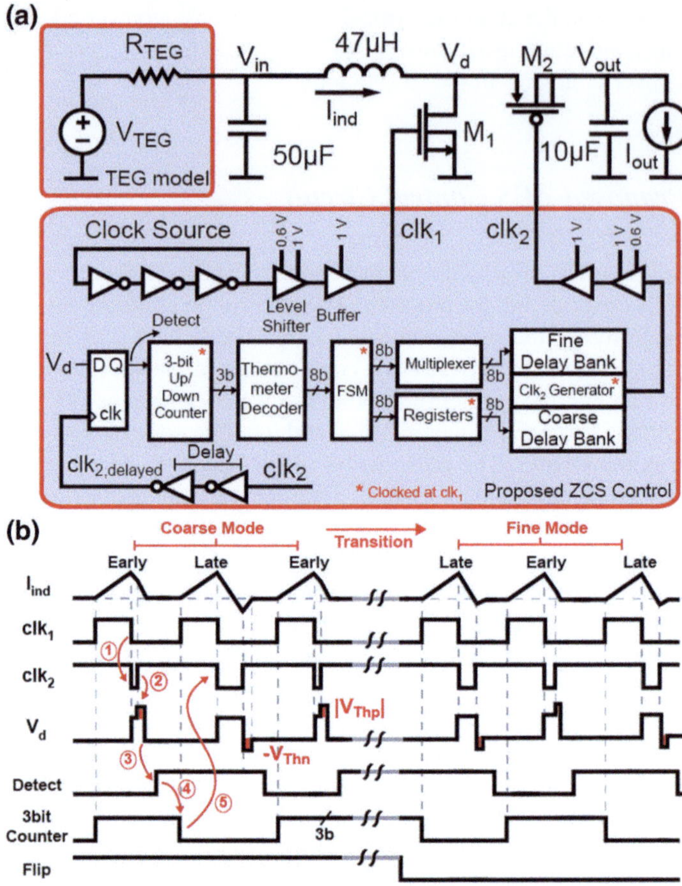

Fig. 15.3 **a** Block diagram of the proposed system **b** Key waveforms of the proposed ZCS

of 7 coarse cycles plus 8 fine cycles and extra cycles for the transition to cover the whole delay range. In one particular condition, when the zero current location does not shift dramatically (within a one fine step), a fixed delay system would be as fast as the proposed system. Finally, note that in the proposed system, there are 8 extra cycles during the start-up (as shown in Fig. 15.4) which are reduced to 1 cycle during normal operation.

In real systems, the noise generated from the switching activity of the converter, especially on the V_d node, might affect the operation of the ZCS control and might cause a faulty evaluation. However, due to the filtration on the V_d node (large L and parasitic capacitance on V_d), the noise on the V_d node is reduced. In addition, the resolution of the coarse and fine steps is selected such that the early and late conditions are correctly evaluated.

Figure 15.5a shows the circuit that generates the clk_2 delay. As depicted in the figure, when clk_1 is high, M_3 pulls up the node mid and charges both C_1 and the

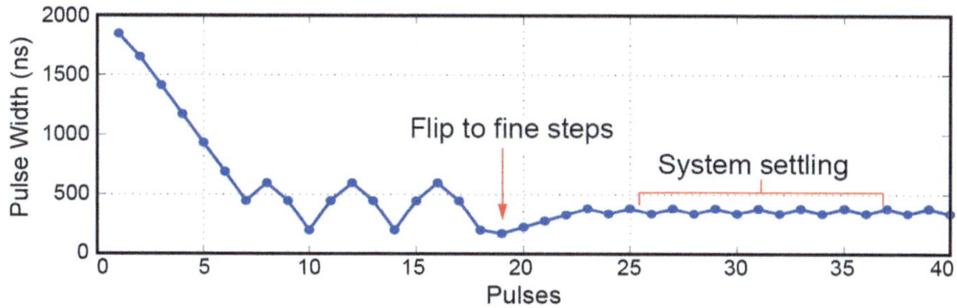

Fig. 15.4 SPICE simulation of system dynamics for coarse and fine delay modes

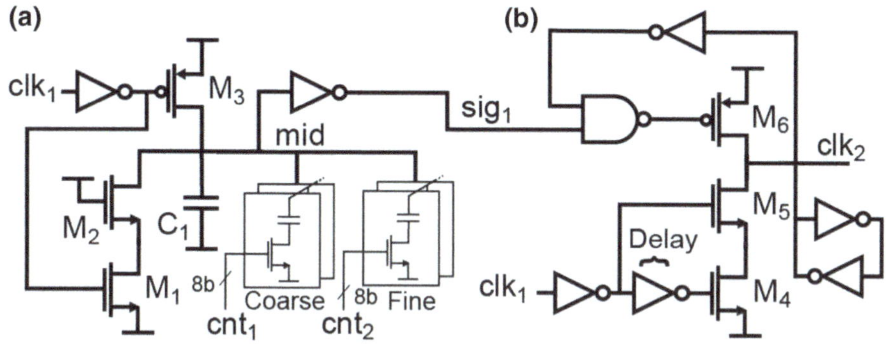

Fig. 15.5 *Clk*$_2$ generation circuit that includes (**a**) Coarse and fine delay circuit (**b**) Push-pull circuit

coarse capacitors since cnt_1 is initially set to logic high; hence, sig_1 is discharged to ground. As such, a maximum delay is generated. Once the evaluation of V_d is detected to be late, cnt_1 bits start to turn off coarse capacitors one by one till the early state is detected. Then, the FSM shifts the state to the fine delay for the final settling. Figure 15.5b shows the pull-push circuit that generates clk_2 signal. As illustrated in the figure, sig_1 from Fig. 15.5a is an input to the circuit in Fig. 15.5b. When clk_1 is high, sig_1 is low and clk_2 is high. In contrast, when clk_1 goes to ground, clk_2 is discharged through M_4 and M_5 and its state is held by the output latch. The state of clk_2 stays the same till sig_1 changes its state depending on coarse and fine delays. Once sig_1 goes high, then clk_2 is pulled high through M_6. Note that M_4 and M_5 can be both ON at the same time due to the delay introduced at the M_4 gate.

15.4 Measurement Results

Figure 15.6 shows the measurement setup of the overall system that was discussed in Fig. 15.3a. A DC power analyzer is used to emulate the voltage generated from a TEG which has a series resistance of 7 Ω (to emulate G2-30-0313 from Tellurex). Further, the DC power analyzer is used to measure the efficiency of the DC-DC converter. In addition, an oscilloscope is utilized to measure the time domain signals. Furthermore, a real TEG (G2-30-0313 from Tellurex) is placed on a human hand for the real case experiment instead of the emulated one.

The bond wires and printed circuit board (PCB) contribute to the total loss of the circuit. For two parallel pads, 10 mm length each and 1 mil diameter bond wire, the resistance is 0.25 Ω [12] which is typically larger than the PCB resistance. As such, with an average input current less than 4 mA, the voltage drop is less than 1 mV.

Figure 15.7a shows the measured efficiency at different loads and TEG voltages. The efficiency is defined as the output power P_{out} divided by the input power from the TEG, P_{in} at the input voltage V_{in} which will be equal to half the V_{TEG} since clk_1 frequency is chosen to achieve the maximum power transfer based on Eq. (15.3). The peak efficiency reaches 81% at V_{TEG} of 50 mV. In addition, Fig. 15.7b shows the output voltage at different V_{TEG} voltages and load currents.

Figure 15.7c shows the relationship between the input and the output voltage. As indicated in the figure, the output voltage is larger than 1 V over the range of 20–50 mV. Note that the output voltage is saturated when the input voltage is larger than 50 mV where no regulation is implemented in the converter. The reason is that the ZCS control circuit is saturated and not able to provide more delay steps for input voltages larger than 50 mV as will be discussed in Fig. 15.9.

Inspection of Fig. 15.7 leads to a key observation regarding the DC-DC converter performance. First, the high efficiency is achieved mainly because of the usage of the high-quality factor inductor with an advanced 65 nm CMOS and low converter losses. However, the proposed ZCS control circuit helps in maintaining the high efficiency as was discussed in Fig. 15.2. Note that this system is part of a biomedical SoC which requires an output voltage from the DC-DC converter to be between 0.9 and 1.3 V [13]. However, Fig. 15.7 shows the performance of the DC-DC converter over a wide range of output currents and voltages for complete characterization.

The measured time domain results of the system in Fig. 15.3a at no load and V_{TEG} of 50 mV are presented in Fig. 15.8. As indicated in Fig. 15.8a, the inductor current

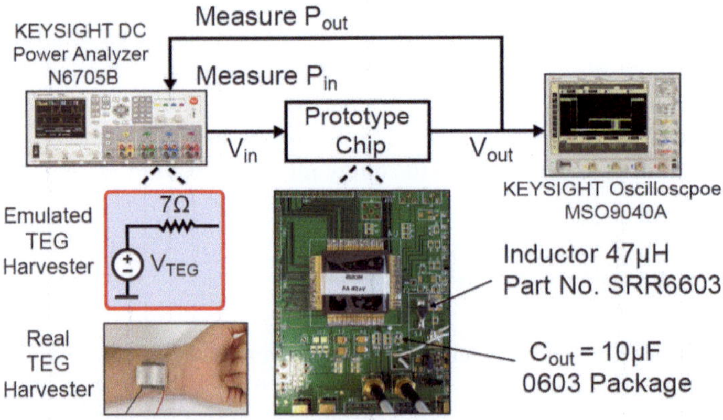

Fig. 15.6 Measurement Setup of the Overall System

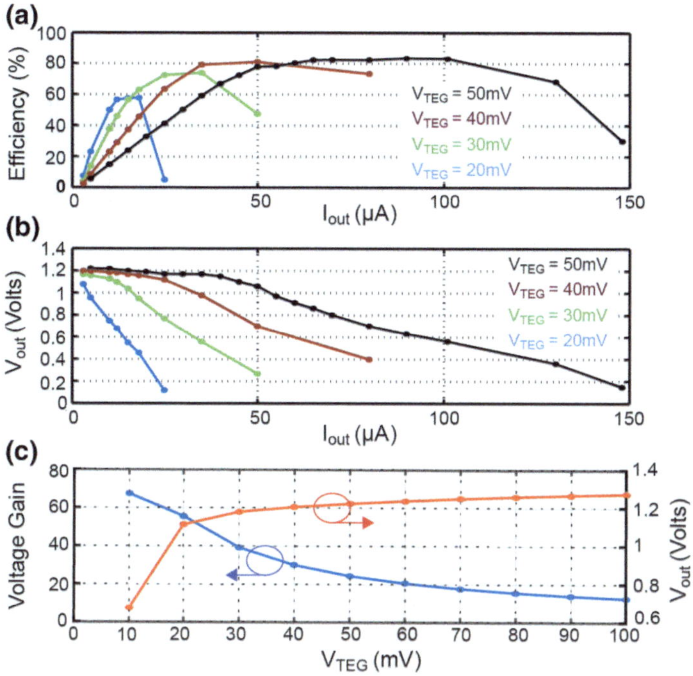

Fig. 15.7 Measured results of (**a**) Efficiency at different output load currents and TEG voltages (**b**) Output voltage at different output load currents and TEG voltages (**c**) Voltage gain at no-load

ramps up during the on-time of clk_1 and then ramps down during the on-time of clk_2. The inductor current is measured by placing a 2.2 Ω in series with the inductor and then the voltage across the resistor is measured. Since the voltage across the resistor is low (∼18 mV), the noise from the oscilloscope affects the measurement. As such, a digital filter inside the oscilloscope is used to filter the noise on the inductor current waveform.

Figure 15.8b, c shows the early and late conditions of the inductor current, respectively. In addition, the pulse width of clk_2 is changed from 270–325 ns (55 ns difference) which represents the resolution of the fine delay. Finally, the 2.2 Ω resistor is inserted only in this setup to measure the current and it is removed in other measurements since it affects the output voltage level. Finally, a negative inductor current is noticed in the early condition which should not be confused with the late condition. This current is not from the output capacitor, but from a transient effect that happens due to the formation of a resonance loop between the inductor and the parasitic capacitance of the NMOS switch.

Figure 15.9a shows the time domain results when the TEG voltage changes from 20–100 mV. Further, Fig. 15.9b shows the value of the 3-bit counter, which is shared between the coarse and fine delay steps. Initially, the TEG voltage is 20 mV and the fine steps toggle between 6 and 7. Once the TEG voltage becomes 50 mV, then the FSM goes into the following states as shown in Fig. 15.9c. First, the fine steps are changed to search for new zero current point. Once the fine steps are saturated,

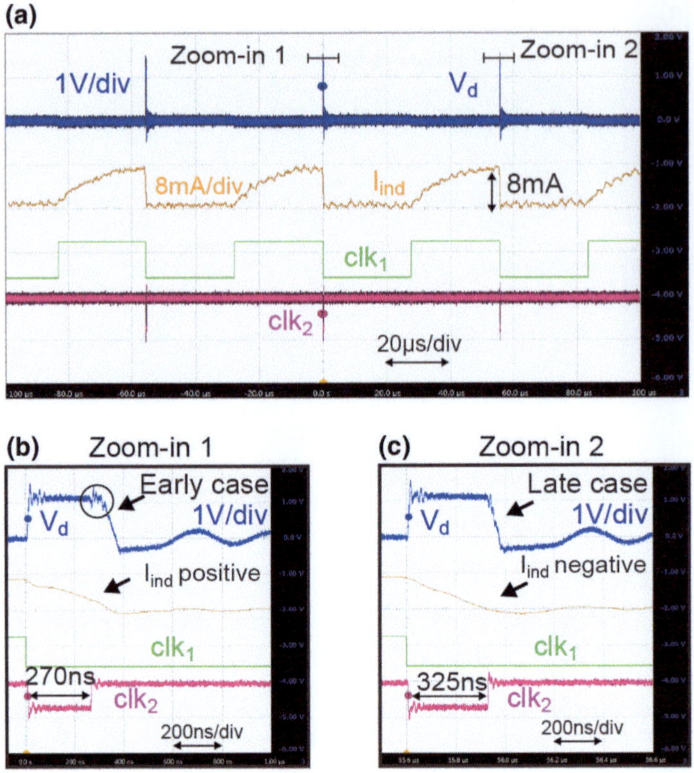

Fig. 15.8 **a** Measured time domain result of the system in Fig. 15.3a **b** Early condition **c** Late condition

then FSM changes to coarse steps. The coarse steps are changed till the zero point is found between the value 5 and 6. Then, the mode is changed to fine steps for the final settling between 4 and 5. One should note that it needs 1 clock cycle to change the mode from coarse to fine. Further, the counter value for input voltage larger than 50 mV is zero. This means that the ZCS circuit is not able to provide more delay steps outside the input voltage range (20–50 mV) and the zero current point is not detected efficiently. This is why we see saturation in the output voltage for TEG voltages larger than 50 mV as shown in Fig. 15.9a, c. However, our range of interest is from 20–50 mV.

Figure 15.10a shows the measured TEG voltage and output voltage for the experiment of real TEG which is placed on a human hand at room temperature as was shown in Fig. 15.6. As depicted in the figure, the TEG input voltage settles at 30 mV and produces a no-load output voltage of 1.1 V. The die photograph is shown in Fig. 15.10b with a total active area of less than 0.04 mm^2. The layout in Fig. 15.10c shows different parts of the chip. To compare the system performance with recent work, Table 15.1 shows that the proposed system achieves a high efficiency at low output power at a range of 20–50 mV and large number of steps with few control bits. Note that the output voltage of the proposed system is not regulated.

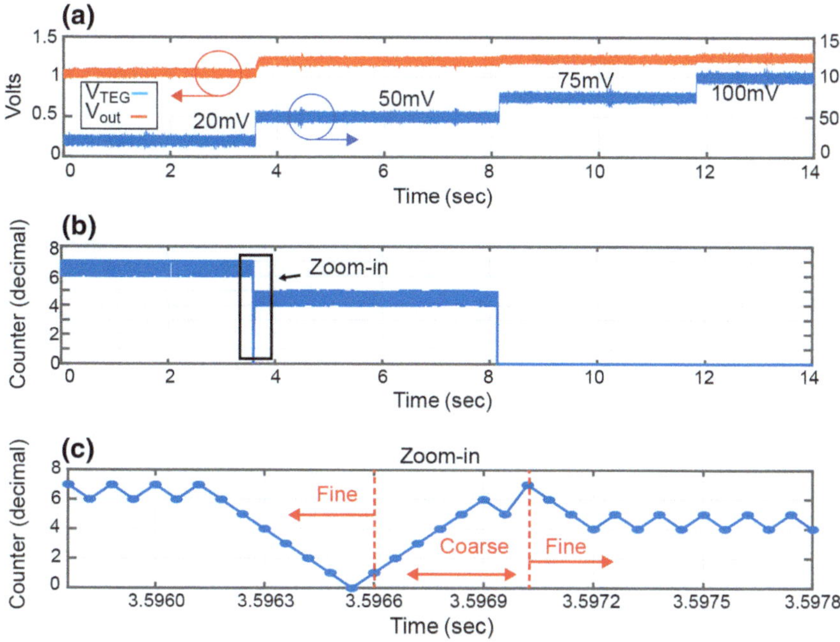

Fig. 15.9 **a** Time domain output voltage at different TEG voltages **b** Counter value that represents the coarse and fine dynamics **c** Zoom-in version of graph b

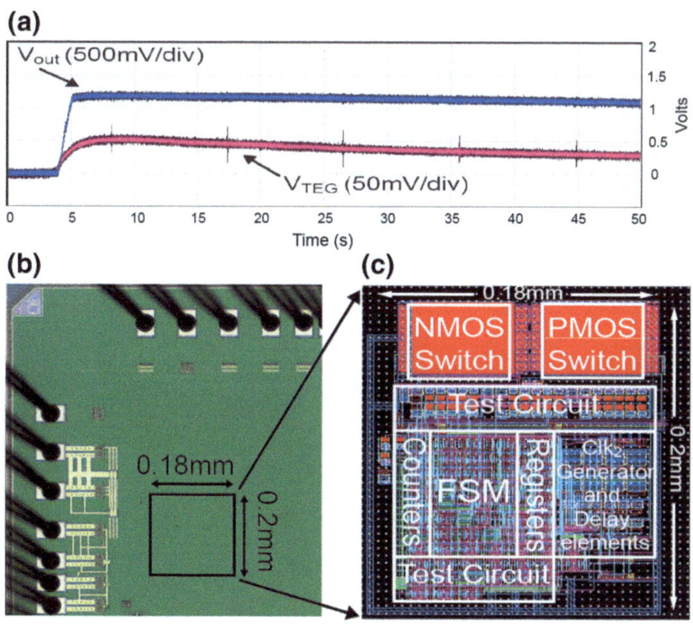

Fig. 15.10 **a** Measured input and output voltage with the real TEG harvester **b** Die photograph of the prototype chip **c** Layout showing different circuit parts

Table 15.1 Comparison with other published work

Parameter	JSSC'11 [3]	JSSC'13 [4]	JSSC'15 [5]	ISSCC'12 [6]	TPEL'15 [7]	TCAS-I'15 [8]	JSSC'10 [11]	This Work
Harvesting	TEG	TEG	TEG/Solar	TEG	Solar	TEG	TEG	TEG
Technology	0.35-μm CMOS	65-nm CMOS	130-nm CMOS	130-nm CMOS	180-nm CMOS	180-nm CMOS	130-nm CMOS	65-nm CMOS
Minimum Input Voltage	25 mV	50 mV	10 mV	40 mV	270 mV	100 mV	20 mV	20 mV
Maximum Output Power	280 μW	1.1 mW	22 mW	2.7 mW	100 μW	1 mW	100 μW	55 μW
Maximum Output Voltage	1.8 V	1.2 V	1.1 V [5]	2 V	1.8 V	0.5 V	1.0 V	1.2 V [5]
Zero Crossing Technique	V_d Method	V_d Method	Latched Comaprator	Diode (asynchronous)	Latched Comaprator	Latched Comaprator	V_d Method	V_d Method
Control bits/ Delay steps	3-bits/ 8 steps	4-bits/ 16 steps	NA	NA	NA	NA	4-bits/ 16 steps	3-bits/ 56 steps
Efficiency at 50mV Input	58% at 50 μW	63% at 100 μW	68% Not Reported	33% Not Reported	NA	NA	65% at 40 μW	81% at 55 μW
Peak Efficiency	58%	73%	83% [1]	61%	86% [4]	83.4% [2]	75%	81% [3]

1- At 300 mV input
2- At 100 mV input and 500 μW output power
3- External 0.6 V and 1 V are used
4- At 500 μW output power
5- Not regulated

15.5 Conclusions

This chapter presented an improved ZCS control circuit that leverages coarse and fine delays. The FSM uses the coarse delay to achieve wide delay range to search and locate zero current point. Then, the fine delay optimizes the zero current point location. As such, the proposed control reduces the number of delay elements while still having many delay steps using few number of control bits. Measured results in 65 nm CMOS confirm 81% peak efficiency of the overall system.

Acknowledgements The authors thank John Pigott, Mark Schlarmann from Freescale and Muhammad M. Khellah, Lilly Huang from Intel for their support. This work has been supported by the Semiconductor Research Corporation (SRC) under the Abu Dhabi SRC Center of Excellence on Energy-Efficient Electronic Systems (ACE^4S), Contract 2013 HJ2440, with funding from the Mubadala Development Company, Abu Dhabi, UAE.

References

1. S. Bandyopadhyay, P. Mercier, A. Lysaght, K. Stankovic, A. Chandrakasan, A 1.1nw energy harvesting system with 544pw quiescent power for next-generation implants, in *IEEE International Solid-State Circuits Conference Digest of Technical Papers (ISSCC)* (2014), pp. 396–397
2. C. Lu, V. Raghunathan, K. Roy, Efficient design of micro-scale energy harvesting systems. IEEE J. Emerg. Sel. Top. Circuits Syst. **1**, 254–266 (2011). Sept.
3. Y. Ramadass, A. Chandrakasan, A battery-less thermoelectric energy harvesting interface circuit with 35 mv startup voltage. IEEE J. Solid-State Circuits **46**(1), 333–341 (2011)

4. P.-S. Weng, H.-Y. Tang, P.-C. Ku, L.-H. Lu, 50 mv-input batteryless boost converter for thermal energy harvesting. IEEE J. Solid-State Circuits **48**(4), 1031–1041 (2013)
5. A. Shrivastava et al., A 10 mV-input boost converter with inductor peak current control and zero detection for thermoelectric and solar energy harvesting with 220 mV cold-start and -14.5 dbm, 915 MHz RF kick-start. IEEE J. Solid-State Circuits **50**, 1820–1832 (2015). Aug
6. J.-P. Im, S.-W. Wang, K.-H. Lee, Y.-J. Woo, Y.-S. Yuk, T.-H. Kong, S.-W. Hong, S.-T. Ryu, G.-H. Cho, A 40mv transformer-reuse self-startup boost converter with mppt control for thermoelectric energy harvesting, in *2012 IEEE International Solid-State Circuits Conference Digest of Technical Papers (ISSCC)* (2012), pp. 104–106
7. R.D. Prabha, G. Rincon-Mora, 0.18- um light-harvesting battery-assisted charger-supply cmos system. IEEE Trans. Power Electron. **31**, 2950–2958 (2016)
8. P.-H. Chen, P.-Y. Fan, An 83.4% peak efficiency single-inductor multiple-output based adaptive gate biasing dc-dc converter for thermoelectric energy harvesting. IEEE Trans. Circuits Syst. I: Regul. Pap. **62**, 405–412 (2015). Feb
9. S. Kim, G. Rincon-Mora, 23.4 dual-source single-inductor 0.18um cmos charger-supply with nested hysteretic and adaptive on-time pwm control, in *IEEE International Solid-State Circuits Conference Digest of Technical Papers (ISSCC)* (2014), pp. 400–401
10. P.-H. Chen, C.-S. Wu, K.-C. Lin, A 50nw-to-10mw output power tri-mode digital buck converter with self-tracking zero current detection for photovoltaic energy harvesting, in *2015 IEEE International Solid- State Circuits Conference - (ISSCC)* (2015), pp. 1–3
11. E. Carlson, K. Strunz, B. Otis, A 20 mv input boost converter with efficient digital control for thermoelectric energy harvesting. IEEE J. Solid-State Circuits **45**(4), 741–750 (2010)
12. https://www.mosis.com/pages/products/assembly/index
13. N. Bayasi et al., Low-power ECG-based processor for predicting ventricular arrhythmia. IEEE Trans. Very Large Scale Integr. (VLSI) Syst. **PP**(99), 1–13 (2015)

Chapter 16
Reconfigurable, Switched-Capacitor Power Converter for IoT

Dima Kilani, Mohammad Alhawari, Baker Mohammad, Hani Saleh and Mohammed Ismail

16.1 Introduction

Advancements in CMOS technology have enabled the development of low-power circuits and devices in many IoT applications including health care, smart buildings and environment monitoring. Energy harvesting is one of the pillars that enable IoTs to achieve energy autonomy [1, 2]. Since the harvested energy depends on the availability of various sources, a power management unit (PMU) is required to monitor the harvested energy level and to manage the power distribution for different system-on-chip (SoC) domains such as the memory and the digital processor. In addition, the PMU is essential to obtain the optimal energy and performance design [3, 4]. Furthermore, since IoT device remains most of the time in sleep mode, reducing the static power or the leakage is very important as it is dominant factor in IoT's power consumption. Therefore, the need of an efficient PMU for IoTs is essential.

D. Kilani · M. Alhawari · B. Mohammad (✉) · H. Saleh · M. Ismail
Khalifa University, Abu Dhabi, UAE
e-mail: baker.mohammad@kustar.ac.ae

D. Kilani
e-mail: dima.kilani@kustar.ac.ae

M. Alhawari
e-mail: mohammad.alhawari@kustar.ac.ae

H. Saleh
e-mail: hani.saleh@kustar.ac.ae

M. Ismail
e-mail: ismail@kustar.ac.ae

Ibrahim (Abe) M. Elfadel and M. Ismail (eds.), *The IoT Physical Layer*,
https://doi.org/10.1007/978-3-319-93100-5_16

16.2 Dynamic Power Reduction Techniques

There are different techniques to reduce dynamic power such as dynamic voltage scaling (DVS) [5], adaptive voltage scaling and frequency scaling. DVS reduces the operating voltage supply to achieve the optimum energy point. Hence, DVS is a key requirement to decrease the power consumption quadratically such that $P \propto CV^2f$ [6]. DVS can be achieved using DC–DC buck converter as discussed in Sect. 16.3. Frequency scaling can be also used in conjunction with DVS to further reduce the power consumption. On the other hand, adaptive voltage scaling reduces the operating voltage based on the process variation or process strength. For example, in a fast fabrication process, the voltage supply can be reduced while maintaining the same performance at the same voltage supply as in the nominal fabrication process.

16.3 DC–DC Buck Converters

In order to achieve DVS, a DC–DC buck converter is utilized which becomes a common building block of the PMU in all today's SoCs. One SoC example that uses a DC–DC converter is the wireless 2.4 GHz IEEE ZigBee as reported in TI [7], STMicroelectronics [8] and Freescale [9]. Some blocks of the ZigBee SoC operate at different voltage levels where the input voltage must be down-converted into multiple voltage levels to reduce the power consumption and increase the efficiency. However, for integrated energy harvesting-based system, the need for a power converter is not only for power efficiency, but also for voltage regulation as the harvested voltage level is variable.

Multiple DC–DC buck converters are common on many SoCs to support independent voltage domains for both analog and digital blocks that run simultaneously at different voltage levels based on performance and functional requirements of minimum voltage [10]. There are three main topologies for DC–DC buck converter, namely low drop-out (LDO) regulator, switched-inductor (SI) and switched-capacitor (SC) converters. LDO regulator provides better integration and fast response to the output load, but the efficiency becomes low at low voltage conversion ratio (V_{out}/V_{in}) [11]. On the other hand, low-frequency SI converter provides high power efficiency [12], but it requires an off-chip inductor [13–15]. However, there are several high-frequency SI converters that utilize very small on-chip and on-package inductors [16–18]. SC converter combines the two advantages of the other two converters where on one hand, it is fully integrated as LDO regulator and its efficiency is relatively high but less than SI converter. In addition, the SC converter becomes a viable solution for ultra-low power applications because of its low power requirement (μW power range) that could achieve high efficiency at a low voltage conversion ratio (V_{out}/V_{in}) [19]. However, the SC converter has inherent losses even if all components are ideal [20]. Detailed comparison between converters is reported in [21, 22].

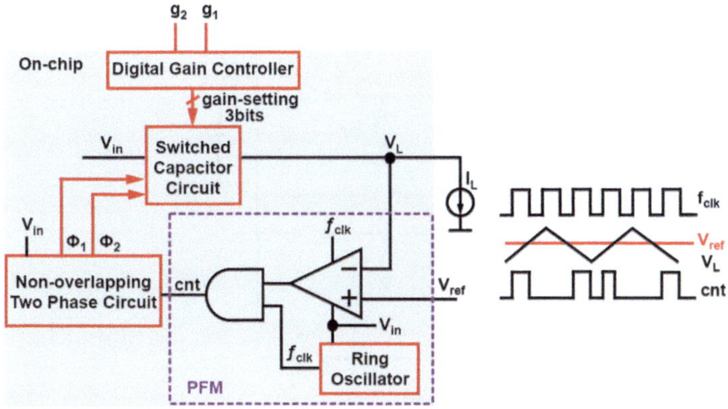

Fig. 16.1 Proposed SC regulator block diagram

16.4 Proposed Switched-Capacitor DC–DC Buck Converter

Figure 16.1 shows the proposed SC regulator block diagram. The design consists of a reconfigurable SC circuit, a digital gain controller, a non-overlapping two-phase circuit and a pulse frequency modulation (PFM) controller. The proposed circuit has been designed to work for an input range from 0.6 V to 1.2 V, which the control circuit powered by the input voltage. The SC converter generates multiple voltage gain levels which are 1, 3/4 and 2/3 based on two digital bits g_1 and g_2 that should be generated from the SoC. However, in this implementation, external bits are used to configure the SC. In addition, a frequency modulation technique is utilized to regulate the output voltage over a wide range of load current. Such technique changes the clock frequency f_{clk} based on the load current. The operation of the PFM is as follows. When the load current increases, V_L becomes lower than V_{ref}. Hence, the PFM feedback loop adjusts the frequency of the SC regulator by forcing the AND gate to pass f_{clk} when the output of the comparator is logic one at the rising edge of the clock. Once V_L exceeds V_{ref}, the output of the comparator becomes logic zero and the AND gate prevents f_{clk} to pass. A non-overlapping two-phase clock ϕ_1 and ϕ_2 is required to create a dead-time in order to prevent any short circuit current during switching.

16.4.1 *Circuit Design*

Figure 16.2a shows the circuit implementation of the SC converter [23]. As indicated in Fig. 16.2a, the SC circuit provides three voltage gain levels controlled digitally by two bits g_1 and g_2. The three combinations of the gain settings are shown in Fig. 16.2b. Furthermore, Fig. 16.2c illustrates the charging and discharging phase configurations

(a)

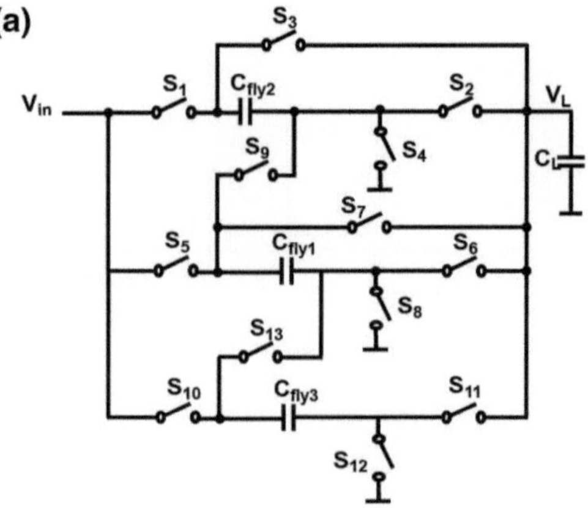

(b)

Gain	g_2g_1	S_1	S_2	S_3	S_4	S_5	S_6	S_7	S_8	S_9	S_{10}	S_{11}	S_{12}	S_{13}
1	00	Φ_1	off	Φ_2	on	Φ_1	off	Φ_2	on	off	off	off	off	off
3/4	01	Φ_1	Φ_1	Φ_2	off	Φ_1	Φ_1	off	off	Φ_2	Φ_1	Φ_1	Φ_2	Φ_2
2/3	10	Φ_1	Φ_1	Φ_2	off	Φ_1	Φ_1	off	Φ_2	Φ_2	off	off	off	off

(c)

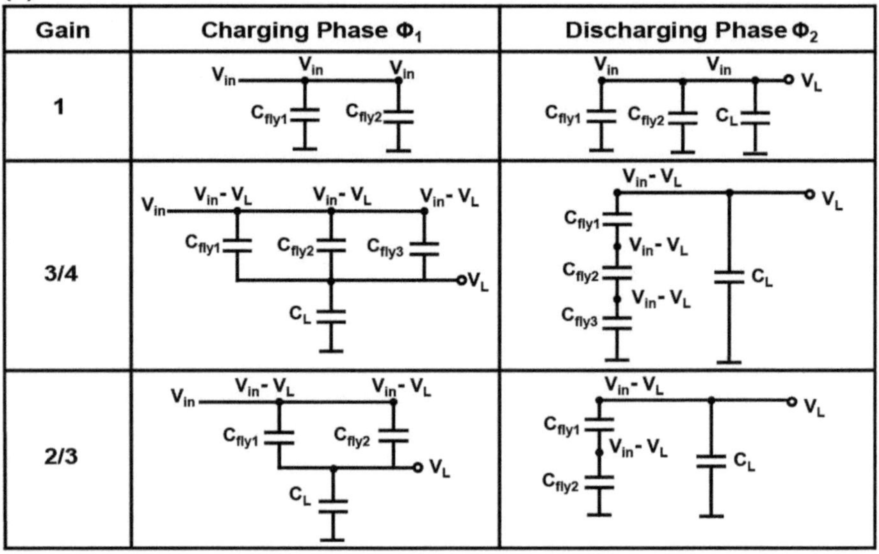

Fig. 16.2 **a** Reconfigurable SC circuit, **b** switch control signal based on gain setting and **c** SC configuration during charging and discharging phases based on g_2g_1

ϕ_1 and ϕ_2 that control the SC converter operation. During ϕ_1, the flying capacitors C_{fly1-3} are charged and during ϕ_2, the charges in C_{fly1-3} are discharged into the load capacitor C_L such that the output voltage V_L is generated. For example, when the gain is set to 1, C_{fly1-2} are connected in parallel and charged to V_{in} during ϕ_1. In contrast, during ϕ_2, V_{in} is disconnected and the stored energy in the flying capacitors is discharged into C_L, which raises V_L to V_{in}. Consider another setting with a gain of 3/4. During ϕ_1, C_{fly1-3} are charged to V_{in}-V_L as they all are connected in parallel between the input and output nodes. The series configuration of C_{fly1-3} discharges the stored energy into C_L during ϕ_2 so that $V_L = 3(V_{in}\text{-}V_L)$ which implies $V_L = 3/4V_{in}$. As shown in Fig. 16.2b, when the gain equals to 2/3, the charging and discharging phase configurations are similar to the voltage gain of 3/4 but with two flying capacitors C_{fly1-2} that are participating in the charge transfer. Therefore, the voltage summation at the output load node is $V_L = 2(V_{in}\text{-}V_L)$ which produces an output voltage of $V_L = 2/3V_{in}$.

Figure 16.3 shows the clocked comparator of the current latched sense amplifier (CLSA). The operation of the CLSA depends on the amount of current passed through M_9 and M_{10} [24]. Nodes *a* and *b* are precharged with the precharge drivers M_1, M_4, M_5 and M_8 when *clk* is low. Once *clk* becomes high, the precharge drivers are turned off and M_9 and M_{10} are turned on. If $V_{ref} > V_L$, then the drain–source current $I_{ds,M_9} > I_{ds,M_{10}}$. I_{ds,M_9} discharges faster into M_{11} decreasing node *a* to 0 V and therefore $cnt_b =$ gnd and $cnt = V_{dd}$. The advantage of this topology is that both V_L and V_{ref} are driving high impedance and no discharge can occur. Voltage latched sense amplifier (VLSA) topology [24] has been evaluated, but due to possible discharge of V_L and V_{ref}, it has not been used. The transistors in CLSA design are implemented with a nominal threshold voltage device, and the sizes are designed to have a valid operation across all

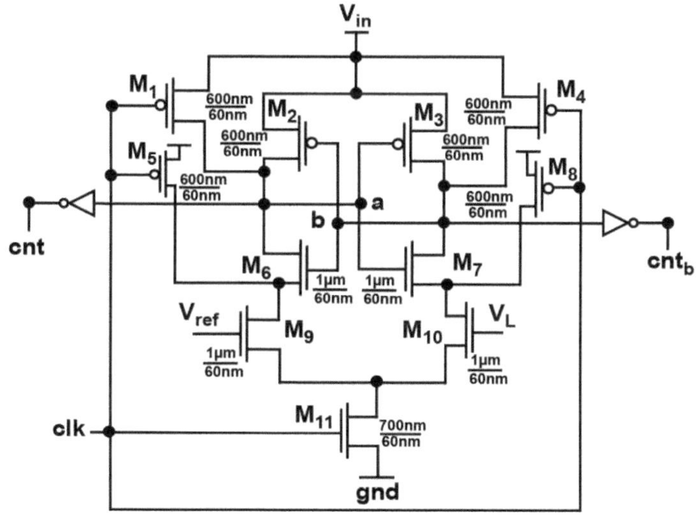

Fig. 16.3 Current latched sense amplifier

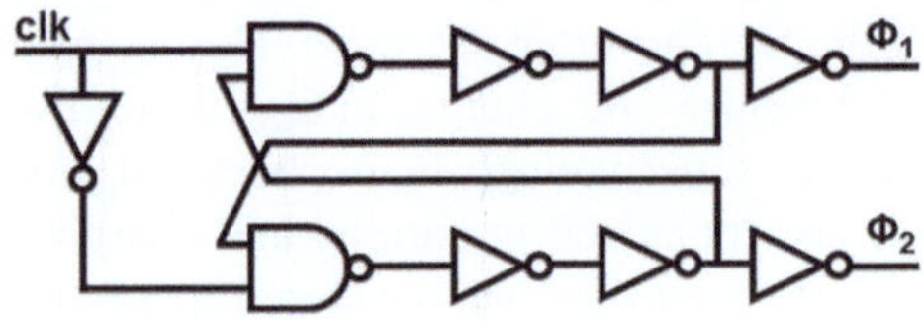

Fig. 16.4 Dead-time circuit

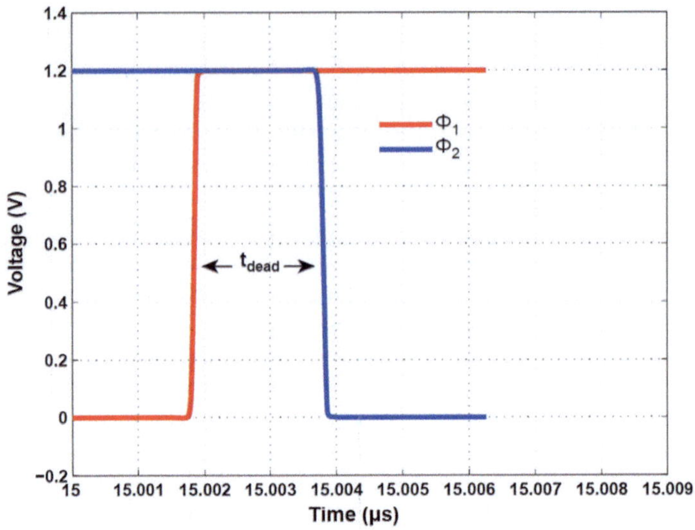

Fig. 16.5 Non-overlapping two-phase clock signals

analysis the corners. During the layout, the transistors are placed close to each other in order to reduce the matching errors while minimizing the interconnection. SPICE across corners has been performed to ensure the operation of the CLSA between 0.6 and 1.2 V. The analysis shows that the CLSA operates down to 0.5 V and, has been verified during the measurement.

A non-overlapping two-phase circuit that is known as a dead-time circuit is shown in Fig. 16.4 [25]. The dead-time circuit creates a time t_{dead} between ϕ_1 and ϕ_2 where all the switches of the SC converter are off for a very short period of time. This is very important to avoid any short circuit current from V_{dd} to ground during the switching. Figure 16.5 shows the output waveform of the dead-time circuit with a dead-time of 1.90 ns.

16.4.2 Loss Analysis of Switched-Capacitor Converter

There are four types of losses associated with the SC converter: conduction losses, switching losses, flying capacitor losses and bottom plate parasitic capacitance losses [26, 27]. Conduction losses P_{cond} are proportional with the drain–source resistance R_{ds} of the transistor and given as:

$$P_{cond} = I_{ds}^2 R_{ds} \tag{16.1}$$

where I_{ds} is the drain–source current (in addition to other parasitic resistance). Note that the conduction loss is dominant at high load current since it is proportional to the quadratic current value as given by Eq. (16.1).

Switching losses are associated with the PFM switching activity of the SC converter. The switching losses are given as (16.2):

$$P_{sw} = C_{ox} V_{in}^2 W f_{sw} \tag{16.2}$$

where C_{ox} is the gate-oxide capacitor per unit width, W is the transistor width and f_{sw} is switching frequency. Both conduction and switching losses increase at heavy load current since P_{cond} is proportional to the load current and P_{sw} is proportional to the switching frequency which is increased by PFM. However, P_{cond} dominates the total losses at heavy load current since it is relatively higher than P_{sw}. On the other hand, both P_{cond} and P_{sw} drop at light load current where the PFM controller lowers the switching activity of the SC converter. P_{sw} is relatively higher than P_{cond} at light load current so that P_{sw} dominates the total losses.

The third loss mechanism is the flying capacitor loss P_{cfly} that is defined as charge loss caused by C_{cfly} during the charge transfer through the charging and discharging phases. Thus, the flying capacitor losses are given as:

$$P_{cfly} = \frac{I_L^2}{C_{fly} f_{sw}} \tag{16.3}$$

where I_L is the load current. Note that P_{cfly} is inversely proportional to the C_{fly} and f_{sw} and proportional to quadratic I_L. This means that P_{cfly} is a load-current-dependent loss. On the other hand, the fourth loss mechanism P_{bp} is caused by the bottom plate capacitor $C_{bp} = \alpha C_{fly}$ where α represents a particular percentage of the total flying capacitance based on the CMOS technology node [28]. C_{bp} is formed due to the PN junction between the N well and the P substrate. P_{bp} is given as:

$$P_{bp} = C_{bp} V_{in}^2 f_{sw} \tag{16.4}$$

16.4.3 Switched-Capacitor Converter Optimization

The power efficiency of the SC converter is not only affected by the design parameters (W, f_{sw}, C_{fly}), but it is also affected by the operating point condition such as V_L and I_L. Depending on both design parameters and operating point conditions, the power losses can be minimized. However, if the operating point condition changes, the SC converter might not maintain a high power efficiency; hence the design parameters have to be adaptive to the output load. In other words, the SC converter that is designed to operate efficiently under heavy load current has different parameters than the SC converter that operates efficiently under light load condition. Thus, the SC converter needs to tune some of the parameters like W, f_{sw} or C_{fly} according to the operating point in order to maintain high power efficiency over a wide range of load currents. PFM and switch width modulation (SWM) controllers are commonly used to enhance the power efficiency [29]. However, an overhead control circuit is added to achieve such adaptive tuning.

In order to design an efficient SC converter, W, C_{fly} and f_{sw} should be optimized so that the power losses are minimized. Figure 16.6 shows the total average power losses of the SC circuit for various switch widths and flying capacitors. The switch width is swept from 50 μm till 1 mm to calculate the total power losses. As depicted in Fig. 16.6, the optimum switch width is achieved at $W = 100\,\mu\text{m}$ where the total losses record the minimum of 100–110 μW depending on C_{fly}. This is because the switching losses are highly affected by the switch size according to Eq. (16.2) so that small transistor width results in low switching losses. However, the total losses increase below $W = 100\,\mu\text{m}$ because conduction losses become dominant. In order to select C_{fly}, Fig. 16.7 studies the effect of the switching frequency upon the total power losses for three different C_{fly} values at a maximum load current. As depicted in the figure, although the minimum power loss at $C_{fly} = 675\,\text{pF}$ is 1.3X less than $C_{fly} = 225\,\text{pF}$, the power density of the latter is 3X higher than the former. Therefore,

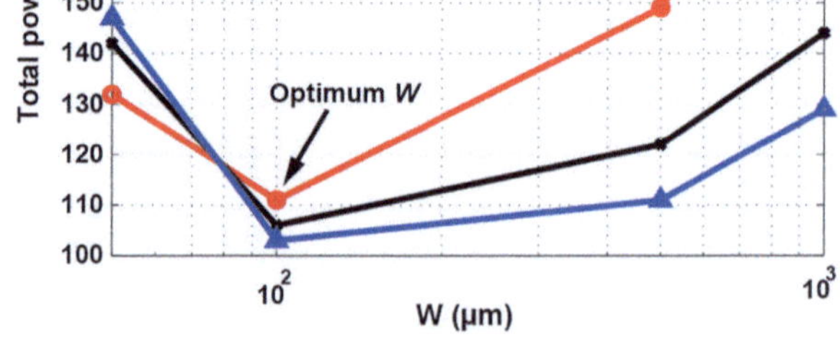

Fig. 16.6 Average power losses of SC circuit against W for different values of C_{fly} at gain=3/4 (worst case) and $I_L = 800\,\mu\text{A}$

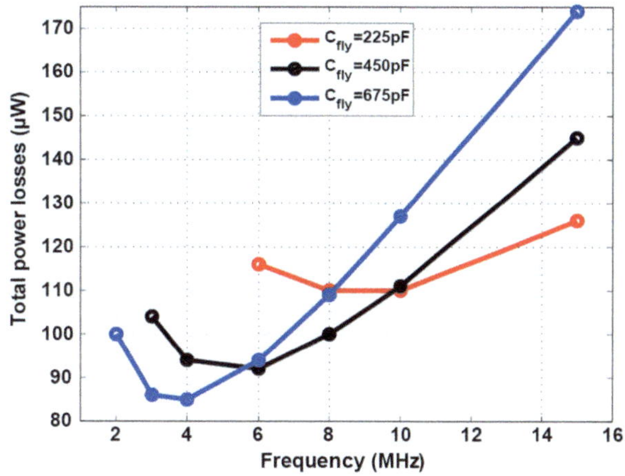

Fig. 16.7 Average power losses of SC circuit against f_{sw} for different values of C_{fly} at gain=3/4, $I_L = 800\,\mu A$ and $W = 100\,\mu m$

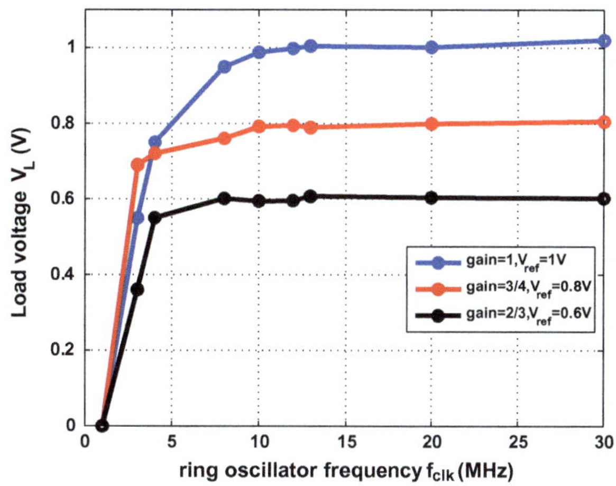

Fig. 16.8 Average load voltage levels when sweeping the ring oscillator frequency at $I_L = 800\,\mu A$

the smallest capacitance of the 225 pF is chosen. In addition, the choice of capacitance used for both flying and load capacitors has been carefully analysed to reduce the bottom plate parasitic losses. Even though MIM capacitor reduces the bottom plate capacitor losses, it has a small capacitance density of 0.92 fF/μm^2. However, MOS capacitor provides higher capacitance density of 2.45 fF/μm^2 but adds more bottom plate parasitic losses. Therefore, MIM capacitor has been selected to implement the three flying capacitors of 675 pF (3*225 pF) because they participate in the charging and discharging every clock cycle, and if the MOS capacitor is used, it will have a negative impact on the efficiency. The load capacitor of 1.67 nF utilizes the MOS capacitance to achieve higher capacitance density. Since MIM capacitor only uses upper layer metals, the active area and lower level metal can be dedicated to the MOS capacitor.

In order to select the appropriate value of f_{clk}, Fig. 16.8 shows the load voltage level while sweeping the clock frequency at a maximum load current of 800 μA. As depicted in the figure, the load voltage levels of 1, 0.8 and 0.6 V are regulated at a clock frequency of 10 MHz but due to PVT variations, 13 MHz has been selected. The PFM of the proposed design adjusts the switching frequency depending on load current levels by the PFM. At heavy load current, the switching frequency is increased, whereas at light load current, the switching frequency is reduced.

16.4.4 Measurement Results in 65 nm CMOS

Figure 16.9 shows the measured average output voltage levels with their corresponding modulated pulses *cnt* at a maximum oscillator frequency of 13 MHz, gain=1, $V_{in} = 1.2$ and at no load condition. Note that the SC converter regulates the output voltage at 1.05, 0.813 and 0.634 V, according to the reference voltage of 1, 0.8 and 0.6 V, respectively. For an output voltage of 0.634 V, the number of switching pulses *cnt* is less than in other two voltage levels.

Figure 16.10 shows the time domain signals of the SC converter when switching from a light load current of 10 μA to a heavy load current of 800 μA. As depicted in the figure, the PFM controller maintains the voltage regulation at approximately 1 V with a voltage drop of 50 mV when switching from a load current of 10–800 μA. This voltage drop is due to the selection of f_{clk} where higher switching frequency is required at higher load current. At a load current of 800 μA, the SC converter operation is pushed to the limit and the ring oscillator cannot support a frequency higher than the selected one. Therefore, the voltage drop can be reduced by increasing f_{clk} or the output capacitance. Note that the switching pulses of *cnt* increase at $I_L = 800$ μA

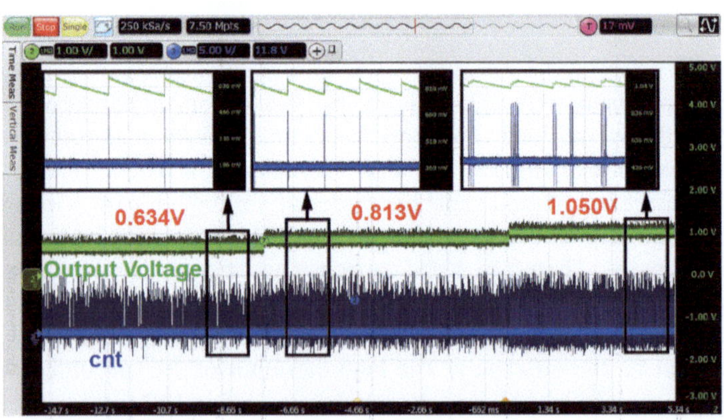

Fig. 16.9 Measured average output voltage level with their corresponding modulated pulses *cnt* at $f_{clk} = 13$ MHz, gain=1, $V_{in} = 1.2$ V and no load

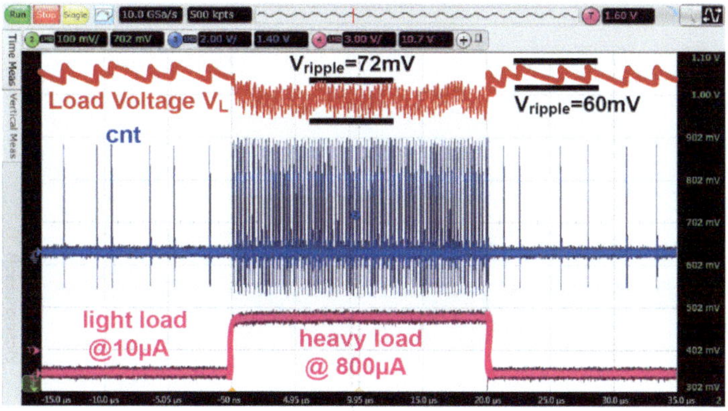

Fig. 16.10 Measured time domain signal when switching from light to heavy load current at $V_L \approx 1\,\text{V}$

in order to maintain a regulated voltage of 1 V. In addition, the voltage ripple at a maximum load current of 800 μA is 72 mV whereas 60 mV at minimum load current of 10 μA.

Figure 16.11 shows the power efficiency of the SC circuit against the variation of the load voltage from 0.4 V up to 1 V. For each gain setting, the voltage reference is changed according to the load voltage in order to achieve voltage regulation. As depicted in Fig. 16.11, the multiple gain settings are important to achieve high power efficiency. For example, the power efficiency at $V_L = 0.8\,\text{V}$ is 67% using a gain of 1 but it is 72% using a gain of 3/4. Figure 16.12 shows the power efficiency of the SC converter for different output voltages from 10 μA up to 800 μA load currents. As shown in the figure, the efficiency reaches the peak of 80% at $V_L = 1\text{V}$ and $I_L = 800\,\mu\text{A}$. Note that higher load power does not lead into higher power efficiency. Indeed, the efficiency at a load power of 10 μW ($V_L = 1\,\text{V}$ and $I_L = 10\,\mu\text{A}$). This is because the SC converter dissipates a power of 40 μW to deliver 10 μW load power, whereas only 8 μW is dissipated to deliver 6 μW load power. This gives an indication that at $I_L = 10\,\mu\text{A}$, it is better to operate the load at the lowest voltage level to achieve power saving rather than dissipation. Note that the voltage regulation of the SC converter as depicted in Fig. 16.12 stops after $I_L = 800\,\mu\text{A}$ at $V_{in} = 1.2\,\text{V}$ as it cannot support higher load current due to the high conduction losses. Note that the quiescent current of the design including the PFM controller, ring oscillator, gain control and non-overlapping two-phase signals has been accounted for the overall measured power efficiency. This count is measured as 3 and 27 μA at a load current of 10 and 800 μA, respectively.

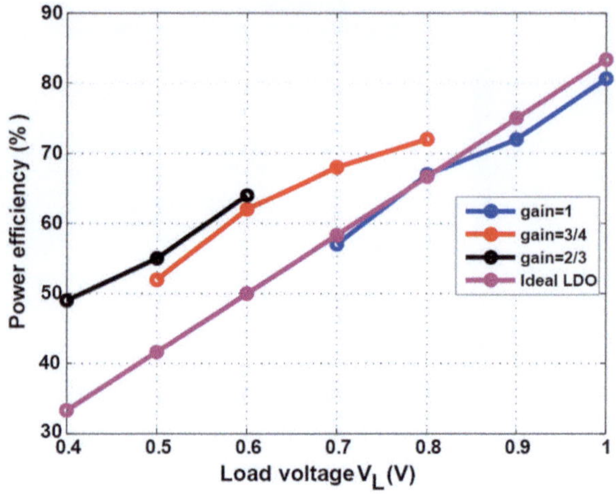

Fig. 16.11 Measured power efficiency of the SC versus load voltage at $V_{in} = 1.2\,\text{V}$ and $I_L = 800\,\mu\text{A}$ with ideal LDO efficiency

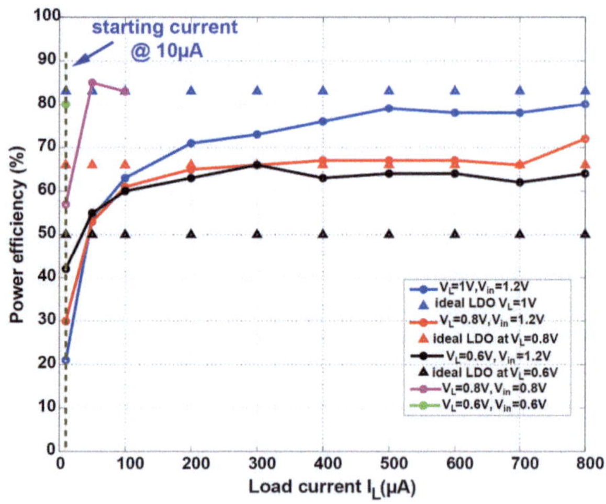

Fig. 16.12 Measured power efficiency of the SC versus the load current at different input voltage levels (regulated output voltage is needed because the input is changing)

16.5 Conclusion

In summary, DVS is an essential power management technique that reduces the dynamic power. DC–DC converter scales down the supply voltage to a particular voltage level that is required by the SoC. The on-chip switched-capacitor DC–DC converter is a viable option for μW power range applications. The presented switched-capacitor converter provides multiple output voltage levels that are 1, 0.8 and 0.6 V supporting load currents of $10\,\mu$A up to $800\,\mu$A. The frequency modulation technique is utilized to optimize the switching losses based on the load current. MIM and MOS capacitor types are used to reduce the active area and bottom plate parasitic losses. The switched-capacitor converter has been fabricated in 65 nm CMOS technology and is able to achieve a maximum power efficiency of 80% at an output voltage of 1 V and load current of 800 μA.

Acknowledgements This work has been supported by the Semiconductor Research Corporation (SRC) under the Abu Dhabi SRC Center of Excellence on Energy-Efficient Electronic Systems (ACE^4S), Contract 2013 HJ2440, with funding from the Mubadala Development Company, Abu Dhabi, UAE.

References

1. C. Lu, V. Raghunathan, K. Roy, Efficient design of micro-scale energy harvesting systems. IEEE J. Emerg. Sel. Top. Circuits Syst. **1**(3), 254–266 (2011)
2. Y. Ramadass, A. Chandrakasan, A battery-less thermoelectric energy harvesting interface circuit with 35 mv startup voltage. IEEE J. Solid-State Circuits **46**(1), 333–341 (2011)
3. P. Luo, S. Zhen, J. Wang, K. Yang, P. Liao, X. Zhu, Digital assistant power integrated technologies for pmu in scaling cmos process. IEEE Trans. Power Electron. **29**(7), 3798–3807 (2014)
4. L.G. Salem, P.P. Mercier, An 85%-efficiency fully integrated 15-ratio recursive switched-capacitor dc–dc converter with 0.1-to-2.2 v output voltage range, in *ISSCC*. (IEEE, 2014), pp. 88–89
5. J. Pyo, Y. Shin et al., 20 nm high-k metal-gate heterogeneous 64b quad-core cpus and hexa-core gpu for high-performance and energy-efficient mobile application processor, in *ISSCC*. (IEEE, 2015), pp. 1–3
6. M. Alioto, Ultra-low power vlsi circuit design demystified and explained. Trans. Circuits Syst. I **59**(1), 3–29 (2012)
7. T. Instruments, A usb-enabled system-on-chip solution for 2.4 ghz ieee 802.15. 4 and zigbee applications (2009)
8. S. STM32W108HB, High-performance, 802.15. 4 wireless system-on-chip, Preliminary data. Doc ID, vol. 16252
9. F. Semiconductor, Mc1322x technical data, rev. 1.3, (2010)
10. B. Mohammad, J. Abraham, A reduced voltage swing circuit using a single supply to enable lower voltage operation for sram-based memory. Microelectron. J. **43**(2), 110–118 (2012)
11. P. Hazucha, T. Karnik, B.A. Bloechel, C. Parsons, D. Finan, S. Borkar, Area-efficient linear regulator with ultra-fast load regulation. IEEE J. Solid-State Circuits **40**(4), 933–940 (2005)
12. C.-W. Chen, A. Fayed, A low-power dual-frequency simo buck converter topology with fully-integrated outputs and fast dynamic operation in 45 nm cmos. IEEE J. Solid-State Circuits **50**(9), 2161–2173 (2015)
13. D. Lu, Y. Qian, Z. Hong, 4.3 an 87%-peak-efficiency dvs-capable single-inductor 4-output dc-dc buck converter with ripple-based adaptive off-time control, in *2014 IEEE International Solid-State Circuits Conference Digest of Technical Papers* (ISSCC). (IEEE, 2014), pp. 82–83
14. C. Tao, A.A. Fayed, A low-noise pfm-controlled buck converter for low-power applications. IEEE Trans. Circuits Syst. I Regul. Pap. **59**(12), 3071–3080 (2012)
15. M. Belloni, E. Bonizzoni, E. Kiseliovas, P. Malcovati, F. Maloberti, T. Peltola, T. Teppo, A 4-output single-inductor dc-dc buck converter with self-boosted switch drivers and 1.2 a total output current, in *IEEE International Solid-State Circuits Conference, ISSCC 2008. Digest of Technical Papers*. (IEEE, 2008), pp. 444–626
16. W. Kim, D.M. Brooks, G.-Y. Wei, A fully-integrated 3-level dc, dc converter for nanosecond-scale dvs with fast shunt regulation, in *2011 IEEE International Solid-State Circuits Conference Digest of Technical Papers (ISSCC)*. (IEEE, 2011), pp. 268–270

17. J. Wibben, R. Harjani, A high-efficiency dc-dc converter using 2 nh integrated inductors. IEEE J. Solid-State Circuits **43**(4), 844–854 (2008)
18. W. Fu, A. Fayed, A self-regulated 588 mhz buck regulator with on-chip passives and circuit stuffing in 65 nm, in *2014 IEEE 57th International Midwest Symposium on Circuits and Systems (MWSCAS)*. (IEEE, 2014), pp. 338–341
19. V. Ng, S. Sanders, A 92%-efficiency wide-input-voltage-range switched-capacitor dc–dc converter, in *2012 IEEE International Solid-State Circuits Conference Digest of Technical Papers (ISSCC)*. (IEEE, 2012), pp. 282–284
20. C. Tse, S. Wong, M. Chow, On lossless switched-capacitor power converters. IEEE Trans. Power Electron. **10**(3), 286–291 (1995)
21. D. Kilani, B. Mohammad, H. Saleh, M. Ismail, Ldo regulator versus switched inductor dc–dc converter, in *2014 21st IEEE International Conference on Electronics, Circuits and Systems (ICECS)*. (IEEE, 2014), pp. 638–641
22. D. Kilani, B. Mohammad, H. Saleh, M. Ismail, Switched capacitor dc–dc converter for ultra-low power applications, in *IEEE International Conference on Electronics, Circuits, and Systems*, vol. 2 (2014)
23. D. Ma, R. Bondade, in *Reconfigurable Switched-Capacitor Power Converters* (Springer, New York, 2013)
24. B. Nikolic, V. G. Oklobdzija, V. Stojanovic, W. Jia, J. K.-S. Chiu, M. Ming-Tak Leung, Improved sense-amplifier-based flip-flop: design and measurements. JSSC **35**(6), 876–884 (2000)
25. R.J. Baker, *CMOS: Circuit Design, Layout, and Simulation*, vol. 1 (Wiley, New York, 2008)
26. H.-P. Le, S.R. Sanders, E. Alon, Design techniques for fully integrated switched-capacitor dc–dc converters. JSSC **46**(9), 2120–2131 (2011)
27. M.D. Seeman, S.R. Sanders, J.M. Rabaey, An ultra-low-power power management ic for energy-scavenged wireless sensor nodes, in *Power Electronics Specialists Conference*. (IEEE, 2008), pp. 925–931
28. Y.K. Ramadass, A.P. Chandrakasan, Voltage scalable switched capacitor dc–dc converter for ultra-low-power on-chip applications, in *PESC*. (IEEE, 2007), pp. 2353–2359
29. Y. Ramadass, A. Fayed, B. Haroun, A. Chandrakasan, A 0.16mm2 completely on-chip switched-capacitor dc–dc converter using digital capacitance modulation for ldo replacement in 45 nm cmos, in *ISSCC*. (2010), pp. 208–209

Chapter 17
Macromodeling of Microbatteries for IoT Micropower Source Integration

Mohammed Shemso Nesro and Ibrahim (Abe) M. Elfadel

17.1 Introduction

The previous two chapters dealt with one important aspect of power management for IoT devices, namely the design of an integrated power conversion and management unit to supply energy to the IoT electronics using a thermoelectrical generator. As pointed out in [17], the amount of power an energy harvester provides is dynamic and unpredictable, and therefore "an energy storage element is needed in IoT devices for uninterrupted operation when ambient power is not available." As is clear from the well-known Ragone diagram shown in Fig. 17.1, battery technology with its high-specific energy is more suitable for IoT devices than supercapacitor technology. Indeed, batteries can provide steady energy over the long-time intervals during which the IoT device is in standby mode. If the battery is recharged using ambient energy, the IoT mode of operation of cycling between short activity intervals and long sleep intervals can in theory be sustained indefinitely. Unfortunately, the scaling of battery technology has very much lagged behind that of microsystems, with the net result being that the battery options of the IoT designer are quite limited. The EnerChip CBC005 thin-film, all-solid-state, rechargeable lithium cell [15] is representative of the battery options that may be available to the IoT designer. This bare-die cell measures 1.7 mm × 2.25 mm × 200 μm and has a rated capacity of 5 μAh at 3.8 V. At 10% depth of discharge, CBC005 is rated at 5000 recharge cycles at 25 °C, which translates into 2.5 mAh of lifetime capacity. If the operating current of the IoT node is 1 μA, the lifetime of this microbattery is from 1 to 2 years. Clearly, such lifetime may not be sufficient for many IoT

M. S. Nesro · Ibrahim (Abe) M. Elfadel (✉)
Masdar Institute at Khalifa University, Abu Dhabi, UAE
e-mail: mohashems@yahoo.com

Ibrahim (Abe) M. Elfadel
e-mail: ielfadel@masdar.ac.ae

Ibrahim (Abe) M. Elfadel and M. Ismail (eds.), *The IoT Physical Layer*,
https://doi.org/10.1007/978-3-319-93100-5_17

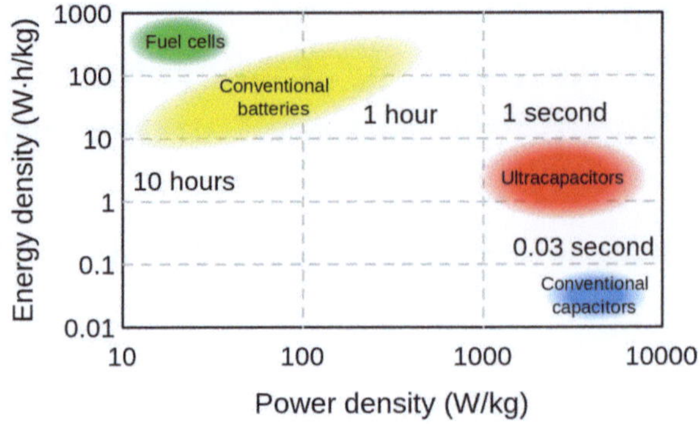

Fig. 17.1 Ragone diagram for batteries and capacitors [16]

applications such as set-and-forget (e.g., home security) or semi-permanent devices (e.g., infrastructure monitoring) [17]. A comprehensive survey of battery technologies and their suitability for IoT node design is provided in [20].

The seamless integration of lithium-ion, solid-state, microbattery with electronics and MEMS has been addressed in several technology areas such as microenergy harvesters, wireless sensor nodes, and biomedical implants [1, 2, 4, 11, 12]. However, the full promise of such integration will not be achieved without integrated microbatteries becoming part and parcel of the process design kits and standard cell libraries so as to include them in the computer-aided design flows of integrated circuits and MEMS. One fundamental component of such design kits and standard libraries is the compact model of the microbattery. The main objective of this chapter is to provide a full methodology for the development of such model.

17.2 Microbattery Modeling

Modeling of batteries is one of the most researched areas. Many of the techniques established for liquid electrolyte batteries can be extended to solid electrolyte microbatteries [18]. Different modeling approaches have been reported targeting different aspects of battery behavior.

Different modeling approaches such as experimental, electrochemical, and equivalent circuit-based are reported in the literature. Equivalent circuit-based models are used by circuit designers to describe electrical input–output behavior of batteries. The simplest equivalent circuit model for a battery could be a voltage source in series with a resister. The model can be made more accurate by adding parallel RC circuits and the Warburg impedance [3]. The identification of all the parameters of this impedance-based model relies on experimental technique called electronic impedance spectroscopy and the parameters, including the Warburg impedance, obtained from this technique have no physical relationship with battery dimension and the physics of its electrochemical processes inside its components and surfaces.

Accuracy, speed, and computational complexity are the main factors for preferring one modeling approach over another. Electrochemical models use partial differential and algebraic equations to describe the electrochemical processes such as the charge transfer kinetics at the electrode/electrolyte interface, diffusion of lithium in the electrodes, and diffusion and migration of ions in the electrolyte [5]. These approaches result in the most accurate battery models whose mathematical complexity and exorbitant transient analysis performance make them unsuitable for use in CAD flows. High-level models such as electrical circuit models, analytical models, and stochastic models have the advantage of good transient simulation speed but at the cost of diminished accuracy.

In [7], a one-dimensional model was used to simulate the performance of all-solid-state lithium-ion batteries. Good agreement with experimental measurements has been attained even for high discharge rates. A similar model, including temperature dependence of parameters, has been developed, and a commercial microbattery manufactured by Front Edge Technology (FET, Baldwin Park, USA) has been used for its verification [6]. Battery capacity and lifetime for an integrated implantable device have also been modeled with the ultimate purpose of designing a microscopic power source [4]. Equivalent circuit models have been developed for microbatteries manufactured at Jet Propulsion Laboratory (JPL) and have subsequently been used to test battery series and parallel interconnection [14].

A flowchart of our modeling methodology is given in Fig. 17.2. One distinguishing feature of our proposed compact modeling approach is the combination of a *behavioral linearization* steps with model order reduction. Unlike Taylor series and similar local expansion methods used in linearizing nonlinear equations, behavioral linearization is based on a global transformation of the original nonlinear model into a linear one that is globally valid across the spatial domain. As illustrated in Fig. 17.2, the starting point of our microbattery flow is the full nonlinear partial differential equations (PDEs) describing the microbattery electrochemistry. The multi-physics simulation of these equations provide the reference output for all subsequent steps in our modeling approach. The main benefit of using the full physical PDE description is to rigorously account for both internal state and boundary condition impact

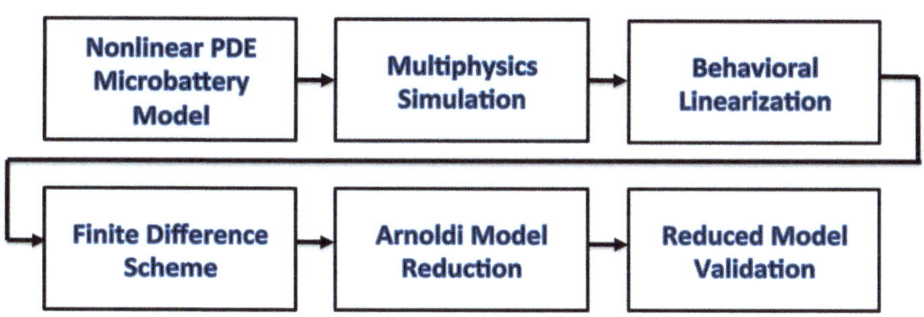

Fig. 17.2 Flowchart for microbattery compact modeling methodology

on microbattery behavior. Other approaches for battery modeling are described in references [5, 6, 14, 18].

The most important step in our approach is arguably the third one, "Model reformulation and linearization." Indeed, the most widespread approach for dealing with nonlinearities is to use local expansions such as Taylor or Volterra series followed by the application of model order reduction techniques to come up with a compact model. Unlike these local approaches, we apply a step that we call *behavioral linearization* that in essence replaces the original nonlinear PDE's with linear ones while ensuring that all the boundary conditions are satisfied. The basis of this behavioral linearization is the a property of the electrolyte that we verify by simulation called *electroneutrality*. Once these linear PDE's are derived, our modeling approach follows the traditional steps of state-space formulation and model order reduction. We use the Arnoldi algorithm [10] to derive the reduced-order model of the microbattery.

17.3 Microbattery Description and Modeling Assumption

Figure 17.3 shows the schematic of a planar all-solid-state lithium-ion cell and typical layer dimensions [9]. Like conventional rechargeable batteries, all-solid-state lithium-ion microbatteries can be operated in charge and discharge modes. Referring to Fig. 17.4, the discharge mode is comprised of the following steps:

1. Oxidation/reduction reaction at the electrolyte/anode interface produces lithium-ions and electrons.
2. Lithium-ions travel through the solid electrolyte (LiPON) to the cathode ($LiCoO_2$).
3. The electrons travel through the external circuit.
4. Lithium-ions are neutralized by reclaiming the electrons.

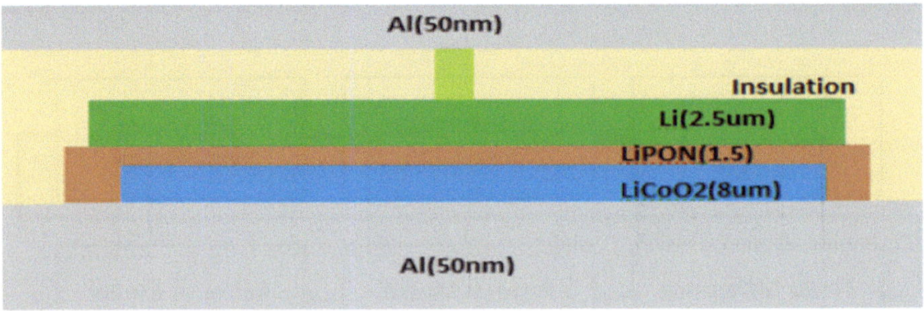

Fig. 17.3 Planar all-solid-state lithium-ion cell

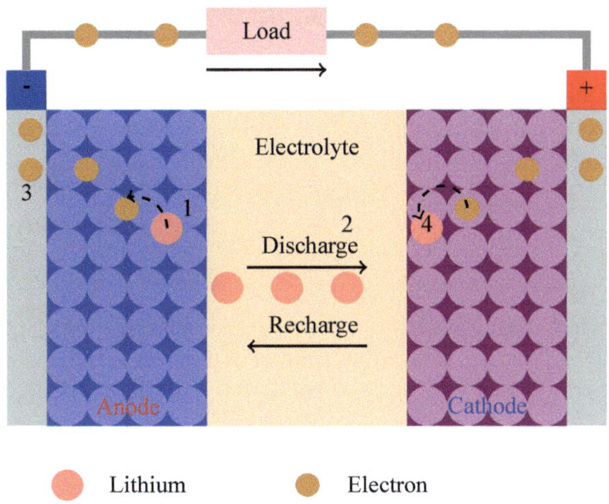

Fig. 17.4 Schematic representation of a thin-film, solid-state lithium microbattery

In the development of the microbattery model, the following assumptions are adopted:

1. Assumption of isotropic transport properties in y- and z-direction will reduce the thin-film battery to 1-D geometry. Figure 17.4 shows the one-dimensional model setup along with the important variables
2. The charge transfer reactions occur only at the surface of the electrodes
3. Lithium is highly reactive and the voltage losses across lithium and at its interface with the electrolyte is ignored
4. The microbattery charge/discharge is assumed to start from relaxed state. This will result in a uniform initial concentration for ions in the electrolyte
5. Electrolyte is assumed to be electroneutral

17.4 Mathematical Formulation

As stated, a simple one-dimensional geometry is considered to simplify the model derivation process. Moreover, surface charge transfer reaction, zero voltage drop at the lithium/electrolyte interface, and uniform initial lithium-ion concentration are assumed. Lithium-ion transport through the electrolyte or electrodes is due to either migration, diffusion, or both. Migration is the transport of a charged body under an electric field, and diffusion is the movement of a chemical species under a concentration gradient.

17.4.1 Electrolyte Domain Equations

Denoting the concentration of species i as c_i, the Nernst–Planck–Poisson (NPP) equation with source term r ($\frac{mol}{m^3 s}$) is used to describe the transport of ions under the influence of concentration and electric potential gradient

$$\frac{\partial c_i}{\partial t} = D_i \frac{\partial^2 c_i}{\partial x^2} + \frac{z_i F D_i}{RT} \frac{\partial (c_i \partial \phi_i)}{\partial x^2} \tag{17.1}$$

where $i = 1$ for lithium and $i = 2$ for negative ion, ϕ ($\frac{J}{C}$) is the electrostatic potential, D_i ($\frac{m}{s^2}$)is the diffusion coefficient for ionic species i, $z_i = 1$ for lithium and $z_i = -1$ for negative ion. F ($96485 \frac{C}{mol}$) is Faraday's constant, R ($8.314 \frac{J}{mol K}$) is gas constant, and T (K) is temperature. The source term r can be described by the "weak electrolyte theory" [7]. According to this theory, the net lithium-ion generation and recombination reaction inside the electrolyte due to thermal agitation can be described by the following equation

$$r = k_d c_T - k_r (c_1 - c_2) \tag{17.2}$$

where k_d ($\frac{mol}{m^3 s}$) is the dissociation rate, k_r ($\frac{mol}{m^3 s}$) is the recombination rate, and c_T ($\frac{mol}{m^3}$) is the total concentration of lithium inside the electrolyte. The Poisson equation is coupled with Nernst–Planck equation to determine the electrolyte potential distribution

$$\frac{\partial^2 \phi}{\partial x^2} = \frac{z_i F}{\varepsilon_r \varepsilon_o} (c_1 - c_2) \tag{17.3}$$

where a relative permeability ε_r of 10 is used and ε_o is the permeability of air. A negative concentration gradient is set at the anode/electrolyte interface for lithium-ion, whereas positive concentration gradient is set at the electrolyte/cathode interface. The negative ions remain inside the electrolyte, and their flux is set to zero at both ends of the electrolyte

$$D_1 \frac{\partial c_1(0, t)}{\partial x} = -\frac{I}{FA}, \; D_1 \frac{\partial c_1(L_e, t)}{\partial x} = \frac{I}{FA} \tag{17.4}$$

where L_e is the thickness of the electrolyte, $I(Amp)$ the current flowing in the electrodes, and $A(\mathrm{m}^2)$ their surface area.

17.4.2 Cathode Domain Equation

Diffusion due to concentration gradient is the main transport mechanism in the cathode and is given by Fick's second law

$$\frac{\partial c_3}{\partial t} = D_3 \frac{\partial^2 c_3}{x^2} \tag{17.5}$$

where c_3 ($\frac{mol}{m^3}$) is the concentration of lithium in the cathode and D_3 ($\frac{mol}{m^2 s}$) its diffusion coefficient. The initial concentration of lithium inside the cathode is c_{3_O}. While a constant negative concentration gradient is set at the electrolyte/cathode interface, it is set to zero at the cathode/current collector interface

$$D_3 \frac{\partial c_3(L_c, t)}{\partial x} = -\frac{I}{FA} \tag{17.6}$$

where L_c is the thickness of the cathode.

17.4.3 Voltage Losses and Terminal Voltage

The analytic expression from [8] is used to find the equilibrium voltage of the cathode. The terminal voltage of the microbattery, V_{bat}, is the difference between the cathode equilibrium voltage, E^+_{eq} and all voltage losses

$$V_{bat} = E^+_{eq} - \eta^{mt} - \eta^{ct} - \eta^d \tag{17.7}$$

where η^{mt} is the mass transfer voltage loss in the electrolyte

$$\eta^{mt} = \phi(L_e) - \phi(0) + \frac{RT}{F} \ln \frac{c_1(L_e, t)}{c_1(0, t)} \tag{17.8}$$

The voltage loss due to charge transfer (η^{ct}) is given by:

$$\eta^{ct} = \frac{R}{\alpha F} \text{arcsinh}(\frac{I_c}{I_{oc}}) \tag{17.9}$$

The diffusion voltage loss in the cathode (η^d) is:

$$\eta^d = E^+_{eq} \left(\frac{c_3(L_c, t)}{c_{3max}} \right) - E^+_{eq} \left(\frac{\int_0^{L_c} c_3(x, t)dx}{L_c c_{3max}} \right) \tag{17.10}$$

17.4.4 Behavioral Linearization

The model can be further simplified by exploiting the electroneutrality property. This property is validated in COMSOL by plotting the charge density across the electrolyte for different values of electrolyte diffusion length λ_D(m) and electrolyte

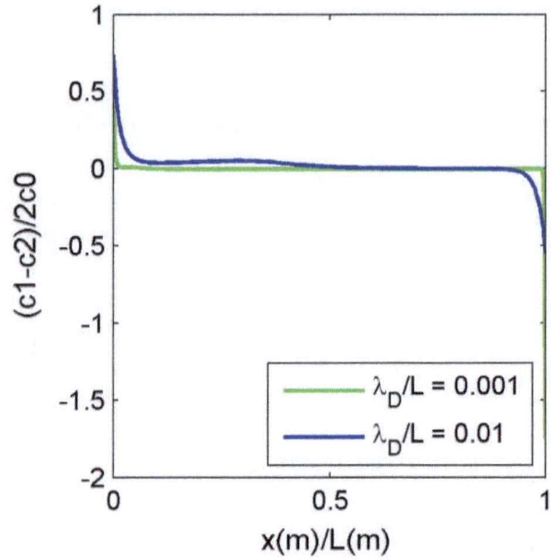

Fig. 17.5 Normalized charge density across the electrolyte

thickness ratios. Figure 17.5 shows that the charge density is close to zero for practical electrolyte thicknesses. For an electrolyte satisfying the electroneutrality property, we have $c_1 = c_2$, and the above nonlinear NPP equation is reduced to a linear diffusion equation with source term [7]

$$\frac{\partial c_1}{\partial t} = D_e \frac{\partial^2 c_1}{\partial x^2} + r \tag{17.11}$$

where $D_e = 2\frac{D_1 D_2}{D_1 + D_2}$. The boundary conditions are also reduced to

$$D_1 \frac{\partial c_1(0,t)}{\partial x} = -\frac{I}{2FA}, \; D_1 \frac{\partial c_1(L_e,t)}{\partial x} = \frac{I}{2FA} \tag{17.12}$$

Equations (17.11) and (17.12) constitute the behavioral linearization of the nonlinear NPP equations in the electrolyte domain. Combined with the linear diffusion equation in the cathode domain, the full electrochemistry of the microbattery can then be described by a set of linear PDE's. The next section will show how this set can be further reduced to a small set of ordinary differential equations using reduced-order modeling.

17.5 Finite Difference Discretization

To apply the macromodeling methodology of model order reduction, we first need to find a state-space representation of electrochemical equations. To do so, we apply finite difference (FD) schemes to the spatial derivatives of (17.11) and boundary conditions of (17.12) for the electrolyte. Similarly, FD is applied to (17.5) and to the boundary condition (17.6) for the cathode. Second-order centered FD is used to approximate the spatial derivatives in the electrolyte and cathode domain. A first-order forward difference scheme is used for the gradient boundary conditions. The discharge rate C_{rate} (unitless) indicates how fast the battery is discharged depending on the current. A C_{rate} of 1 corresponds to discharging the microbattery in one hour, and $\frac{1}{2}$ C_{rate} corresponds to discharging it in two hours. Generally, it is the ratio between the actual discharging current and the 1 C_{rate} current. For an M-point discretization, the electrolyte state-space system is given by

$$\frac{dc}{dt} = \mathbf{A}c + \mathbf{b}u \tag{17.13}$$

where the $M \times M$ system matrix $\mathbf{A}$ and the $M \times 1$ input vector $\mathbf{b}$ are given by

$$\mathbf{A} = \begin{bmatrix} \lambda + \beta & \beta & \cdot & 0 & 0 \\ \beta & \lambda & \cdot & 0 & 0 \\ 0 & \cdot & \lambda & 0 & 0 \\ 0 & 0 & \cdot & \lambda & \beta \\ 0 & 0 & 0 & \beta & \lambda + \beta \end{bmatrix}, \mathbf{b} = \begin{bmatrix} \gamma + \sigma \\ \gamma \\ 0 \\ \gamma \\ \gamma - \sigma \end{bmatrix} \tag{17.14}$$

in which $\lambda = -(\frac{2D_e M^2}{L_e^2} - 2k_r c_o - k_d)$, $\beta = \frac{2D_e M^2}{L_e^2}$, $\sigma = \frac{IC_{crate}}{FD_e}$ and $\gamma = -k_d c_T + k_r c_o^2$

A similar discretization can be applied to the cathode equation in (17.5) to get a state-space system similar to (17.13).

17.6 Arnoldi Model Order Reduction

Model order reduction (MOR) allows us to generate a compact model that is much smaller in dimension but still preserves the input–output behavior of the original model with reasonable accuracy [13]. The common approaches for MOR of linear systems are based on iterative methods such as Arnoldi iteration [19]. A slightly modified version of the Arnoldi algorithm of [10] is used to generate our compact model using the following procedure:

1. Translate the origin of the state space $\tilde{c} = c - c_0$ where $\tilde{c}$ is the homogeneous state vector with zero initial condition
2. Start Arnoldi iteration by setting $\mathbf{M} = -\mathbf{A}^{-1}$ and $r = -\mathbf{A}^{-1} b_1$ where $b_1 = \mathbf{A}c_0 + b$
3. Generate the Arnoldi projection matrix $\mathbf{Q}$ and find the reduced system matrix $\mathbf{A_r}$ and input vector $\mathbf{b_r}$ using the following relationships

$$\mathbf{A_r} = \mathbf{Q^T M Q}, \mathbf{b_r} = \mathbf{Q^T b_1}$$

4. Solve the reduced system using the equation

$$\frac{dc_r}{dt} = \mathbf{A_r} c_r + \mathbf{b_r} u \tag{17.15}$$

5. Recover the original concentration state variable c using

$$c = \mathbf{Q} c_r + c_o \tag{17.16}$$

17.7 Numerical Comparisons

The main microbattery parameters used for the full and reduced model are listed in Table 17.1. Both the full and reduced models are solved using MATLAB ODE solver.

Figure 17.6 shows the relevant lithium concentration plots with respect to electrolyte position and time during 1 C_{rate} for the full and reduced models. The original nonlinear model is also solved using COMSOL's batteries and fuel cells module. The discrepancy between the reduced-order model, full model, and nonlinear original model is negligible. The diffusion equation for the cathode is the same for the three models, and they generate exactly the same lithium concentration profile in the cathode. Equation (17.7) is used to find the microbattery's terminal voltage. The total voltage loss due to diffusion, migration, and charge transfer is plotted in Fig. 17.7. Despite their low values, the voltage losses due to electrolyte diffusion and migration are dominant at the initial stage of discharge curve. The charge transfer and cathode diffusion losses are dominant at the later stage of the discharge curve while the cathode diffusion voltage loss is responsible for the increased voltage drop at the end of the discharge curve.

Table 17.1 Microbattery parameters used

Symbol	Value	Description
L_e	1.5 μm	Electrolyte length
L_c	320 nm	Cathode length
Q_{max}	10 μAh	Rated capacity
D_1	$0.9*10^{-15}$	Lithium-ion diffusion coefficient (electrolyte)
D_2	$5.1*10^{-15}$	Negative-ion diffusion coefficient (electrolyte)
D_3	$1.76*10^{-15}$	Lithium-ion diffusion coefficient (cathode)
c_T	$60100 \frac{mol}{m^3}$	Total concentration of lithium (electrolyte)

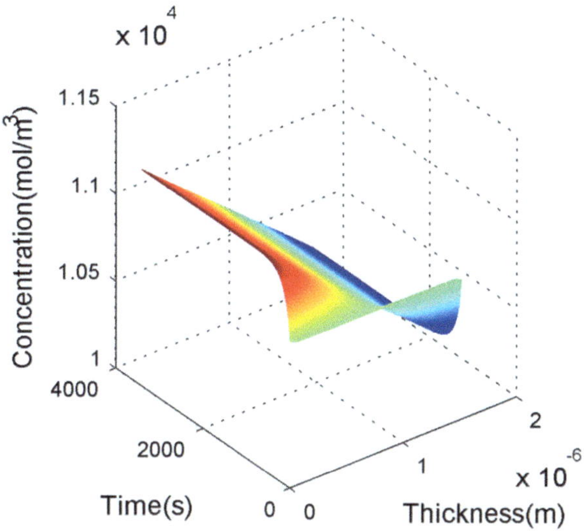

Fig. 17.6 Electrolyte lithium-ion concentration for the full and reduced models at $1C_{rate}$

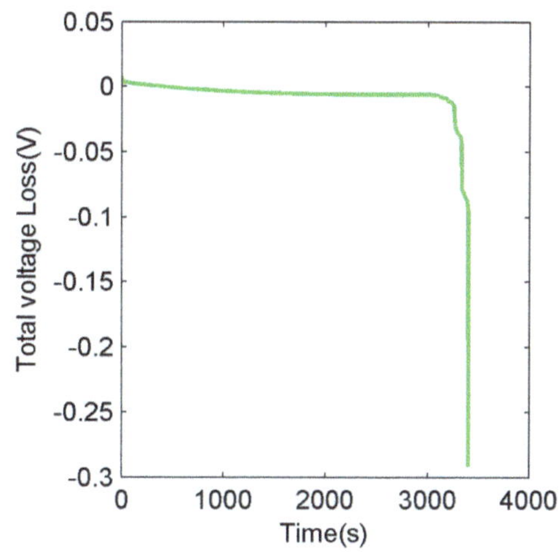

Fig. 17.7 Total voltage losses during $1C_{rate}$ discharge

Figure 17.8 shows the discharge curve for different multiples of the 1 C_{rate} current. The result shows that the microbattery can be discharged at a rate as fast as that of some capacitors. The predictions of the original nonlinear, linear, and reduced models at the terminal voltage are virtually the same.

Figure 17.9 shows the terminal voltage for the linear and reduced models. They are almost identical, which shows the accuracy of the reduced-order model. The advantage of the reduced-order model is a significant reduction of transient simulation time. Table 17.2 summarizes the simulation run times for different reduced model sizes. The original linear system has a state-space dimension of 1000. Run-time improvements of 10–30X are observed while the relative error between the two models is negligible.

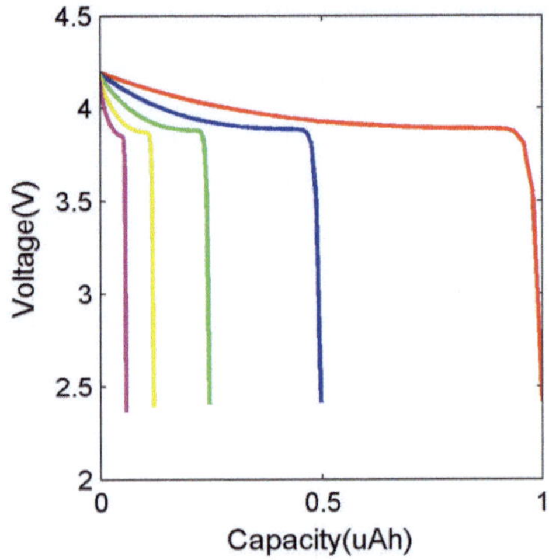

Fig. 17.8 Battery terminal voltage at different discharge rates

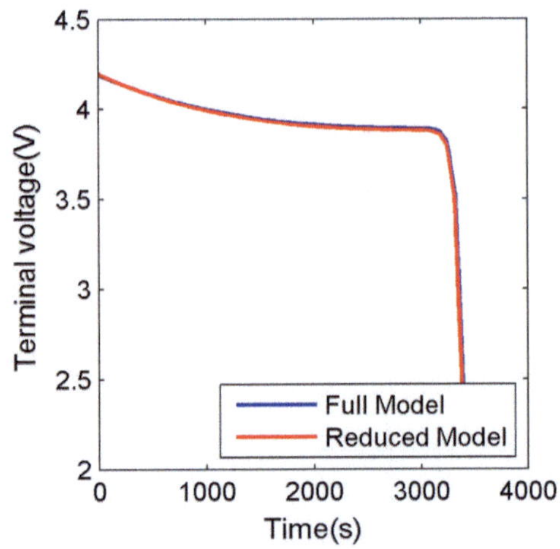

Fig. 17.9 Terminal voltage comparison for the full and reduced models

Table 17.2 Transient simulation speed comparison for the full and reduced models

Full model order	Reduced model order	Full model simulation time (s)	Reduced model simulation time (s)	Speed improvement (X)
1000	200	2.47	0.23	10
1000	100	2.06	0.14	15
1000	10	1.676	0.07	24
1000	5	1.66	0.05	33

17.8 Conclusions

In this chapter, a compact model for a lithium-ion solid-state microbattery has been developed. The electrolyte of the original system is described by nonlinear PDEs. The full model is solved using a commercial finite element solver. The electroneutrality assumption is first validated and later used to replace the nonlinear PDEs with linear ones. The linear system output agrees with the finite element output of the original system. The Arnoldi-based model order reduction is able to replicate the input–output behavior of the full model with negligible error while achieving a speedup of at least 30x in transient simulation with respect to the nonlinear PDE model.

Acknowledgements We would like to acknowledge our SRC industrial liaison, Dr. Lizhong Sun from Applied Materials, for his guidance and input throughout this project. This work was supported by the Semiconductor Research Corporation (SRC) under contract 2012-VJ-2336 with customized funding from the Mubadala Investment Company, Abu Dhabi, UAE.

References

1. F. Albanoa, A fully integrated microbattery for an implantable microelectromechanical system. J. Power Sources **185**, 1524–1532 (2008)
2. M. Balkanski, Solid-state microbatteries for electronics in the 21st century. Solar Energy Mater. Solar Cells **62**, 21–35 (2000)
3. H.J. Bergveld, W.S. Kruijt, P.H.L. Notten, Electronic network modeling of rechargeable batteries. J. Electrochem. Soc. **145**, 3764–3778 (1998)
4. D. Blaauw et al., A fully integrated microbattery for an implantable micro-electro-mechanical system. J. Power Sources **185**, 1524–1532 (2008)
5. G.G. Botte, V.R. Subramanian, R.E. White, Mathematical modeling of secondary lithium batteries. Electrochim. Acta **45**, 2595–2609 (2000)
6. P. Bouillon et al., Charge/discharge simulation of an all solid-state thin film battery using a one-dimensional model. J. Electrochem. Soc. **159**, A104–A115 (2012)
7. D. Danilov, R.A.H. Niessen, P.H.L. Notten, Modeling all-solid-state Li-ion batteries. J. Electrochem. Soc. **58**, A215–A222 (2011)
8. T.S. Dao, C.P. Vyasarayani, J. McPhee, Simplification and order reduction of lithium-ion battery model based on porous-electrode theory. J. Power Sources **198**, 329–337 (2012)
9. N. Dudney, Y. Jang, Analysis of thin-film lithium batteries with cathodes of 50nm to 4um thick LiCoO2. J. Power Sources **22**, 119–121 (2003)
10. I.M. Elfadel, D.L. Ling, A block rational Arnoldi algorithm for multipoint passive model-order reduction of multiport RLC networks, in *International Conference on Computer Aided-Design* (San Jose, CA, 1997), pp. 66–71
11. M.H. Gahaed et al., Circuits for a cubic-millimeter energy-autonomous wireless intraocular pressure monitor. IEEE Trans. Circuits Syst. **60**(12), (2013)
12. L.M. Goncalves, J.F. Ribeiro, M.F. Silva, M.M. Silva, J.H. Correia, Integrated solid-state film lithium battery. Procedia Eng. **5**, 778–781 (2010)
13. K. Henrik A. Olsson, J. McPhee, Model order reduction with rational Krylov methods, Doctoral thesis in Numerical Analysis, KTH, Stockholm (2005)
14. K.B. Herbert et al., Switch array system for thin film lithium microbatteries. J. Power Sources **136**, 401–407 (2004)
15. http://www.cymbet.com/pdfs/DS-72-41%20v6.pdf

16. https://commons.wikimedia.org/wiki/File:Supercapacitors_chart.svg
17. H. Jayakumar et al., Powering the internet of things, in *Proceedings of the 2014 International Symposium on Low Power Electronics and Design* (La Jolla, California, USA, 2014), pp. 75–380
18. M.R. Jongerden, B.R. Haverkort, Which battery model to use? IET Softw. **3**(6), 445–457 (2010)
19. Y. Saad, *Iterative Methods for Sparse Linear Systems*, 2nd edn. (Society for Industrial and Applied Mathematics, Philadelphia, 2003)
20. J. Sather, Battery technologies for IoT, in *Enabling the Internet of Things*, ed. by M. Alioto (Springer, Berlin, 2017)

Part V
Systems and Security

"Always design a thing by considering it in its next larger context: a chair in a room, a room in a house, a house in an environment, an environment in a city plan."
Eliel Saarinen, Finnish-American Architect

Chapter 18
Self-Powered SoC Platform for Wearable Health Care

Mohammad Alhawari, Dima Kilani, Temesghen Habte, Yonatan Kifle, Nourhan Bayasi, Ismail Elnaggar, Nicholas Halfors, Baker Mohammad, Hani Saleh and Mohammed Ismail

18.1 Overview of the System

The Internet of things (IoTs) is generating a level of connectivity never seen before. The ability of devices to constantly monitor and report data for remote analysis could bring a new era of home health care services. Constant monitoring systems (CMS) could measure cardiac activity in patients with heart problems, providing 24 hour data to physicians using software to recognize potential issues before they even happen [1]. This type of data could be used to prevent heart attacks, to send rapid response medical assistance in the event of a stroke, or to adjust medications based on

M. Alhawari · D. Kilani · T. Habte · Y. Kifle · N. Bayasi · I. Elnaggar · N. Halfors
B. Mohammad · H. Saleh (✉) · M. Ismail
Khalifa University, Abu Dhabi, UAE
e-mail: hani.salef@kustar.ac.ae

M. Alhawari
e-mail: mohammad.alhawari@kustar.ac.ae

D. Kilani
e-mail: dima.kilani@kustar.ac.ae

T. Habte
e-mail: temesghen.habte@kustar.ac.ae

Y. Kifle
e-mail: yonatan.kifle@kustar.ac.ae

N. Bayasi
e-mail: nourhan.bayasi@kustar.ac.ae

N. Halfors
e-mail: nicholas.hallfors@kustar.ac.ae

B. Mohammad
e-mail: baker.mohammad@kustar.ac.ae

M. Ismail
e-mail: ismail@kustar.ac.ae

Ibrahim (Abe) M. Elfadel and M. Ismail (eds.), *The IoT Physical Layer*,
https://doi.org/10.1007/978-3-319-93100-5_18

changing heart activity. CMS could not only increase effectiveness of patient care, but also significantly lower the long-term cost of care as the majority of analysis could be performed by software, only to be reviewed by a physician as necessary. Integration of CMS into the IoTs could be the future of health care, ensuring constant monitoring and rapid response in the event of a cardiac episode.

The implantable cardioverter-defibrillator (ICD) has been considered as the best protection against sudden death from ventricular arrhythmias in high-risk individuals. However, most sudden deaths occur in individuals who do not have high-risk profiles. Long-term ECG monitoring is the criterion standard for the diagnosis of ventricular arrhythmia. If the clinical situation permits, the 12-lead ECG is obtained before conversion of the rhythm and analyzed to detect any changes in the characteristics of the ECG signal. By extracting information about intervals, amplitudes, and waveform morphologies of the different P-QRS-T waves, the onset of the ventricular arrhythmia can be detected.

The proposed system shown in Fig. 18.1 is a potential life savior for patients who are susceptible to ventricular arrhythmia by alerting them of immediate attention to their medical condition. The system aims to design and develop a self-powered integrated biomedical processor chip that is capable of acquiring the ECG signal and processing it using noninvasive methods and sending it wirelessly to a point of care. The system employs the state-of-the-art power reduction design techniques to extend the operational lifetime. In addition, the system is fully integrated on the same chip and combines different computational engines which are necessary for performing the signal processing tasks of the ECG and other sensors. The top-level design of the system connecting all the sub-modules is shown in Fig. 18.1. As depicted in the figure, an energy harvesting system that harvests the thermal and vibrational energy of the human body is utilized to power all the modules in the system. A graphene-based dry ECG sensors are designed to acquire the ECG signal from the patient. An analog front-end and an analog-to-digital converter amplifies and digitizes the ECG signal to be analyzed by the processor. A low-power biomedical processor extracts the ECG feature and performs classification based on an accurate prediction algorithm. A low-power wireless communication module is added to the mobile wearable device which can even alert a medical center of a life-threatening situation that they need to attend or enable remote logging of patient info to a cloud-based medical database. In the following sections, each module in Fig. 18.1 is highlighted with brief explanation.

18.2 Reduced Graphene Oxide Coated Fabric ECG Sensor

The requirements of sensors in long-term and wearable health monitoring systems include patient comfort, low-power consumption, excellent reliability/durability, and high signal quality [2]. The sensor acts as the interface between the CMS hardware and the patient, by directly measuring surface electrical potential at the skin. This signal is passed on to the analog front-end (AFE) and analog-to-digital converter

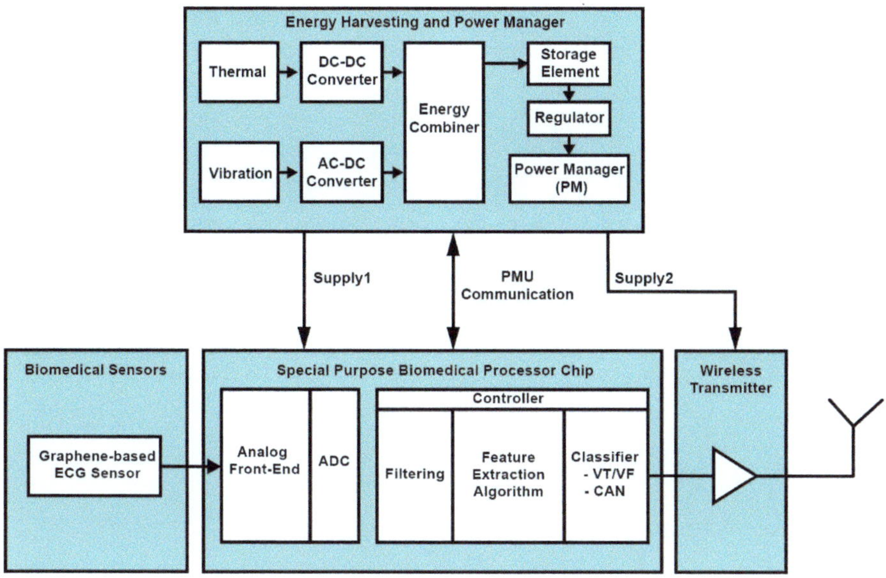

Fig. 18.1 Overall block diagram of self-powered ECG system

(ADC) where it is translated into a digital signal and sent along to the processing unit for filtration and analysis.

The current standard for ECG electrodes is based on the Ag/AgCl gel-type electrode, and nearly all clinical ECG collection is done with this type of electrode due to its excellent signal quality and good resistance to motion artifacts [3]. However, the use of this type of electrode is restricted to a clinical setting, typically with a patient confined to a bed. These electrodes require an electrolyte gel and adhesive patch, both of which may cause irritation and patient discomfort over long-term use, and both must be regularly reapplied as they dry out. Due to the inconvenience and discomfort associated with gel-type electrodes, they are not suitable for CMS [2].

To address the requirements of a CMS electrode, a reduced graphene oxide (rGO_x) coated fabric ECG sensor has been illustrated in Chap. 11 of this volume. By coating a piece of fabric (Nylon) in rGO_x, as shown in Fig. 18.2, a sensor is produced that maintains the intrinsic softness, flexibility, and comfort associated with a woven textile, while also gaining the electrical properties of graphene [4]. The resulting coated fabric may be easily incorporated into clothing or accessories to be comfortably worn by the patient. rGO_x coated fabric can be produced with very low electrical resistance, providing an excellent signal transduction interface. The coating is durable, flexible and stretchable, does not affect the comfort of the fabric or cause any additional irritation while worn [5], and does not show any evidence of loss of signal integrity over extended periods of use.

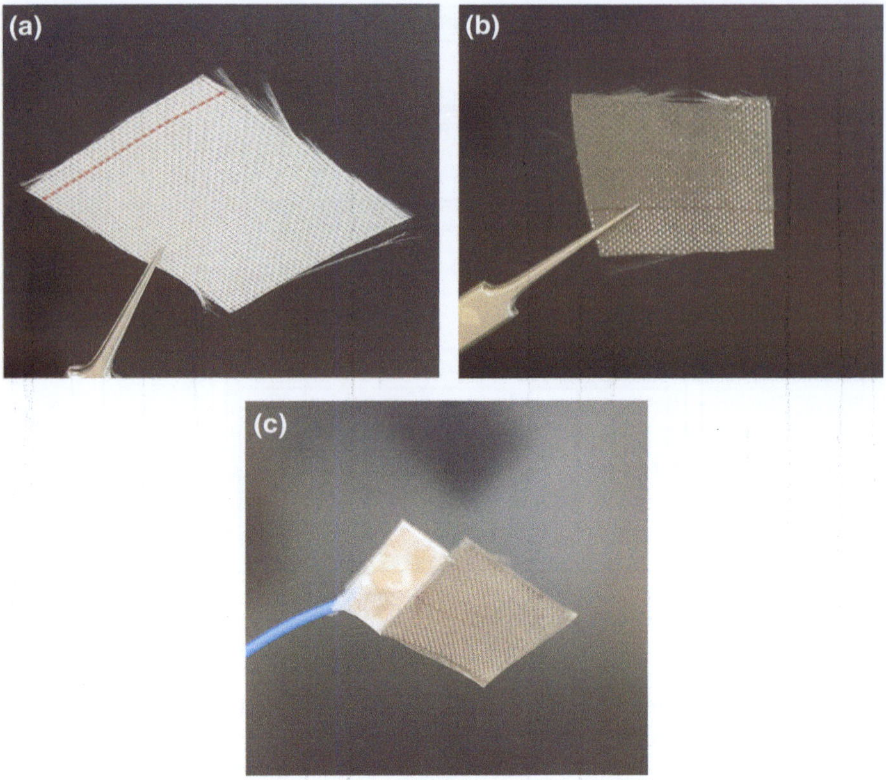

Fig. 18.2 Uncoated, plain nylon fabric (**a**) followed by coating in rGO_x (**b**) and incorporated into a prototype ECG electrode (**c**) consisting of copper tape and a soldered wire lead

18.3 Energy Harvesting and Power Management Unit for IoTs

A thermoelectric generator (TEG) which converts a temperature gradient into an electrical energy has been an excellent candidate as an energy harvester for IoT devices [6]. The basic principle of the TEG is based on the creation of an electron-hole pair in a semiconductor due to a temperature gradient. The electric potential generated across the TEG is related to the temperature difference (ΔT) by the equation:

$$V_{TEG} = S\Delta T \tag{18.1}$$

where S is the Seebeck coefficient of the material used [7, 8]. To utilize the energy harvested from the TEG, efficient interface circuits are required to convert and store the harvested energy in order to power the load. This includes DC-DC converters and a power management unit (PMU). DC-DC converters are necessary to convert the low TEG voltage into a higher level. PMU is an essential block for the SoC since it controls the power generation, storage, and delivery. In addition, PMU can have different techniques to reduce the power consumption such as dynamic voltage scaling (DVS).

18.3.1 Energy Harvesting Stage

The main function of the energy harvesting stage is to extract the maximum power from the energy source and then charge the output storage. A typical TEG device generates 20–50 mV when placed on a human skin at room temperature [8]. In addition, the series resistance of the TEG is typically between 3–100 Ω. The output voltage should be higher than 1 V since the load in our system needs two regulated voltages, 0.6 and 1 V. As such, the power stage should have a high conversion gain to achieve a high output voltage. In our case, the output voltage from the power stage reaches 1.2 V.

Figure 18.3 shows the block diagram of the power conversion stage. An inductor-based converter boosts the small TEG voltage into an output voltage of 1.2 V. The controller is essential for the converter operation as well as achieving maximum power transfer from the TEG. In addition, the controller includes zero current switching circuit that is essential for the converter to operate. Finally, the output capacitor is needed for reducing the output ripple of the converter.

18.3.2 Power Management Unit

A power management unit (PMU) is required to monitor the harvested energy level and to manage the power distribution for different domains in an SoC such as a memory and a processor. In addition, the PMU is essential to obtain the optimum energy and performance point [9, 10]. This is done through enabling DVS [11] and power gating [12]. Power gating allows the SoC to switch between active and sleep modes

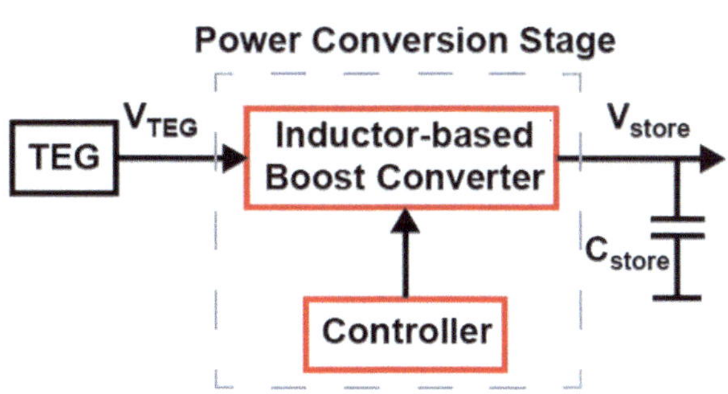

Fig. 18.3 Block diagram of the power conversion stage

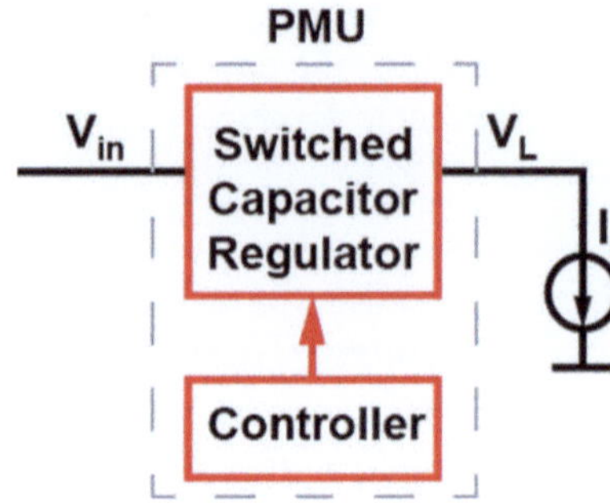

Fig. 18.4 Block diagram of the PMU

based on the harvested input voltage level and functional requirements. Moreover, DVS reduces the operating voltage supply to achieve the optimum energy point. Hence, DVS is a key requirement to decrease the power consumption quadratically such that $P \propto CV^2 f$ [13]. Frequency scaling can also be used in conjunction with DVS to further reduce the power consumption.

Figure 18.4 shows the block diagram of the PMU. It consists of switched capacitor (SC) regulator and a controller. The controller helps the SC to generate two regulated output voltages, 0.6 and 1 V. Further, it manages different power modes based on the available harvested energy from the power conversion stage. In addition, it controls how frequent the transmitter is turned on to send the processed data in the SoC.

18.3.3 Integrated Energy Harvesting IC

Before explaining the proposed PMU architecture, the load current should be analyzed. The energy harvesting IC (EHIC) is designed to drive a μWatt biomedical SoC that samples a real-time ECG signal [14]. This SoC extracts different ECG features to predict arrhythmia in the heart. The SoC acquires 4 s of data (∼ three heartbeats) and then runs an algorithm for 30 ms at 1 MHz and 100 µW power consumption. As such, a sleep mode model should be implemented such that the SoC is active during execution and then goes to sleep when acquiring a new data.

Figure 18.5 shows the PMU architecture with the various control signals. As indicated in the figure, the SoC power consumption is emulated using the control signal *cnt* generated from a microcontroller that drives two current sources. Note that the sleep current is estimated to be 10 µA which includes leakage and power consumption of the always-on blocks in the SoC.

For the switches S_{boost} and S_{SC}, one of them is ON at a time depending on the energy available from the TEG. When the TEG provides enough energy for the SC to drive the SoC, then the S_{SC} will be ON. However, during low-energy mode, S_{SC} will be off and the SC will be bypassed using S_{boost} such that the SoC is driven directly through the boost converter. In addition, bypassing the SC could also be used when higher efficiency is needed. For example, the efficiency of the EHIC at 50 mV input,

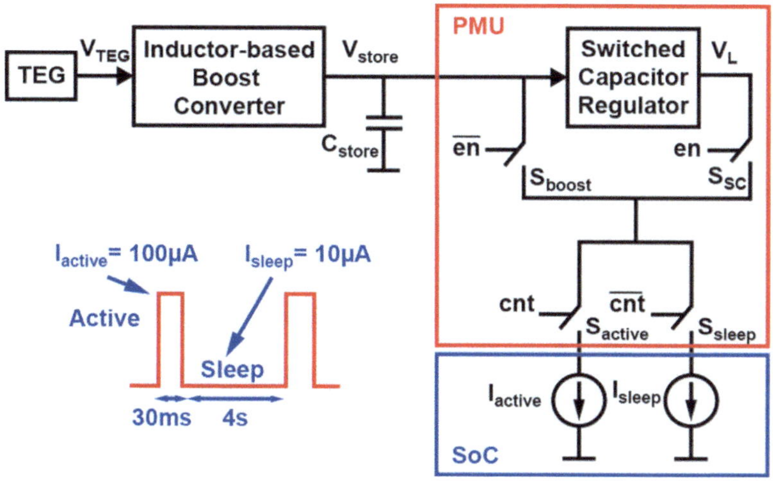

Fig. 18.5 Architecture of the proposed PMU in μWatt SoCs

50 μA and 0.6 V output voltage is 45%. In contrast, the boost converter efficiency at the same condition is around 75% with output voltage of 1.1 V. One should note that the boost converter output is not regulated; however, as long as the boost converter can support the SoC to perform its task, then this scheme can be used.

Typically, during the active period of the SoC, the current required for the SoC is high such that it is supported mainly from the storage capacitor. As such, C_{store} should be chosen carefully such that the droop in V_{store} is low enough to achieve the regulation by the switched capacitor. For our system, C_{store} is chosen to allow a droop on V_{store} by a maximum of 100 mV. As such, C_{store} can be calculated using the equation:

$$C_{store} \geq \frac{100\mu A \times 30\,ms}{100\,mV} \rightarrow C_{store} \geq 30\mu F \tag{18.2}$$

As such, C_{store} is chosen to be 47 μF. At this value, the droop is 64 mV. The time required to recover the droop in the voltage is shown in equation:

$$t_{charging} = \frac{47\mu F \times 64mV}{(50\mu A - 10\mu A)} \rightarrow t_{charging} = 75.2\,ms \tag{18.3}$$

which is much smaller than the available time in sleep period. Note that the available charging current from the boost converter at V_{TEG} of 50 mV is 50 μA.

Figure 18.6 shows the time domain signals for the setup shown in Fig. 18.5. As depicted in the figure, *cnt* signal is activated for 30 ms. During this period, V_{store} has a droop of approximated 60 mV while V_L maintains a regulated 1 V output voltage.

Figure 18.7 displays the end-to-end efficiency for the EHIC at different load current, input voltages V_{TEG} and output voltages V_L. Inspection of Fig. 18.7 leads to several key observations regarding the EHIC performance. First, the efficiency of the

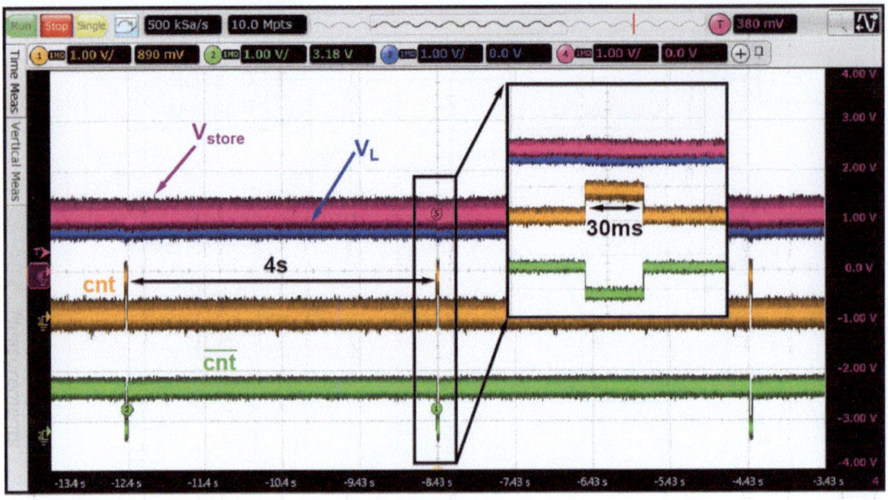

Fig. 18.6 Measured time domain signal in sleep mode model for the SoC

EHIC increases linearly as the load current increases since the output power increases linearly while the input power is kept constant at fixed V_{TEG} .Further, a wide range of load current is achieved with maximum of 300 μA at multiple output voltages, namely 0.6, 0.8, and 1 V. In particular, at V_{TEG} of 35 mV, the EHIC is capable only of supporting 0.6 V at 30 μA which is considered a low-energy mode for the SoC. As such, the SoC should adapt to different interruptions such that it goes to sleep or low-energy mode when necessary. Note that the end-to-end peak efficiency is 71% at output power of 182 μW, V_L of 0.8 V, and V_{TEG} of 100 mV.

Finally, Table 18.1 compares the performance of the proposed EHIC to recent prior work. As shown in the table, the prototype presented in this paper supports DVS due to the switched capacitor regulator. In addition, the maximum power efficiency achieved by this work is 71% at lower input voltage compared to [15]. As such, the implementation of DVS do not compromise much the efficiency compared to [15].

18.4 Analog Interface for the ECG Sensors

The first on-chip circuit block responsible for acquiring and digitizing the raw ECG data is the analog front-end (AFE). There are some general AFE specifications that are followed in the literature especially when the application is ECG. The amplitude of the ECG signal is too small, which sets a design challenge on the system's noise requirement. Further, a DC offset of ±300 mV is created due to the electrode–skin interface. Moreover, the AFE is interfaced directly with the ECG electrodes. The commercial electrodes have few mega ohms contact impedance; hence, it is evident that the input impedance of the AFE should be much higher.

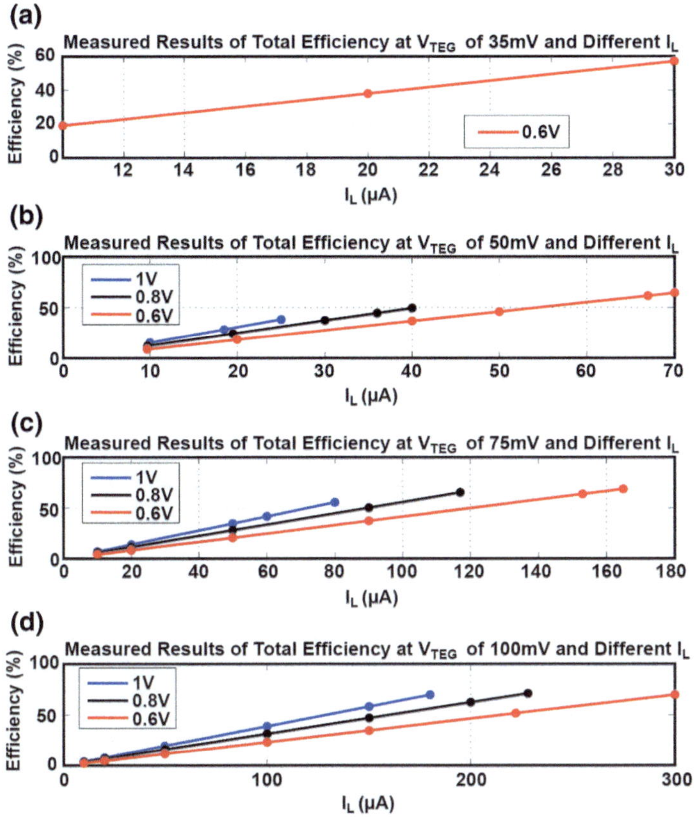

Fig. 18.7 Measured end-to-end efficiency of the EHIC at **a** $V_{TEG} = 35\,\text{mV}$ **b** $V_{TEG} = 50\,\text{mV}$ **c** $V_{TEG} = 75\,\text{mV}$ **d** $V_{TEG} = 100\,\text{mV}$, and regulated output voltage of 0.6, 0.8 and 1.0 V

Table 18.1 Comparison table between the proposed EHIC and the state of the art

Comparison	JSSC, 2010 [18]	JSSC, 2011 [11]	JSSC, 2013 [10]	This work
Process	0.13µm CMOS	0.35µm CMOS	65nm CMOS	65nm CMOS
TEG-based converter structure	L-based boost converter	L-based boost and L-based buck converter	L-based boost converter	L-based boost and switched-capacitor buck converter
Minimum input voltage V_{TEG}	20mV	25mV	30mV	35mV
Regulated output voltage V_L	1V	1.8V	1.2V	0.6V, 0.8V, 1V
DVS implementation	No	No	No	Yes
Peak efficiency @ V_{TEG} =50mV	66% @ P_L=40µW	46% @ P_L=50µW	63% @ P_L=100µW	65% @ P_L=42µW
Peak efficiency	79% @ P_L=100µW V_L=1V, V_{TEG}=240mV	58% @ P_L=295µW V_L=1.8V, V_{TEG}=100mV	73% @ P_L=1.75mW V_L=1.2V, V_{TEG}=135mV	71% @ P_L=182µW V_L=0.8V, V_{TEG}=100mV
Area	0.117mm^2	1.6mm^2	0.997mm^2	0.453mm^2

Table 18.2 AFE Performance Summary

Parameter	ASSCC 2010 [19]	JSSC 2014 [20]	ISCAS 2008 [21]	JSSC 2009 [22]	RFIT 2010 [23]	Miller OTA Chop. outside	Miller OTA Chop. inside	Folded Casc. Chop. inside	Folded Casc. Chop. outside
Technology	65nm CMOS	65nm CMOS	350nm CMOS	350nm CMOS	350nm CMOS	65nm CMOS	65nm CMOS	65nm CMOS	65nm CMOS
Supply Voltage	1V	0.6V	1V	1V	2.7V	1.2 V	1.2 V	1.2 V	1.2 V
Current (μA)	2.1	-	0.26	0.33	9.9	1.6	1.6	0.858	0.858
Vrms (μV)	6.7	5.8	2.1	2.5	0.97	0.22	2	4	1.2
CMRR (dB)	110	80	67	71.2	114	65	42	70	164
Input Impedance	80M	10M	-	-	1G	5M	10	10	5M
Gain (dB)	40	36.92	45.5	40	-	40	39.7	39.5	39.5
BW (Hz)	-	0.117 - 547	0.05 - 200	0.04 - 460	-	0.5 – 10K	0.6 – 2K	0.2 – 4K	0.3 – 9K

The AFE consists of the preamplifier with noise cancelation scheme and the analog-to-digital converter (ADC). A chopping technique is implemented to mitigate the DC offset and noise. The preamplifier is followed by GM-C filter to filter out the chopping ripples. A programmable variable gain amplifier (PGA/VGA) then follows to guarantee maximum amplitude fed to the ADC. Finally, the ADC digitizes the analog signal to be further processed by the SoC.

The whole system performs at a minimum dynamic range of 40 dB and SNDR of 32 dB so that it is able to deal with the ECG signal of as low as 100 μV. The placement of the choppers with two amplifier topologies is investigated. The spice simulation results of the different orientation of the input and output choppers for folded cascode and Miller amplifiers are summarized in Table 18.2. Applying chopper technique outside the feedback loop for both amplifiers enhances the noise performance. A chopper technique outside the feedback loop gives better CMRR and noise performance, but worse input impedance. Folded cascode is recommended over the Miller amplifier, as it is more robust across process corners and consequently has better CMRR on top of the lower power consumption. The table also shows the comparison with other work in the literature [16–20].

There are many ADC topologies that are reported in the literature. The fact that the ADC does not need to operate at a fast clock makes the SAR topology an optimum choice. After performing several behavioral analysis, the requirement for our system is an 11 bit SAR ADC to digitize the ECG signal. The ECG is sampled at 1KS/S where the digitization is done at a frequency of sampling frequency multiplied by the number of bits, namely 11 kHz.

18.5 SoC Implementation

Prediction systems would aid in reducing deaths due to sudden heart failures such as sudden cardiac death (SCD). SCD originates from ventricular arrhythmia (VA), and VA detecting systems have been reported in [21, 22]; however, there are only few systems that do prediction. A prediction system that can predict VA upto 3 hour prior its onset is reported in [14]. In order to perform detection or prediction, the first step is to extract ECG features and define parameters that could generate accurate classification results. ECG features include P, QRS, and T waves that could be utilized to define factors and intervals for prediction or detection systems. Various techniques have been developed for extracting such features where some of them use discrete wavelet transform (DWT) [23] and time domain analysis [24]. Efficient processing platforms are crucial to enable ultralow-power operation for wearable devices which requires optimization at the algorithm level and at architectural level.

18.5.1 ECG Processing Architecture

The proposed ECG system architecture is shown in Fig. 18.8. It consists of ECG feature extraction, feature selection, and VA prediction. The first stage is ECG feature extraction that delineates the ECG fiducial points. The second stage evaluates the features required for the classification system. The final stage is a VA prediction based on naive Bayes-based classifier. A custom designed FSM controls all the parts of the architecture and coordinates the timing of each stage. A data bus width of 16 bits is used and scaling is done whenever necessary during the arithmetic operations such as multiplication. Clock gating is applied as a power reduction technique especially for the ECG feature extraction block.

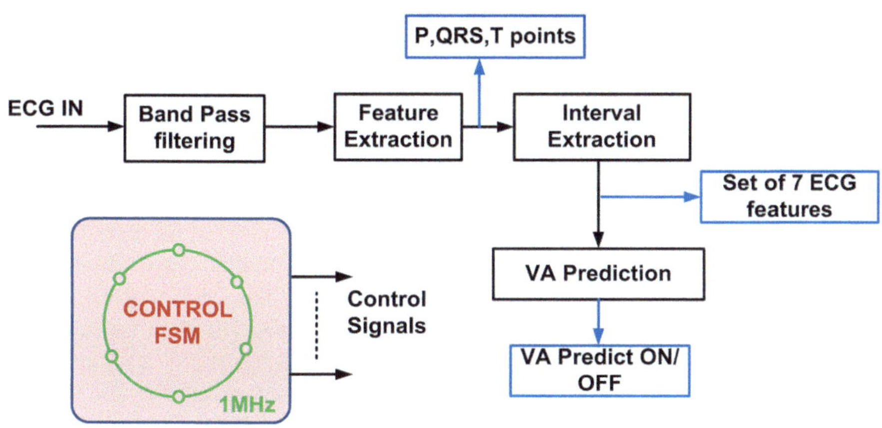

Fig. 18.8 Block diagram of the ECG processing architecture

18.5.2 ECG Feature Extraction Architecture

The architecture for feature extraction is constructed based on an algorithm described in [25]. Based on this algorithm, an architecture is developed and implemented. In the implemented architecture, the ECG processing comprises preprocessing, transformation, and delineation. The ECG preprocessing is realized using a band-pass filter in which it enhances the signal-to-noise (SNR) of the ECG wave. Moreover, the filter removes low-frequency artifacts such as baseline wander and motion artifacts as well as high-frequency interference. In the succeeding stage, a transformation using curve length transform (CLT) is applied which enhances QRS complex. QRS peaks are detected using a threshold in the transformed ECG signal. The last step in the feature extraction is delineating T and P wave. The delineation is accomplished using DWT. All the ECG feature extraction operations are controlled by a custom FSM.

18.5.3 ECG Feature Selection for VA Prediction

Extracted ECG features could be utilized for detecting and predicting various cardiac problems. Here, a VA prediction system is described that utilizes ECG wave intervals. A set of seven intervals, namely RR, PQ, QP, RT, TR, PS, and SP intervals, are selected that give high accuracy for VA prediction. These intervals are input to the VA prediction system. This system is constructed using naive Bayes classifier and is integrated along with the feature extraction and feature selection. The Bayesian classifier uses Bayes theorem to find out the probability of data belonging to a particular class of observations. The naive Bayes classifier being a linear classifier has advantages in hardware implementation over other complicated classification techniques.

In the classifier architecture, the main operations are finding the mean, standard deviation, and conditional probability. The arithmetic operations required are addition, subtraction, and division. For each of the features used the conditional probability is evaluated, and the commutative log score is utilized for classification.

Overall the proposed architecture is computationally efficient. If we look at the feature extraction, it is performed using CLT and DWT. The CLT is pipelined and requires only one square, one square root, and summation for each sample. Further, the DWT is implemented as cascade of filters. The filter coefficients are 2 and 1/2 which are implemented using shift and do not require multiplication or division. Besides, the DWT is performed on a small length of the signal on a window which is the probable location of TP waves. Only 256 bytes of memory are required for the DWT since only a single scale is used for the delineation process and the maximum window size does not exceed 256. The feature selection and feature classification were implemented heavily using combinatorial circuits which aid in reducing the overall power consumption. The VA prediction requires finding conditional probabilities, and logarithmic scores were used which avoids the need for

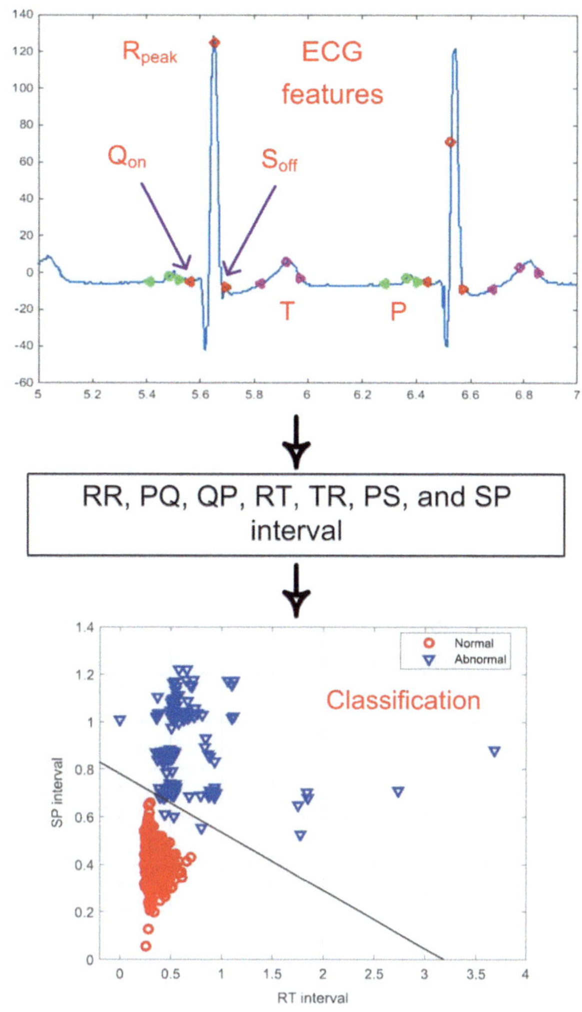

Fig. 18.9 ECG feature extraction, feature selection, and VA classification demonstration

complex multiplication and division operations. Results shown in Fig. 18.9 demonstrate ECG features extracted using the test chip, features selection, and VA prediction indicator.

18.6 System Level Integration

Most IoT devices utilize different modes of operation depending on the available power budget. Figure 18.10a shows a general scheme where the device is active for a small period of time and then goes to sleep to reduce power consumption. Typically, the power consumption is dominated by leakage. Hence, the device should

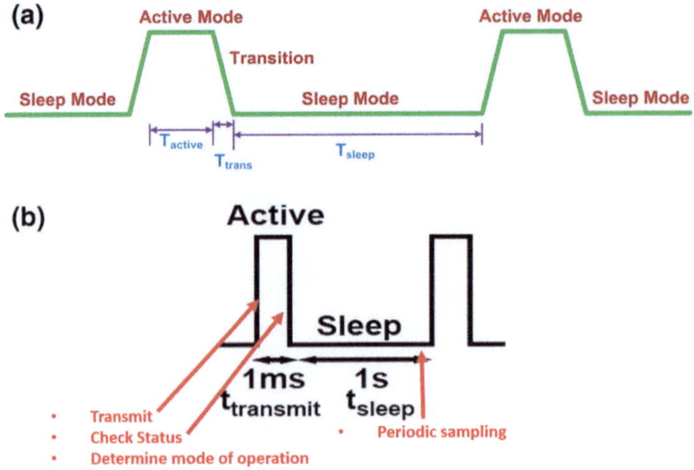

Fig. 18.10 **a** A general sleep model in IoT devices **b** The sleep model implementation for the system

be carefully designed to reduce the overall leakage current. Figure 18.10b shows the scheme adopted in our SoC. The wireless transmitter is active for 1 ms to send the data outside the chip, and then it is turned off for 1 s. During the sleep mode, the AFE and the ADC are active to acquire the ECG signal. Further, the SoC executes the ECG data at the end of the sleep mode so that the data is ready for transmission before the active mode. Note that the sleep mode is also important for the energy harvesting module since it allows the storage capacitor to recover the charge that was consumed during the active mode.

Figure 18.11 shows a visualization of a possible prototype that includes all the system components inside a necklace. There will be two ECG electrodes on front

Fig. 18.11 A visualization concept of the overall system integrated as part of a necklace

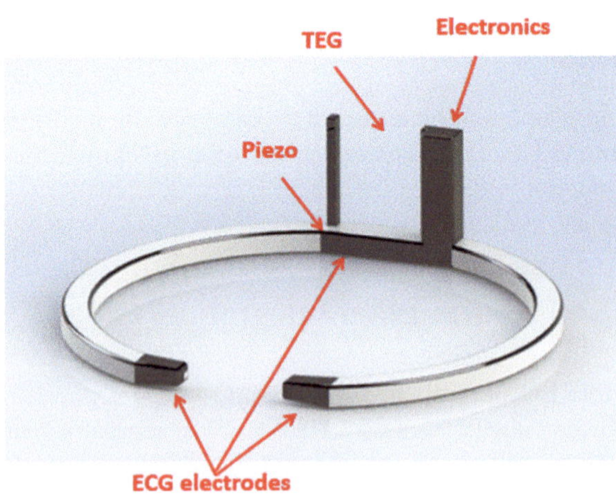

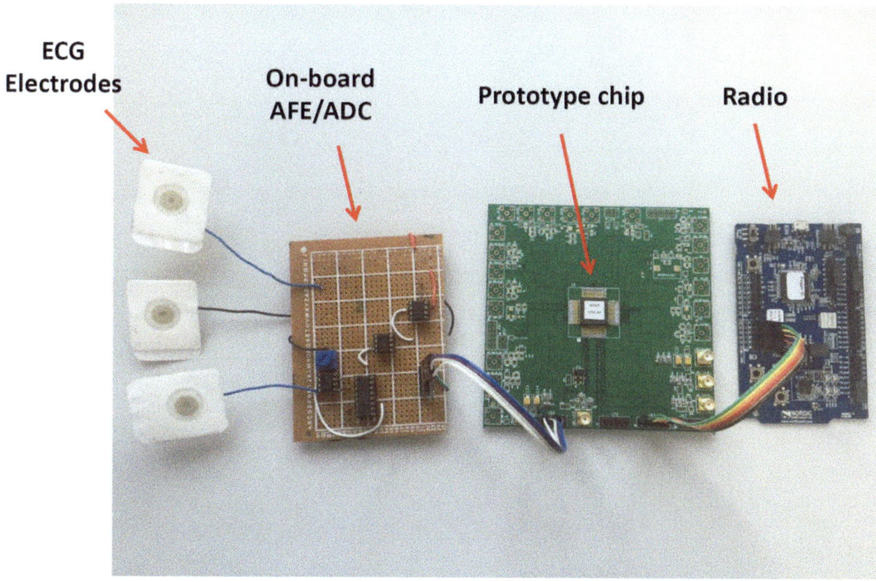

Fig. 18.12 A prototype of the overall system

side of the user's neck and a reference point on the back of the neck. The SoC will be placed in a compartment on the back of the neck along with a TEG to harvest energy from the body heat. A possible improvement in this design is to create the necklace using an elastic material to accommodate people with different neck circumferences.

To prove the concept of the overall design concept, a prototype system has been implemented. Figure 18.12 shows the complete system that includes the ECG electrodes, off-the-shelf-AFE/ADC, SoC, and a radio. In the AFE module, a TI instrumentation amplifier (INA116) is used, along with the INA116 and TI op-amp (UA741) to amplify the signal. Further, a right-leg drive circuit is implemented to cancel 50 Hz noise which allows a less-noisy ECG output signal to be produced. Once the signal is amplified, an ADC is used to convert the signal to digital to be processed by the SoC. The SoC extracts all the necessary features from the ECG signal and sends it serially as a 16-bit signal through an SPI to an Arduino board as in Fig. 18.13. The Arduino along with a SeeedStudio Bluetooth shield is then able to transmit the signal via Bluetooth to a mobile device. From this mobile device, a user can see the ECG signal in real time.

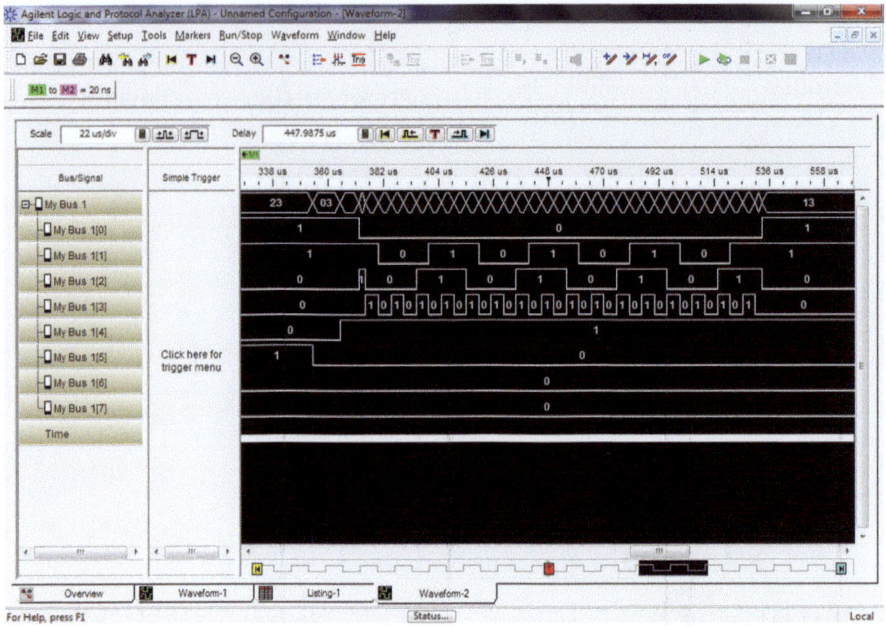

Fig. 18.13 SPI interface results between the SoC and the wireless interface

Figure 18.14a shows a detailed block diagram of the self-powered biomedical SoC. The figure presents in details the top-level design of each block and the interface between various modules. The energy harvesting block provides two regulated voltages, V_{low} and V_{high}. V_{low} is used to power the AFE and the processor while V_{high} powers the transmitter. The biomedical processor transfers the data to the transmitter serially which will be transmitted wirelessly using on–off keying modulation. The complete system shown in Fig. 18.14a is fabricated on one chip as shown in Fig. 18.14b. In addition, Fig. 18.14b shows a summary of performance of each block in the biomedical SoC.

(a)

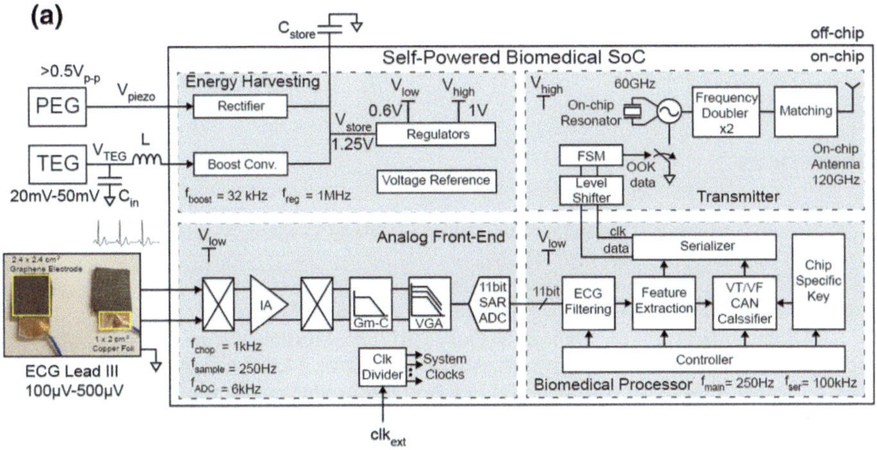

(b)

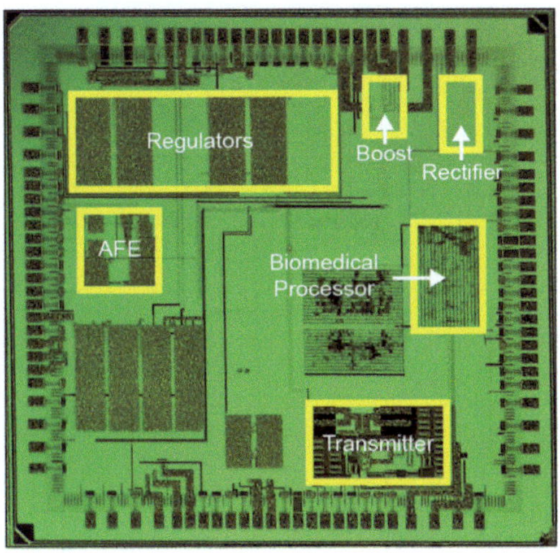

Technology		65nm
Die area		3 x 3 mm^2
EH	V_{in}	20 mV-50 mV
	V_{out}	1.25 V
	η_{max}	70%
	Start-up	Vibration @>0.5 V_{p-p}
Supp Reg.	V_{in}	>1.05 V
	V_{out}	0.6 V,1 V
	η_{max}	71%
	$I_{quiescent}$	3-6µA
AFE/ ADC	Supply	0.6 V
	Power	1 µW
	Gain	28-48dB
	BW	320 Hz
	CMRR	80 dB
	ADC bits	11 bits
	ADC Sampling	6 kHz
	Input Referred Noise	80nV/√Hz ,0.957µV (rms)
DSP	Supply	0.6 V
	Power	500 nW
	Clock Frequency	250 Hz
	Total Memory	2kByte
TX	Supply	1 V
	Power	2.4 mW
	Data rate	up to 10Mbps
	Output Power	-25 dBm
	Modulation	OOK
	Band	120 GHz

Fig. 18.14 **a** Detailed block diagram of the complete system **b** Die photo and performance table

18.7 Conclusions

This chapter presented a top-level design of a self-powered SoC system for biomedical application. The system acquires ECG signals and performs feature extraction and classification to accurately predict ventricular arrhythmia. The complete system has been demonstrated using off-the-shelf components to prove the concept, and then all the modules have been fabricated on-chip. The proposed platform could monitor a patient's condition continuously for long time with minimum maintenance and low cost.

Acknowledgements This work has been supported by the Semiconductor Research Corporation (SRC) under the Abu Dhabi SRC Center of Excellence on Energy-Efficient Electronic Systems

(ACE^4S), Contract 2013 HJ2440, with funding from the Mubadala Development Company, Abu Dhabi, UAE.

References

1. E. Nemati, M.J. Deen, T. Mondal, A wireless wearable ECG sensor for long-term applications. IEEE Commun. Mag. **50**(1), 36–43 (2012)
2. Y.M. Chi, T.-P. Jung, G. Cauwenberghs, Dry-contact and noncontact biopotential electrodes: methodological review. IEEE Rev. Biomed. Eng. **3**, 106–119 (2010)
3. J.-Y. Baek, J.-H. An, J.-M. Choi, K.-S. Park, S.-H. Lee, Flexible polymeric dry electrodes for the long-term monitoring of ecg. Sens. Actuators A: Phys. **143**(2), 423–429 (2008)
4. Y.A. Samad, Y. Li, S.M. Alhassan, K. Liao, Non-destroyable graphene cladding on a range of textile and other fibers and fiber mats. RSC Adv. **4**(33), 16935–16938 (2014)
5. M.K. Yapici, T. Alkhidir, Y.A. Samad, K. Liao, Graphene-clad textile electrodes for electrocardiogram monitoring. Sens. Actuators B: Chem. **221**, 1469–1474 (2015)
6. C. Lu, V. Raghunathan, K. Roy, Efficient design of micro-scale energy harvesting systems. IEEE J. Emerg. Sel. Top. Circuits Syst. **1**, 254–266 (2011)
7. P.-S. Weng, H.-Y. Tang, P.-C. Ku, L.-H. Lu, 50 mv-input batteryless boost converter for thermal energy harvesting. IEEE J. Solid-State Circuits **48**(4), 1031–1041 (2013)
8. Y. Ramadass, A. Chandrakasan, A battery-less thermoelectric energy harvesting interface circuit with 35 mv startup voltage. IEEE J. Solid-State Circuits **46**(1), 333–341 (2011)
9. P. Luo, S. Zhen, J. Wang, K. Yang, P. Liao, X. Zhu, Digital assistant power integrated technologies for pmu in scaling cmos process. IEEE Trans. Power Electron. **29**(7), 3798–3807 (2014)
10. L.G. Salem, P.P. Mercier, An 85%-efficiency fully integrated 15-ratio recursive switched-capacitor dc-dc converter with 0.1-to-2.2 v output voltage range, in *ISSCC* (IEEE, 2014), pp. 88–89
11. J. Pyo, Y. Shin et al., 20nm high-k metal-gate heterogeneous 64b quad-core cpus and hexa-core gpu for high-performance and energy-efficient mobile application processor, in *ISSCC* (IEEE, 2015), pp. 1–3
12. H. Singh, K. Agarwal, D. Sylvester, K.J. Nowka, Enhanced leakage reduction techniques using intermediate strength power gating. IEEE Trans. Very Large Scale Integr. (VLSI) Syst. **15**(11), 1215–1224 (2007)
13. M. Alioto, Ultra-low power VLSI circuit design demystified and explained. Trans. Circuits Syst. I **59**(1), 3–29 (2012)
14. N. Bayasi, T. Tekeste, H. Saleh, B. Mohammad, A. Khandoker, M. Ismail, Low-power ecg-based processor for predicting ventricular arrhythmia. IEEE Trans. Very Large Scale Integr. (VLSI) Syst. **99**, 1–13 (2015)
15. E. Carlson, K. Strunz, B. Otis, A 20 mv input boost converter with efficient digital control for thermoelectric energy harvesting. IEEE J. Solid-State Circuits **45**(4), 741–750 (2010)
16. Q. Fan, F. Sebastianen, H. Huijsing, K. Makinwa, A 2.1 μw area-efficient capacitively-coupled chopper instrumentation amplifier for ecg applications in 65 nm cmos, in *2010 IEEE Asian Solid State Circuits Conference (A-SSCC)* (IEEE, 2010), pp. 1–4
17. Y.-P. Chen, D. Jeon, Y. Lee, Y. Kim, Z. Foo, I. Lee, N.B. Langhals, G. Kruger, H. Oral, O. Berenfeld et al., An injectable 64 nw ecg mixed-signal soc in 65 nm for arrhythmia monitoring. IEEE J. Solid-State Circuits **50**(1), 375–390 (2015)
18. X. Zou, X. Xu, J. Tan, L. Yao, Y. Lian, A 1-v 1.1-μw sensor interface ic for wearable biomedical devices, in *2008 IEEE International Symposium on Circuits and Systems* (IEEE, 2008), pp. 2725–2728
19. X. Zou, X. Xu, L. Yao, Y. Lian, A 1-v 450-nw fully integrated programmable biomedical sensor interface chip. IEEE J. Solid-State Circuits **44**(4), 1067–1077 (2009)

20. X. Pu, L. Wan, Y. Sheng, P. Chiang, Y. Qin, Z. Hong, A wireless 8-channel ecg biopotential acquisition system for dry electrodes, in *2012 IEEE International Symposium on Radio-Frequency Integration Technology (RFIT)* (IEEE, 2012), pp. 140–142
21. C.-Y. Chiang, H.-H. Chen, T.-C. Chen, C.-S. Liu, Y.-J. Huang, S.-S. Lu, C.-W. Lin, L.-G. Chen, Analysis and design of on-sensor ecg processors for realtime detection of vf, vt, and pvc, in *2010 IEEE Workshop on Signal Processing Systems (SIPS)* (2010), pp. 42–45
22. Q. Li, C. Rajagopalan, G. Clifford, Ventricular fibrillation and tachycardia classification using a machine learning approach. IEEE Trans. Biomed. Eng. **61**, 1607–1613 (2014). June
23. J. Martinez, R. Almeida et al., A wavelet-based ecg delineator: evaluation on standard database. IEEE Trans. Biomed. Eng. **51**(4), 570–581 (2004)
24. N. Bayasi, T. Tekeste, H. Saleh, A. Khandoker, B. Mohammad, M. Ismail, Adaptive technique for p and t wave delineation in electrocardiogram signals, in *2014 36th Annual International Conference of the IEEE Engineering in Medicine and Biology Society (EMBC)* (2014), pp. 90–93
25. T. Tekeste, N. Bayasi, H. Saleh, A. Khandoker, B. Mohammad, M. Al-Qutayri, M. Ismail, Adaptive ecg interval extraction, in *2015 IEEE International Symposium on Circuits and Systems (ISCAS)* (IEEE, 2015), pp. 998–1001

Chapter 19
Toward an Integrated, Low-Power Platform for Continuous Congestive Heart Failure Monitoring

Shahzad Muzaffar, Ayman Shabra and Ibrahim (Abe) M. Elfadel

19.1 Introduction

Heart failure is one of the leading causes of death and morbidity in developed countries and is expected to reach endemic levels by the year 2040 [2]. Patients with heart failure are prone to frequent hospitalizations, most commonly due to edema, which is the buildup of fluids in the body. This was the major contributor to the total $37.2 billion annual cost of caring for these patients in 2009 in the USA [2]. Normal care in the form of regular physician visits does not provide the necessary safeguards to detect nor prevent the rapidly deteriorating conditions associated with decompensation experienced by the patients in the days or weeks leading up to hospitalization [5]. Early detection is essential for maintaining the patient's well-being.

Among chronic conditions, cardiac and pulmonary disorders will profit the most from the development of wearable, accurate remote patient monitoring systems [19]. There are currently large-scale patient studies for developing algorithms to detect the onset of heart failure decompensation with focus on sensitivity and specificity [2]. To be implemented in hardware, such algorithms require the use of energy-efficient sensing front ends, which will include the non-invasive monitoring of body fluids, heart rate, electrocardiogram (ECG), respiration, and activity level [2, 5]. Patients with decompensation conditions also suffer from difficulties in maintaining posture and balance and therefore prone to falling. It is, therefore, desirable to also include sensors that detect abnormalities in posture, balance, and gait. Furthermore, it is

S. Muzaffar · Ibrahim (Abe) M. Elfadel (✉)
Masdar Institute at Khalifa University, Abu Dhabi, UAE
e-mail: ielfadel@masdar.ac.ae

S. Muzaffar
e-mail: smuzaffar@masdar.ac.ae

A. Shabra
MediaTek, Woburn, MA, USA
e-mail: aushabra@yahoo.com

Ibrahim (Abe) M. Elfadel and M. Ismail (eds.), *The IoT Physical Layer*,
https://doi.org/10.1007/978-3-319-93100-5_19

critical to detect subtle but consistent changes in the activity levels of heart failure patients, including lower ambulatory activity and longer sleep patterns. Such patterns are very challenging to monitor using traditional techniques such as pedometers, which are known to provide erroneous readings in patients moving at slow speeds [8].

Several heart failure monitoring platforms have been developed in recent years for detecting acute decompensation. The low-power sensing, processing, and communication technologies underlying such platforms have also been under intense development over the past decade. Several methodologies are now available for outpatient congestive heart failure monitoring. One of them is the Home Automated Telemanagement (HAT) system, which queries the patient daily about their health condition, monitors their weight, and provides her with feedback on her health status [4, 9]. Not only do these systems need a central server but also they require the visits of a care management team to educate the patient. They are also very prone to the human error resulting from the patient entering inaccurate information. Another source of grave error is due to the irregular intervals at which queries and measurements are taken each day. Such irregularity is detrimental to good care management. Game systems such as Xbox are used as user-friendly interface, with centralized cardiac monitoring systems [10], but the underlying issues remain the same, with no observed improvement in healthcare outcomes. Another cardiac monitoring system consists of chest-mounted bipolar ECG electrodes and acoustic transducers to detect CHF based on the phase shifts of cardiac and thoracic acoustics that are also coupled with ECG signals [12]. The measured signals require additional spectral analysis and result in heavy post-processing loads and power hungry units. Microwave systems for early-stage CHF detection scan vertically the rear side of the human torso and are similar in operation and acquisition to monostatic radar [20]. The system is comprised of a microwave transceiver, equipped with a wideband folded antenna to scan the patient, and a computer system to analyze and visualize the information statically. However, the system requires a special location for its setup, and patients need to visit it regularly to get themselves scanned. Another intelligent mHealth home monitoring system, MyHeart [1], has been recently proposed and is currently under evaluation at Linda University Medical Center. This system also supports the CHF-care by providing a means of communication between the experts and the patient through a mobile application and a wireless health device. The hazard associated with this kind of system is the patient's non-expert observation and reporting on her own conditions which may lead medical experts to provide an inappropriate diagnosis. Yet another approach [7] is the blue-scale system that measures a variety of vital parameters including weight, body impedance, electrocardiography (EKG), and photoplethysmography (PPG). Experiments have shown that the measurements need additional signal processing to achieve acceptable accuracy. Furthermore, activities and conditions prior to measurements may cause severe variability in the collected data sets which, in turn, lead to uncertainties in healthcare outcomes.

Impedance, optical absorption, and weight measurements are some of the approaches that can be used to determine changes in body fluids. Current medical practice relies on weight measurements to determine changes in body fluid decompensation. Therefore, shoe-mounted sensors are most appropriate for designing wearable CHF system. Not only can they be used for measuring continuously the actual weight but also their information can be post-processed to get body fluid decompensation, thus generating all the vital signals used for CHF prediction. Recent medical studies support the use of changes in weight to achieve the following:

- Detect patterns preceding hospitalization [6]
- Predict imminent hospitalization due to the deterioration in heart conditions [23].

The integration of the medical information on CHF pattern detection and hospitalization prediction in hardware requires configurable platforms that can easily incorporate design improvements in its individual modules such as sensors, processor, power, communication, and storage. To make CHF monitoring based on weight changes measurement, wearable, it has to satisfy several requirements, including:

- Ultra-low power consumption
- Comfortable wearability
- Accuracy so as to avoid false alarms about hospitalization
- Robustness with respect to activity and lifestyle changes.

To help evaluate the figures of merit of a wearable CHF system based on the above metrics, we present, in this chapter, a full system architecture and proof-of-concept FPGA platform for prototyping and analysis. The platform is comprised of three main subsystems:

- Sensory subsystem with two sensor nodes, one for each foot to measure body weight
- Back-end system-on-chip, to analyze weight measurements
- Communication subsystem that relies on body channel transmission to send data from the sensors to the back-end.

To exercise this platform, we have implemented two walk cycle emulators. The first emulator is comprised of a potentiometer which generates the two complementary sinusoid signals using a network of resistors and capacitors. The second emulator is based on the analysis of real-world human walk cycles and incorporates physics-based concepts for analyzing weight and gait. The system composed of the FPGA platform, and the walk cycle emulator has been tested and shown to be able to provide continuous information on walk cycle (and therefore weight measurements) with a sampling rate of 1MSPS, a configurable buffer size of up to maximum of 16 bits, 1024 words, average data rate of 4.1 Mb/s, and low power consumption.

Such platform can also be reused to integrate other measurements such as posture, speed, and acceleration. Such measurements can help in determining gait which can also contribute to improving the accuracy of CHF prediction.

This chapter is organized as follows. Section 19.2 describes the proposed CHF monitoring system architecture with a detailed explanation of its main building blocks such as sensor nodes, digital assistant, body communication channel (BCC), and transmission protocol. The weight measurement methodology is discussed in Sect. 19.3. Section 19.4 focuses on the walk cycle demonstrator prototype. The walk cycle models used in the prototype are described in Sect. 19.5. Section 19.6 highlights the FPGA implementation and the experimental results. A development methodology in Sect. 19.7 shows the ease with which the proposed system architecture can be prototyped. Finally, the conclusions are given in Sect. 19.8.

19.2 System Architecture

The proposed system architecture, shown in Figs. 19.1 and 19.2, divides the system into four main parts: footwear sensor nodes as a measurement mechanism, body skin medium as a body channel communication (BCC), ultra-low-power communication protocol, and digital assistant. The footwear sensor nodes continuously sample and store the weight measurements which later are transmitted to the digital assistant for further processing. Transmission is carried out through the human body using BCC transceivers and an appropriate single-wire communication protocol. Received information at digital assistant is then processed for weight estimation. Based on the analysis results, the system detects the CHF conditions, and the patient is notified accordingly in an appropriate visual and audio formats. The results can also be transmitted to the healthcare team for an expert advice and for the record. Details on the system's building blocks are given in the following subsection.

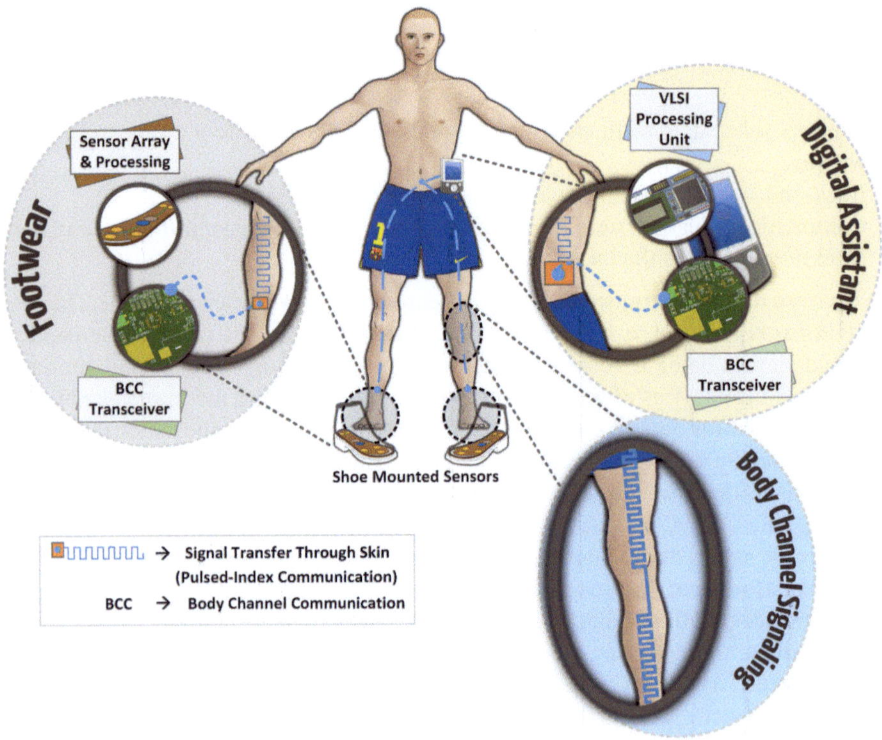

Fig. 19.1 Continuous CHF monitoring system architecture

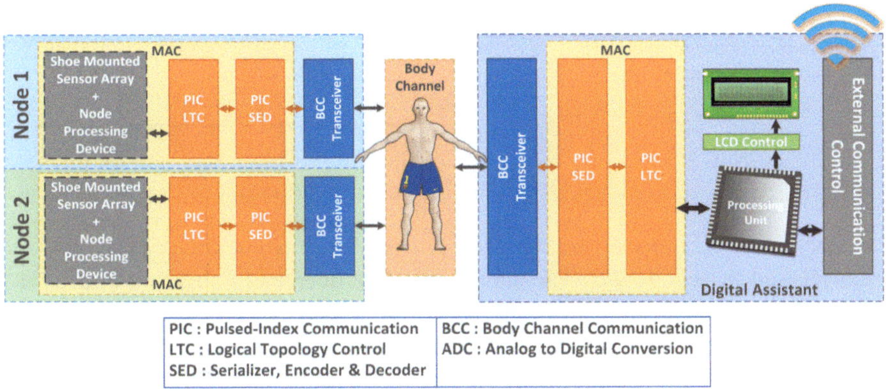

Fig. 19.2 Block diagram of continuous CHF monitoring system architecture

19.2.1 Weight Measurement Subsystem

Sensor nodes are part of the sensing and measurement subsystem that determines weight changes by sampling pressure changes on footwear-mounted sensors. These sensors are also used to measure activity levels. The subsystem is divided into two parts, the sensors and the processing device, which are discussed in the following subsections.

Weight Sensors

Weight is being measured using, for instance, shoe sole-mounted piezoresistive pressure sensors, as shown in Fig. 19.1. A preliminary evaluation of such sensors is already presented in Chap. 2. Due to the flexibility, cost effectiveness, comfort, and easy insole application, the semi-conductive polymer composite material is typically preferred over strain gauges for the design of weight sensors although it suffers from severe history and memory effects. The more the pressure is applied, the more the resistance increases. This increase in resistance is used to convert pressure into weight. The details on extracting the weight information from the measured force (in newton) are presented in Sect. 19.3. The proposed platform is meant to use pressure/force measurements to certify a variety of weight and gait analysis algorithms irrespective of the used sensing mechanism. Depending on the sensing system, the sensing and processing algorithm for the sensor node processing device can be updated to support a range of available sensor subsystems. The sensor nodes processing device isolates the sensing mechanism from the rest of the system, and therefore, allows the use of various kinds of available sensor insoles and shoes.

Sensor Node Processing Device

To measure weight accurately, the sensor nodes are equipped with a processing device. The processing device is connected to the body using appropriate electrodes in order to transmit the measurements through body channel to the digital assistant

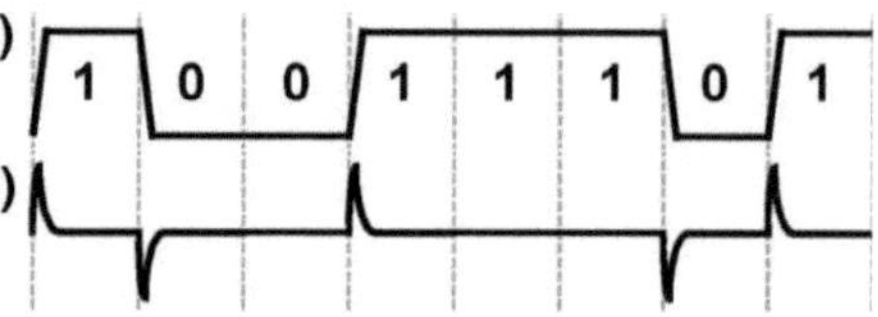

Fig. 19.3 Digital signal through human body: **a** input signal and **b** output signal

for analysis. BCC transceivers are used to establish this communication link and are discussed in Sect. 19.2.3. The node processing device is comprised of a variety of modules including memory to store the measurements, Power Management to wirelessly recharge the batteries and to save power, signal processing (DSP), control system (FSM and User Detect) to monitor the shoe-mounted sensor arrays via analog front end (AFE), and a Time Base to facilitate the synchronization with the digital assistant.

19.2.2 Body Communication Channel

The ability to transmit and receive data at a very low energy level per bit is an important characteristic of the CHF monitoring system. Body channel communication (BCC) based on the use of the human skin as a communication channel has demonstrated impressive figures of merit [21]. This mode of communication is being used primarily between the sensor and the digital assistant. The human body behaves as a band-pass filter (BPF) in the frequency domain [21]. Therefore, as shown in Fig. 19.3, for a square wave input the positive and negative pulse signals are generated at the output of the human body channel. BCC transceivers are responsible of dealing with signal deformations to recover the transmitted data successfully. BCC transceivers are discussed in Sect. 19.2.3.

19.2.3 Digital Assistant

The digital assistant is the main block of the CHF monitoring system which is responsible not only for the analysis of the patient's conditions but also for controlling the information transmission to and from sensor nodes by means of a single-channel communication protocol. It also helps in sending the information to the medical experts for an advice based on the detected edema conditions. The digital assistant can be divided into four main building blocks:

- Processing unit
- Body channel communication control
- External communication control
- BCC transceivers.

Processing Unit

The processing unit focuses on analyzing the data received through the measurement and communication subsystems. The processing unit is also responsible for the control of all the building blocks and manages the communication both with the external world and the footwear sensor nodes. It synchronizes the system flow in order to carry out a successful operation. The processing unit also notifies the patient in visual and audio formats, especially to hear alert about critical conditions.

Channel Communication Control

Channel communication control, which lies between the processing unit and the BCC transceivers, manages the outgoing and incoming data transfers employing a body channel communication protocol, described earlier in Chap. 11 and also briefly in this Chapter in Sect. 19.2.4. The module takes care of the successful collection of data from each node and provides the data next to the processing unit for further processing.

External Communication Control

This module is responsible for establishing a communication link with the external world to provide connectivity to the experts at healthcare centers. Using this module, the system connects to the wireless service providers through wireless transceivers which can be interfaced using standard interfacing protocols such as I2C and SPI.

BCC Transceiver

The BCC transceiver scheme is based on the one described in Chap. 8, see also [21, 22]. The scheme is meant to use a wideband signaling (WBS) digital transceiver chip by exploiting a direct digital transmitter and an all-digital clock and data recovery (CDR) circuit. To further reduce power consumption, this scheme incorporates a low-voltage digitally-controlled oscillator and a quadratic sampling technique for the CDR. However, such a BCC transceiver consumes several milliwatts and is the main contributor to the total power consumption for a sensor node, as shown in Fig. 19.4. The communication power consumption can be reduced significantly (mW to uW) if a single-channel communication technique is used such as the Dallas Single-Wire [13]. However, such protocols have very low data rates that are not appropriate for the system under discussion. The CHF platform needs to transfer 1000 words, at least 16 bits wide, from each foot's sensor node to the digital assistant in a very short interval of time (10–20 ms) as required by the duty cycling of the BCC transceivers. Such duty cycling is essential to further reduce power consumption. To overcome the challenge of high date rates ($\approx$8 Mb/s) and ultra-low power consumption, the Pulsed-Index Communication (PIC) protocol described in Chap. 11 is used that is also reviewed briefly later in this chapter. PIC is a novel single-channel signaling technique that greatly simplifies the design of BCC transceivers. Power consumption can be reduced even more by applying additional source coding techniques to compress the data prior its transmission via PIC.

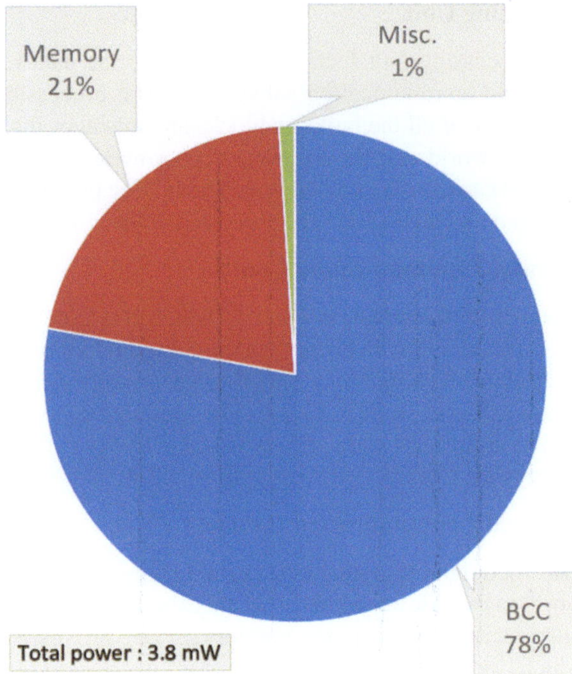

Fig. 19.4 Sensor node power consumption using traditional BCC transceivers

19.2.4 Body Channel Communication Protocol

The human body channel behaves as a single wire and therefore any of the single-channel communication techniques would require a clock and data recovery (CDR) for the successful and reliable data transfer. Previous work on body channel communication reveals that efficient and low-power CDR was one of the primary goals. During the system operation, the measured information has to be reliably communicated in real time without draining the energy resources of the digital assistant and the sensor nodes. One way to save such resources is to code efficiently the data burst, use single-channel communication, and adopt ultra-low-power communication circuit techniques. CDR circuits are typically significant consumers of energy on traditional single-channel communication protocols. Pulsed-Index Communication (PIC) [14–17], on the other hand, presents a novel single-channel protocol that does not require any CDR circuitry. The protocol is based on the novel concept of a pulsed index where data is encoded to minimize the number of ON bits, move them to the LSB end of the packet, and transmit the ON bit indices in the form of a pulse stream. The pulse count is equal to the index of the ON bit. Beside the elimination of CDR, the implementation of PIC is very area-efficient, low-power and highly tolerant of clocking differences between transmitter and receiver. Moreover, PIC supports dynamic data rates in the range of 3.1–8.5 Mb/s (an average of 4.1 Mb/s) with a 25-MHz clock. For an ASIC implementation on a 65nm technology, PIC can reduce area

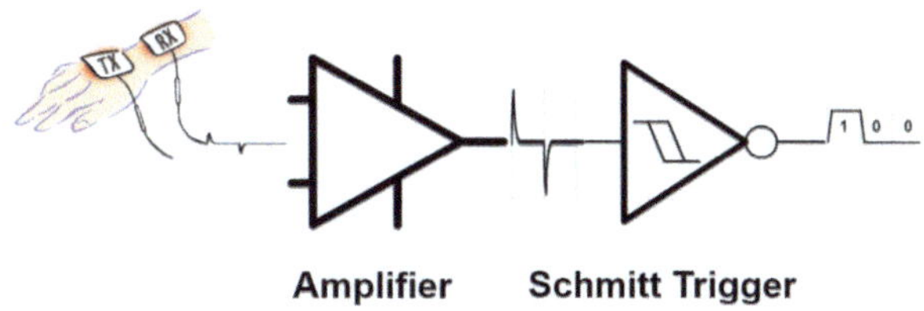

Fig. 19.5 A simple BCC transceiver

by more than 80% and power by more than 70% in comparison with a CDR-based serial bit transfer protocol. More details on PIC are presented in Chap. 11.

Based on the improved figures of merit, PIC is selected as a communication backbone of the presented system. Because the protocol's main idea is to use pulses to detect and infer data bits, the BCC transceiver circuitry gets simplified by using an amplifier and a schmitt trigger, as shown in Fig. 19.5, hence reducing the transceivers power consumption.

19.2.5 System Integration

Figure 19.2 shows the overall system integration incorporating all the modules required to realize the CHF monitoring system. PIC-SED (serializer, encoder, and decoder) and PIC-LTC (logical topology control) are part of PIC protocol. In brief, the LTC is the topology control unit and allows to implement PIC communication according to the application requirements (i.e., master–slave, ring, star). BCC transceivers, on both sides, lie in between the body channel and PIC-SED. Appropriate electrodes are used to connect transceivers to the body parts. The LTC end of PIC module is connected to the processing unit at the digital assistant end and to the node controller at the sensor node end. For the presented work, the LTC is configured in master–slave mode where the digital assistant is a master, and the sensor nodes act as slaves. Digital assistant with the help of PIC and BCC transceivers requests the nodes individually to send the measured information. The nodes answer back the request with a burst of measured weight values (typically 1000 measurements/node).

19.3 Weight Measurement

To monitor the weight change patterns, it is necessary to incorporate the walk cycle information accurate wight estimation. Additionally, the gait analysis of the subject through such a walk cycle information helps to improve the decision accuracy of CHF detection algorithms.

A walk cycle can be divided into two phases, one for each foot. As shown in Fig. 19.6, each phase is a half walk cycle that is comprised of four body poses:

Fig. 19.6 One walk cycle

contact, recoil, passing, and high-point. Among these four poses, the contact and high-point are the appropriate candidates to extract the weight information. The recoil and passing poses provide the same information as the high-point pose and will not be included in the analysis.

The shoe-mounted sensor array measures the distributed force the foot applies on the ground. Such force measurement is used to determine the patient's weight. As shown in Fig. 19.7, the total applied force F_T is decomposed into a y-component due to the weight W, and an x-component F due the friction between the shoe and the ground. The latter horizontal force is responsible for the forward motion of the body. The vector representation of these components is shown in Fig. 19.7, where θ specifies the direction of the applied force with respect to the horizontal axis. Subscripts 1 and 2 will be used in the analysis to represent the first and second feet, respectively.

19.3.1 Weight During High-Point Pose

During the high-point pose, one foot is in contact with the ground and bears alone the total force (F_T) while the other foot has no contact as shown in Fig. 19.7a. F_T is given by

$$F_T = F_{TF1} + F_{TF2} \tag{19.1}$$

where F_{TF1} and F_{TF2} are the total forces applied on the ground by the first and second feet, respectively, with $F_{TF2} = 0$. The weight component (in Kg) of the applied force is determined using the equation

$$W = \frac{W_{F1}}{g} = \frac{F_{TF1} \sin \theta_1}{g} \tag{19.2}$$

where g is the gravitational acceleration constant, W_{F1} is the weight component (in *Newton*) of the total applied force on the first foot, and θ_1 is the direction of the total applied force that is determined by the sensor node circuitry.

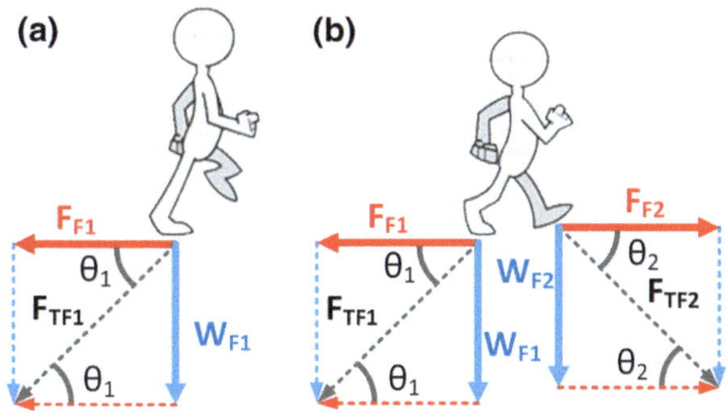

Fig. 19.7 Force distribution during a half walk cycle: **a** high-point position and **b** contact position

19.3.2 Weight During Contact Pose

During the contact pose, the total force is distributed among both feet as they are both in contact with ground, as shown in Fig. 19.7b. Moreover, during motion, the force on one foot increases while the force on the other foot decreases proportionally. The total applied force, F_T, is given by

$$F_T = F_{TF1} + F_{TF2} \tag{19.3}$$

The weight component (in Kg) of the applied force is determined by

$$\begin{aligned} W &= \frac{W_{F1} + W_{F2}}{g} \\ &= \frac{F_{TF1}\sin\theta_1 + F_{TF2}\sin\theta_2}{g} \end{aligned} \tag{19.4}$$

where W_{F1} and W_{F2} are the weight components (in *Newton*) of the total forces applied on first and second feet, respectively, and θ_1 and θ_2 are the directions of the forces.

19.4 CHF Monitoring Prototype

A prototype of the CHF monitoring system is developed using the Xilinx Virtex-7 FPGA platform to showcase the end-to-end operation of the proposed system architecture and to support the implementation, analysis, and comparison of various pattern detection algorithms for early CHF detection. As shown in Fig. 19.8, the prototype system block diagram is composed of three integrated modules: the digital assistant block, the sensor block, and the communication block. The BCC transceivers are replaced with PIC-PHY models, and the processing unit is

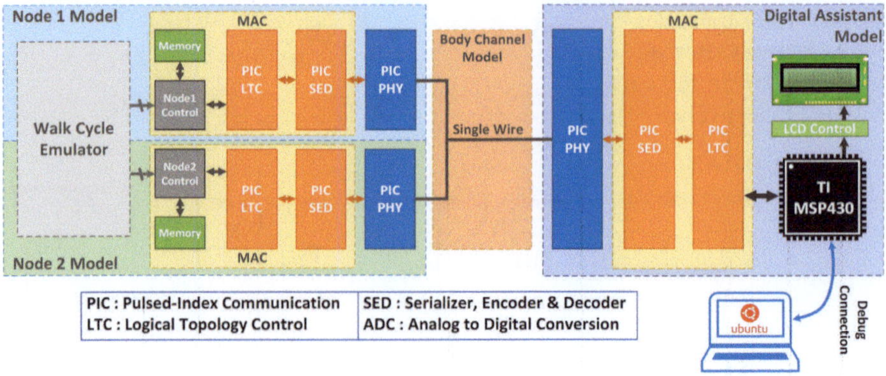

Fig. 19.8 Prototype system block diagram

implemented using an opens-source IP of TI MSP430. A single wire model is used as a body channel. Nodes are replaced with their models which are composed of node controllers, memory storage, and a walk cycle model to mimic the behavior of human walk. A more detailed description of these integrated modules will follow.

19.4.1 Digital Assistant

The digital assistant model employs Verilog IP core of TI MSP430 (OpenMSP430) microcontroller [18] to implement the processing unit. The software development environment uses Eclipse, and MSP430 C libraries are used to implement the microcontroller programs. Programming and debugging tools, running on Linux, are used to load and test the program. The digital assistant sends a measurement request every second to both nodes using the PIC modules (MAC and PHY). It then receives the node-provided information and uses it to detect human movement and estimate weight. Depending on the analysis results, the digital assistant signals the LCD (display) control module to display the status on the LCD. The MSP430 hardware and software, PIC-MAC, PIC-PHY, and LCD module are now discussed:

MSP430 Hardware

- *Processor Configuration*: OpenMSP430 can be configured to a variety of members of the of MSP430 processor family. This is done by changing the parameter settings in configuration provided with the IP core. For our prototype, the MSP430F2330 processor configuration is used.
- *I/O Ports*: A total of four 8-bit MSP430 ports are available. The processing unit also needs to control the PIC and LCD modules, which requires at least 24 I/O ports. The port requirement is fulfilled through the core's peripheral wrapper, which is customized in Verilog HDL to meet the I/O needs of the prototype.

- *Memory*: The MSP430 core needs two kinds of memory: program memory and data memory. Both are created using the Xilinx IP Core generator and connected internally with the core. The program and data memories are of size 8KB and 1KB, respectively.

MSP430 Software

- *Program Development and Compilation*: The eclipse IDE development environment is used to code, verify, and compile the MSP430 program. MSPGCC compiler version 20120406-p20120502 is configured under eclipse. Additionally, to work with the compiler, MSP430 plug-in for eclipse is necessary to generate the proper compile settings and let the IDE know the compiler location on the computer file system. The compiler generates an executable file (.elf) which is loaded into the processor program memory and executed using the debugging tool described in the next bullet point.
- *Debugging Tool*: Ubuntu Linux is used to run the Tool Command Language (TCL) debugging scripts; the scripts are provided with the OpenMSP430 IP core package [18]. The debugging tool, openmsp430-minidebug, is a minimalistic debugger with a simple GUI and used to select and load the eclipse-generated executable files (.elf). The minidebug loads the program into program memory and waits for the run command to start executing. The tool also provides a small assembly language debugging functionality.

LCD, PIC-MAC, and PIC-PHY Layers

The LCD module is implemented in Verilog HDL and connected to the processing unit. The LCD receives the decision continuously from the processing unit and displays either the weight or the activity status: Is the person walking or standing? The PIC-MAC layer, which is composed of the SED (serializer, encoder, and decoder) and LTC (logical topology control unit), is also implemented in Verilog HDL and helps the processing unit regularly request updated measurements from the nodes. The PIC-LTC is implemented in a master–slave topology in which the digital assistant behaves as a master. The PIC SED operates exactly the same as described in Chap. 11; see also [17]. Because the system under discussion is a prototype, it utilizes a simple single-wire model in place of the real body channel. Therefore, the tri-state buffers, shown in Fig. 19.9, are enough to mimic the behavior of a BCC transceiver.

19.4.2 Sensor Subsystem

The sensor subsystem is composed of a walk cycle model, two sensor node models, PIC-MAC layer, and PIC-PHY layer, as shown in Fig. 19.8. Two walk cycle models are implemented: An Activity Detector (AD), and a Weight Analyzer (WA). The AD model tests only the movement activity of the walk cycle (i.e., walking or standing). The WA model determines also the weight using the force on each foot as described in Sect. 19.3. Both of the models are described in Sect. 19.5. Within the sensor subsystem, an important component is node control sub-block.

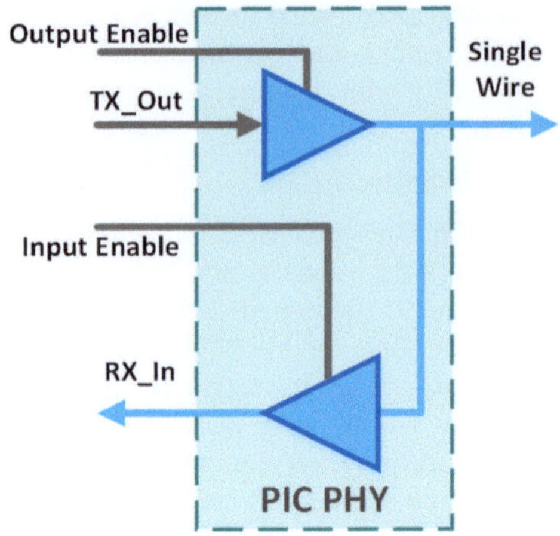

Fig. 19.9 PIC-PHY layer

Node Control

Node control block reads up to 1024 (configurable) sampled force measurements of 12 bits each and stores these into the memory. The node memories are generated using the Xilinx Core generator. When the request is received from the digital assistant, the node's unique ID is verified and transmission of the stored measurements is initiated with the help of PIC-MAC layer. Node controllers are implemented in Verilog HDL.

19.4.3 Body Channel Model

As mentioned previously, the body behaves like a single channel and therefore it is replaced by a single-wire model in the prototype setup. The single-wire model is shown in Fig. 19.10.

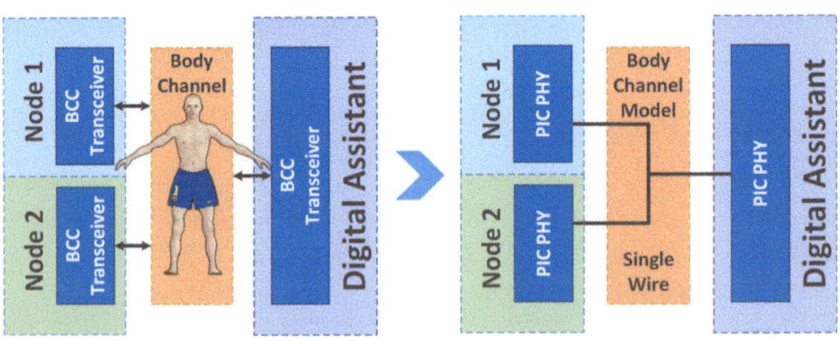

Fig. 19.10 Single-wire model for BCC

19.5 Walk Cycle Emulator

The presented system architecture is just a starting point of a flexible platform that will facilitate the exploration of monitoring and diagnosis algorithms for CHF. Of course, the CHF databases are needed to fully validate the approach and the clinical requirements. To test the functionality of the platform and validate the weight monitoring algorithm, we use Parkinson's disease database. For the algorithm testing, the healthy subjects have been selected as they provide similar to those pressure/force distributions, on each foot during a walk cycle, that are needed for the CHF monitoring.

19.5.1 Activity Detector

As shown in Fig. 19.11, a human walk cycle is similar to that of two complementary sinusoid waves. If these waves are generated, they can be sampled and used as human movement emulators. This is used in the CHF prototype where a single potentiometer and a network of resistors and capacitors are used to generate the complementary sinusoid signals. Rotating the potentiometer in one direction followed by an equal rotation in the opposite direction generates a full walk cycle.

Signals generated by the walk cycle model are digitized using Virtex 7 on-chip ADC. The digitized signals are then stored in memory through the node controllers. The stored information is later transmitted to the digital assistant via the PIC layers as requested. PIC-LTC's at both node ends are configured in a slave mode and, therefore, wait for the request command from the master digital assistant. All the modules in the sensor subsystem are implemented in Verilog HDL.

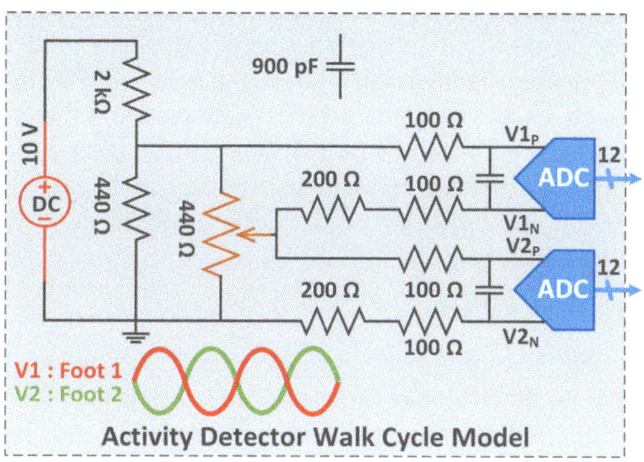

Fig. 19.11 Activity Detector walk cycle model

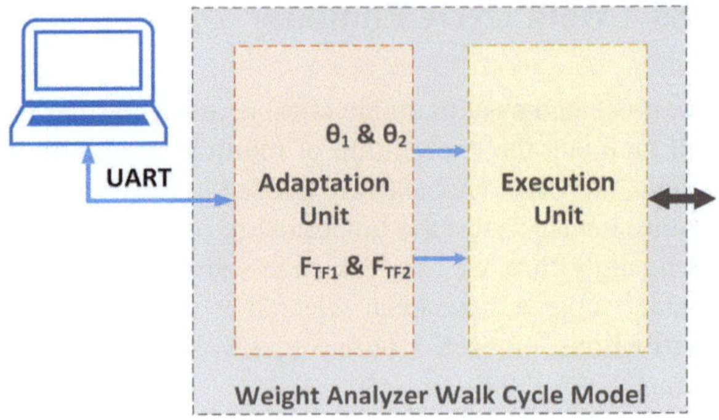

Fig. 19.12 Weight Analyzer walk cycle model

19.5.2 *Weight Analyzer*

The Weight Analyzer (WA) model is developed using the human walk cycle data collected for several healthy control subjects during a study of gait patterns in Parkinson's disease patients [11]. The database provides the force measurement, N, taken every 10 ms from foot-mounted sensors for each of the feet. Additionally, the actual weight (Kg) is included in the data for each of the subjects, which helps validate the predictive value of the WA model.

As shown in Fig. 19.12, the WA model is comprised of two parts: the adaptation unit and the execution unit. In the adaptation part, the angles θ_1 and θ_2 are calculated using the information provided by [11] on the weight (Kg) and total force N applied on each feet. In an actual system, the angles θ_1 and θ_2 would be measured using angle gyros or accelerometers. If these measurements are made available, then (19.4) provides the weight estimate. Since the focus in this chapter is the platform itself rather than the sensing system, the walk cycle database is used to determine the angles and illustrate how these angles can be used to recover the weight estimates.

In the execution part, the force measurements and calculated angles are fed to the node controllers, and are then stored in memory by these controllers. Later, the stored information is transmitted, through the various PIC layers, to the digital assistant for pattern detection. Both parts of the model are implemented in Verilog HDL, and the walk cycle information samples are transmitted via UART from the PC to the FPGA hardware. As described earlier, two walk cycle poses are included in the analysis. As a result, the adaptation part is further divided into two phases based on these poses as described below.

High-Point Phase

During the high-point pose, only one foot is on the ground and therefore, the total force is equal to the force applied by that one foot on the ground, as shown in Fig. 19.13. Rearranging (19.2), one can derive the high-point pose angle as follows:

$$\theta_1 = \sin^{-1}\left(\frac{W\,g}{F_{TF1}}\right) \tag{19.5}$$

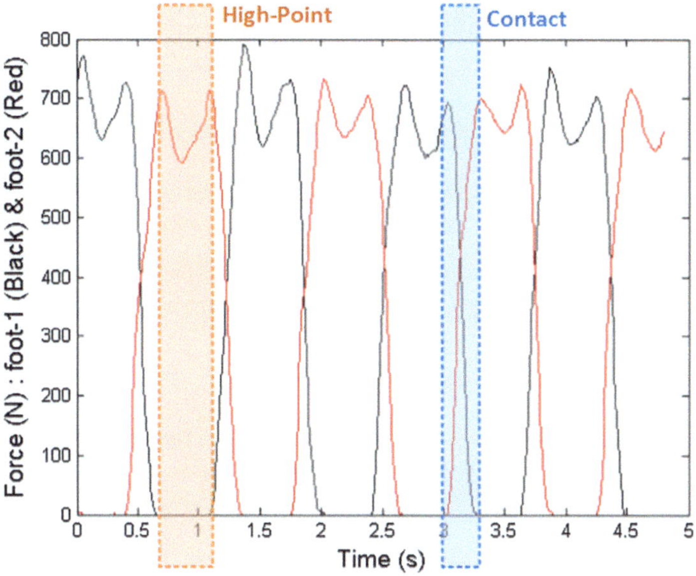

Fig. 19.13 Forces measured at each foot (56 Kg human)

where W is the measured weight of the human subject in Kg. It must be noted that there are two high-point poses in a walk cycle, one for each foot. (19.5) can be used for both high-point poses with the foot touching the ground being Foot 1. The angle θ_1 for both feet, calculated using (19.5), is shown in Fig. 19.14a.

Contact Phase

During the contact pose, both feet are in contact with the ground and therefore, the total force is distributed among them, as shown in Fig. 19.13. The force on Foot 1 decreases gradually as it separates from the ground while the force on Foot 2 increases. For this phase, θ_1 is determined using (19.5) and θ_2 is calculated incorporating the θ_1 information. Rearranging (19.4), we get:

$$\theta_2 = \sin^{-1}\left(\frac{W\,g - F_{TF1}\sin\theta_1}{F_{TF1}}\right) \tag{19.6}$$

where W is the measured weight of the human subject in Kg. Note that there are two contact poses in a walk cycle alternating between the two feet separating from the ground. (19.5) and (19.6) can be used for both contact poses with the convention that the foot separating from the ground is Foot 1 while the one approaching the ground is Foot 2. The angles θ_1 and θ_2 for both feet, calculated using (19.5) and (19.6), are shown in Fig. 19.14.

Fig. 19.14 Pose angles θ_1, θ_2 for a 56 Kg Human subject: **a** θ_1 for Foot 1 and Foot 2 and **b** θ_2 for Foot 1 **c** θ_2 for Foot 2

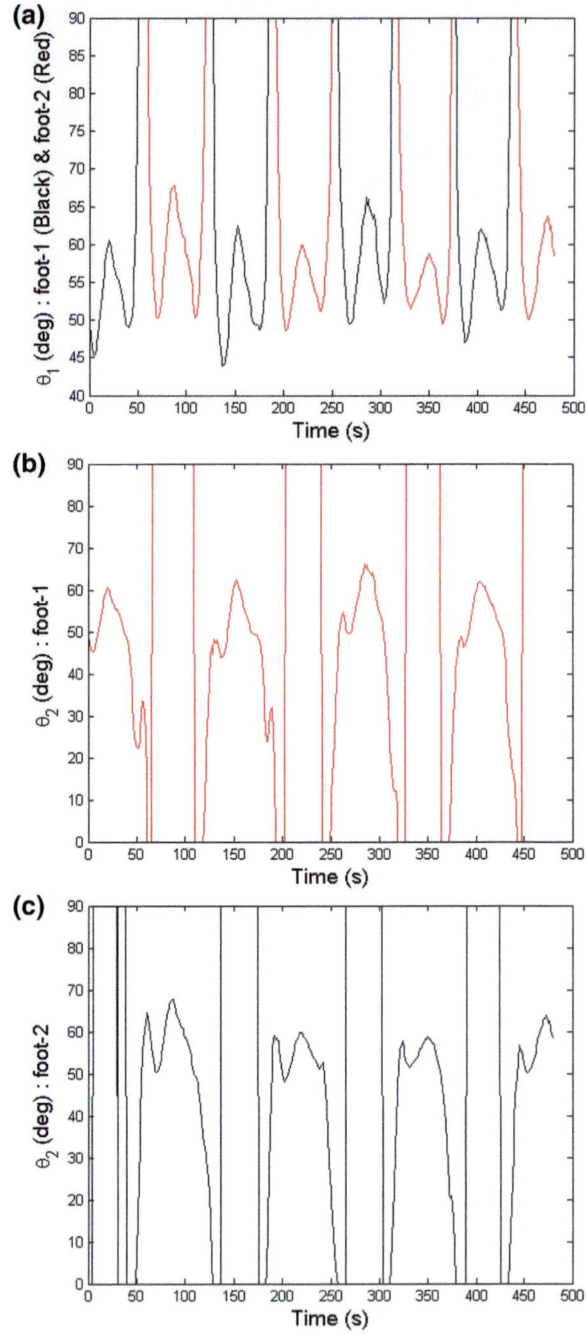

19.5.3 Weight Analyzer Test Algorithm

To exercise the platform with the weight analysis (WA) model, a simple test algorithm is developed in embedded C to execute on the TI MSP430 microcontroller. The TI MSP430 receives the angles and the applied force information from both sensor nodes and applies (19.4) to determine the patient's weight and analyze the gait pattern.

19.6 Testing and Results

The developed prototype system is tested rigorously through both software simulations and on Virtex-7 real-time hardware. First, to test the complete communication link, the potentiometer of the AD walk emulator is used to create test walk cycles which are detected successfully, and the results are displayed on the LCD module showing whether the person is walking or standing. The prototype works with a sampling rate of 1*MSps*, a configurable buffer size of up to maximum of 16 bits, 1024 words, and an average signaling rate of 4.1 Mb/s. A simplified high-level system flow is shown in Fig. 19.15a, c.

Next, the WA walk cycle emulator is used to test the functionality of the system. Data of 30 human subjects from [11] are used where each of the data files (one file per subject) contains an average of 110 walk cycles and each walk cycle has an average of 130 data samples. Data samples include the time stamp, force measurements of all the eight sensors per foot, and the total force for both of the feet. Data samples of all subjects are transmitted from the PC to the FPGA via UART. The angles θ_1 and θ_2 are calculated by the adaptation unit and fed to the execution unit along with the total forces measured on both feet. The data flow proceeds exactly as described in Sect. 19.5.2. The digital assistant receives the data, applies the algorithm of weight measurement, and analyzes the gait. All the input data from the Parkinson database have resulted in accurate weight measurements with the values appropriately displayed on the LCD module. The measurements for a 56-Kg subject are shown in Fig. 19.16. A simplified high-level system flow is shown in Fig. 19.15b, c.

In Table 19.1, the system presented in this paper is compared with existing CHF systems using wearability, footprint, and power as the main figures of merit. The wearability of our system is due to the use of shoe-mounted sensor arrays instead of using traditional scales for weight measurement. Another aspect contributing to wearability is the use of body channel communication for transmitting data between the sensors and the digital assistant [22]. The digital assistant itself can be made no larger than the commercially available MSP430 development boards which are of approximately credit card size (85.60×53.98 mm). As for low power, the proposed system benefits from significant power savings due to the use of energy-efficient communication protocols, low duty cycles, and ultra-low-power transceivers adapted to the body communication channel. The automated and reliable recording of weight contributes to the reduction of human expert interventions as well as human errors.

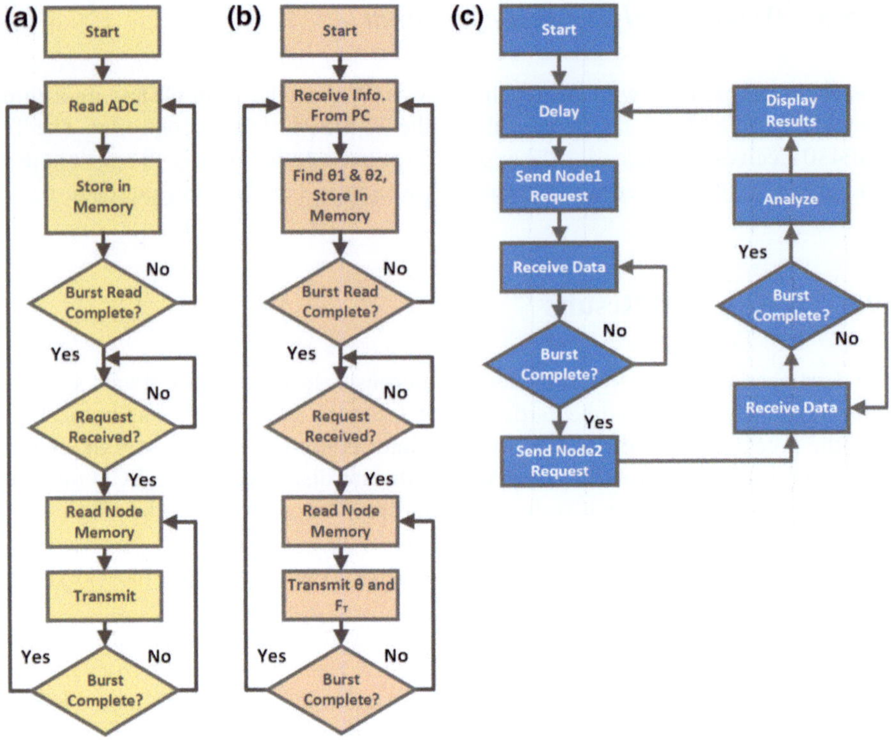

Fig. 19.15 High-level system flow: **a** activity monitor flow, **b** weight analysis flow, **c** digital assistant flow

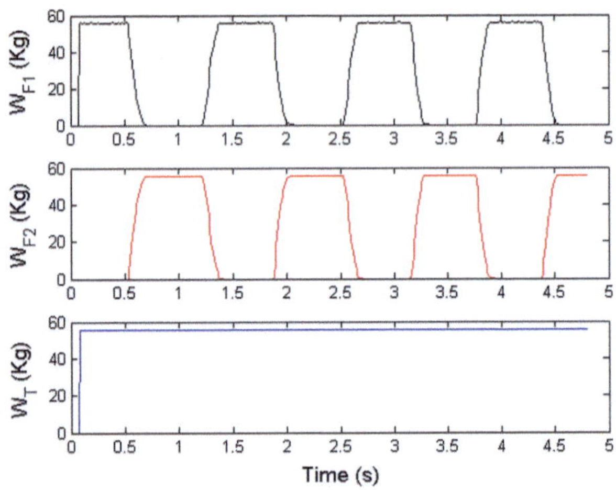

Fig. 19.16 Weight measurement results for a 56 Kg human subject

Table 19.1 Proposed solution versus existing solutions

Solution	Footprint	Need of Experts/Visits	Wearability	Power	Human error
This work	Small	No	Yes	Low	Low
[9]	Big	Yes	No	High	High
[4]	Big	Yes	No	High	High
[10]	Big	Yes	No	High	High
[12]	Big	Yes	No	High	Low
[3]	Big	Yes	No	High	Low
[20]	Big	Yes	No	Moderate	Low
[1]	Medium	Yes	Partially	Moderate	High
[7]	Medium	Partially	No	High	High

19.7 Prototype Development Methodology

The proposed CHF platform architecture is amenable to seamless and fast prototyping. Not only does the proposed system readily adapt to the specific test cases but also the various pattern recognition algorithms can be analyzed without any change in the hardware. All the prototype modules are either implemented in Verilog or in instantiated IPs. Some IPs can effortlessly plug into the prototype while others are easily generated using Xilinx IP core generator. This section is meant to describe the platform design methodology and to illustrate the fast prototyping ramp-up even for novices in FPGA and HDL. It is also meant to spell out the methodology steps, using a variety of user-friendly IP's, well-known CAD tools, and glue logic HDL. These steps are shown in Fig. 19.17 and described below.

19.7.1 Prepare PIC Layers

The developed Verilog IP for the PIC protocol in Chap. 11 is a plug-and-play module that is easy to configure. In particular, it does not require a specific clock rate and can thus work with any of the clock domains provided by the system. On the other hand, the clock rate impacts the PIC data rate linearly as was shown in Chap. 11. While the PIC default settings guarantee a good level of immunity to timing and clocking differences between transmitter and receiver, a more refined analysis can be performed following [15]. Additionally, the PIC protocol coefficients such as inter-symbol delay, pulse width, inter-symbol delay threshold, and error threshold can be rigorously selected to achieve the required immunity to clock jitter and skew as well as clock rate discrepancy between transmitter and receiver. These coefficients can be updated in the PIC module settings. As shown in Fig. 19.9, PIC-PHY is easy to implement in Verilog using tri-state buffers. If power capping is a system

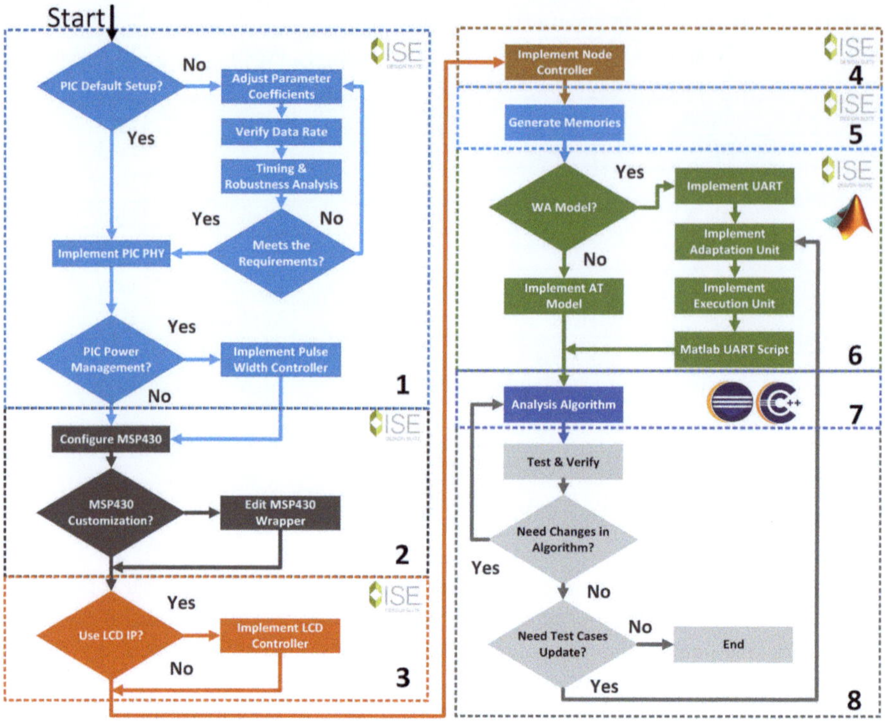

Fig. 19.17 Prototype development flow

requirement, additional control on power consumption can be achieved using pulse width modifications. [14]. Note that the PIC layers on both ends of the channel are identical and must therefore share the same IP. In our flow, Xilinx ISE is used to edit, synthesize, and implement the PIC layers. Other CAD tools such as Synopsys Synplicity can also be used.

19.7.2 Configure Digital Assistant Processing Module

As described in previous sections, the TI MSP430 Verilog IP is also easy to configure and can support a large number of microcontrollers in 430 family. Moreover, the IP wrapper can be updated to support custom microcontroller pin-outs or added functionality. In our design, Xilinx ISE is used to edit, synthesize, and implement the digital assistant processing module. However, any other CAD tool such as Synopsys Synplicity can also be used.

19.7.3 Add a Notification Module

The on-board LCD module of an FPGA development board is the most convenient way to implement a notification feature. In our design, the LCD controller is implemented in Verilog using Xilinx ISE. An open-source IP can also be employed to save development time.

19.7.4 Develop Node Control Module

The main task of the node controller for the prototype system is to collect information from the walk cycle model, store it in memory, and transmit it using PIC in a burst of 1000 samples when requested by the digital assistant. Another task of the node controller is to manage the PIC layer on the sensor's end. This can be done simply by placing the data at the PIC input and signaling PIC to start processing. This control module can also be altered to support another format of data collection. Both node controllers are identical and can share the same IP.

19.7.5 Generate Memories

The memory blocks for both MSP430 and the node controller are generated using the Xilinx IP Core generator, which provides an accessible path to the management of the available on-chip block memories. One can also use the Verilog memory constructs.

19.7.6 Select and Develop Walk Cycle Emulators

- Activity Detector (AD): There is one such detector on each foot. The identical designs employ the Virtex 7 on-chip ADC along with the interface IP generated by the Xilinx IP Core generator. The controllers are simple to instantiate and work with. When using a simpler FPGA board, an external ADC can be added using a Verilog interface module. In this case, there is no need to use the Xilinx ADC IP block.
- Walk Analysis (WA) Model: The walk analysis module receives the walk cycle information from the PC and calculates the angles, θ_1 and θ_2. The calculation of angles is necessary because they emulate the functionality of an external circuit dedicated to accurately measuring the angle of the applied force. To transmit the walk cycle samples from the PC, a MATLAB script is used. At the FPGA end, the IP of an UART receiver is used to receive and store the incoming samples into memory. The latter memory is generated by the Xilinx IP Core generator.

To calculate the angles, Xilinx IPs to calculate the sine function, floating point multipliers, and subtracters are generated and instantiated into the other modules. Another module that is easy to implement, the execution unit, fetches the angles and force information from memory. The execution unit can be considered a complex tester, generating test patterns for the verification of the digital assistant algorithm.

19.7.7 Implement Weight and Gait Analysis Algorithm

The final step is to write the MSP430 embedded C code for the analysis as per the targeted algorithm. The code need also to handle the PIC master mode transfer requests as well as the LCD control.

19.7.8 Test and Verify

The developed prototype is tested using the flow described in the previous section. If the algorithm is not working according to the requirements, go back to Step 7 and update the algorithm. If the test cases need an update, go to Step 6. Only the last two steps need frequent updates in order to analyze a variety of CHF pattern detection algorithms.

19.8 Conclusions

An integrated low-power platform for CHF monitoring using body channel communication has been proposed. For the early detection of CHF conditions, the system continuously measures the patient weight using shoe-mounted wearable sensors to analyze the decompensation of body fluids. An energy-efficient digital assistant analyzes the weight changes to detect and report the CHF conditions. The body channel enables efficient communication and eliminates the requirement of wired and wireless connections. Data is transferred reliably over a human body with the help of a newly introduced low-power and high data-rate single-channel communication protocol, which does not require any clock and data recovery and is robust toward skews, jitters, and clock inaccuracies between the transmitter and receiver. A proof-of-concept prototype, developed on an FPGA platform, tests the end-to-end feasibility of the proposed platform. The highly modular architecture enables plug-and-play integration of sensor nodes, BCC transceivers, and custom VLSI digital assistants. The latter may embody several options for pattern recognition algorithms dedicated to the early prediction of CHF.

Acknowledgements This work has been supported by the Semiconductor Research Corporation (SRC) under the Abu Dhabi SRC Center of Excellence on Energy-Efficient Electronic Systems (ACE^4S), Contract 2013 HJ2440, with funding from the Mubadala Development Company, Abu Dhabi, UAE.

References

1. N. Alnosayan, E. Lee, A. Alluhaidan, S. Chatterjee, L. Houston-Feenstra, M. Kagoda, W. Dysinger, Myheart: an intelligent mhealth home monitoring system supporting heart failure self-care, in *e-Health Networking, Applications and Services (Healthcom)* (October 2014), pp. 311–316
2. I. Anand et al., Design of multi-sensor monitoring in congestive heart failure (music) study: prospective trail to assess the utility of continuous wireless physiologic monitoring in heart failure. J. Card. Fail. **17**(1), 11–16 (2011)
3. M. Arad, S. Abboud, Electrical impedance tomography versus whole thoracic impedance for monitoring lung fluid content in congestive heart failure patients, in*Computing in Cardiology Conference (CinC)* (September 2013), pp. 465–466
4. S. Bakhshi, X. Li, N. Semenov, J. Apodaca-Madrid, M. Mahoor, K. Newman, C. Long, C. Neuman, Congestive heart failure home monitoring pilot study in urban denver. Eng. Med. Biol. Soc. EMBC, 3150–3153 (2011)
5. S.I. Chaudhry et al., Telemonitoring for patients with chronic heart failure: a systematic review. J. Card. Fail. **13**(1), 56–62 (2007)
6. S.I. Chaudhry, Y. Wang, J. Concato, T.M. Gill, H.M. Krumholz, Patterns of weight change preceding hospitalization for heart failure. Healthc. Innov. Conf. (HIC) **116**(14), 1549–1554 (2007)
7. J. Chen, S. Quadri, L. Pollonini, S. Naribole, J. Ding, Z. Zheng, E.W. Knightly, C.C. Dacso, Blue scale: early detection of impending congestive heart failure events via wireless daily self-monitoring, in *Healthcare Innovation Conference (HIC)* (October 2014), pp. 63–66
8. E. Cyarto et al., Pedometer accuracy in nursing home and community dwelling older adults. Med. Sci. Sports Exerc. **36**(2), 205–209 (2004)
9. J. Finkelstein, C. Dennison, A pilot study of home automated telemanagement (hat) system in african americans with congestive heart failure, in *eHealth, Telemedicine, and Social Medicine* (February 2010), pp. 90–94
10. J. Finkelstein, J. Wood, Implementing home telemanagement of congestive heart failure using xbox gaming platform. Eng. Med. Biol. Soc. EMBC, 3158–3163 (2011)
11. Gait in parkinson's disease: https://physionet.org/pn3/gaitpdb/
12. A. Kaiser, C. Pocock, P. Sharma, N. Browdues, K. Newman, F. Barnes, Towards a method for early detection of congestive heart failure with an electrocardiogram and acoustic transducers, in *Signal Processing in Medicine and Biology Symposium (SPMB)* (December 2012), pp. 1–5
13. MAXIM. One Wire Viewer User's Guide, Version 1.4. AN3358 (2009)
14. S. Muzaffar, I.M. Elfadel, Power management of pulsed-index communication protocols, in *IEEE International Conference on Computer Design (ICCD)* (October 2015), pp. 375–378
15. S. Muzaffar, I.M. Elfadel, Timing and robustness analysis of pulsed-index protocols for single-channel IoT communications, in *IFIP/IEEE International Conference on Very Large Scale Integration (VLSI-SoC)* (October 2015), pp. 225–230
16. S. Muzaffar, I.M. Elfadel, Automatic protocol configuration in single-channel low-power dynamic signaling for IoT devices, in *IFIP/IEEE International Conference on Very Large Scale Integration (VLSI-SoC)* (September 2016)
17. S. Muzaffar, A. Shabra, J. Yoo, I.M. Elfadel, A pulsed-index technique for single-channel, low power, dynamic signaling, in *Design, Automation and Test in Europe (DATE)* (March 2015), pp. 1485–1490

18. http://opencores.org/project,openmsp430
19. J. Pare, M. Jaana, C. Sicotte, Systematic review of home telemonitoring for chronic diseases: the evidence base. J. Signal Inf. Process. **1**(3), 269–277 (2007)
20. S. Rezaeieh, K. Bialkowski, A. Abbosh, Effects of antenna characteristics on performance of microwave based system designed for early stage congestive heart failure detection, in *Microwave Symposium (AMS)* (June 2014), pp. 35–36
21. W. Saadeh, Y. Yonatan, J. Yoo, A hybrid ofdm body coupled communication transceiver for binaural hearing aids in 65nm cmos, in *2015 IEEE International Symposium on Circuits and Systems (ISCAS)* (IEEE, 2015), pp. 2620–2623
22. S.-J. Song, N. Cho, H.-J. Yoo, A 0.2-mw 2-mb/s digital transceiver based on wideband signaling for human body communications. J. SolidState Circuits **42**(9), 2021–2033 (2007)
23. J. Zhang, K.M. Goode, P.E. Cuddihy, J.G. Cleland, Predicting hospitalization due to worsening heart failure using daily weight measurement: analysis of the trans-european networkehome-care management system (ten-hms) study. Eur. J. Heart Fail. **11**(4), 420–427 (2009)

Chapter 20
Hardware Security and Trust: Logic Locking as a Design-for-Trust Solution

Muhammad Yasin, Bodhisatwa Mazumdar, Jeyavijayan Rajendran and Ozgur Sinanoglu

20.1 Introduction

Integrated circuits (ICs) are ubiquitous and an essential component of electronic systems ranging from home appliances and smartphones to satellites and military equipment. ICs serve as the root-of-trust for computing systems. The software computations can only be trustworthy if the underlying hardware is reliable and trustworthy. IC supply chain has evolved drastically over the years. Traditionally, companies such as Intel and IBM would use their own foundries, and design and fabricate ICs in-house, in trusted settings. However, the evolving complexity of integrated circuits (ICs), the ever-shrinking time-to-market, and the skyrocketing costs of building or maintaining a semiconductor foundry have propelled the globalization of IC design and fabrication flow [4].

20.1.1 Globalization of IC Supply Chain and Trust Issues

As of today, many semiconductor companies, such as Apple, operate fabless and outsource the fabrication of ICs to *offshore foundries*. Apart from the fabrication of

M. Yasin (✉)
New York University, Brooklyn, NY 11201, USA
e-mail: yasin@nyu.edu

B. Mazumdar · O. Sinanoglu
New York University, Abu Dhabi 129188, UAE
e-mail: bm105@nyu.edu

O. Sinanoglu
e-mail: ozgursin@nyu.edu

J. Rajendran
University of Texas at Dallas, Richardson, TX 75080, USA
e-mail: jv.ee@utdallas.edu

Ibrahim (Abe) M. Elfadel and M. Ismail (eds.), *The IoT Physical Layer*,
https://doi.org/10.1007/978-3-319-93100-5_20

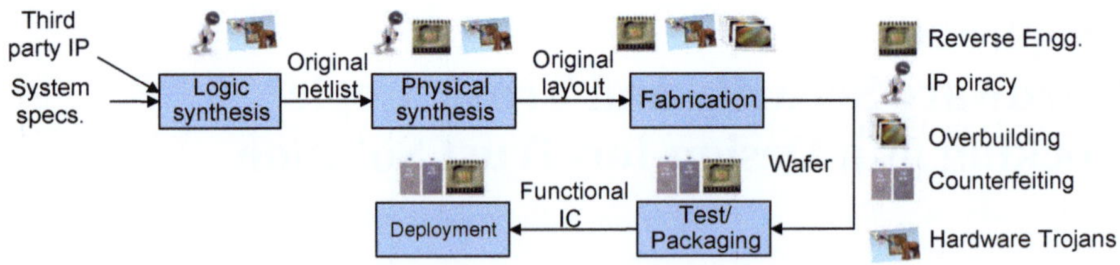

Fig. 20.1 Security threats at different stages in the IC design flow

ICs, test, assembly, and packaging services may also be outsourced to *outsourced assembly and test (OSAT)* companies such as ASE [27] and Amkor [2]. Even *design houses* may purchase and use third-party intellectual property (3PIP) cores to reduce their design effort and meet strict time-to-market constraints. In a globalized and distributed IC supply chain, several (potentially untrusted) agents have access to the valuable IP or the physical IC. This increased access to critical assets opens up opportunities for rogue elements to exploit valuable information and/or jeopardize the trust in the IC design flow. Hardware security vulnerabilities and threats that have emerged at various stages of the IC supply chain, as a consequence of the globalization, are highlighted in Fig. 20.1 and are described below.

Reverse Engineering

Reverse engineering of an IC refers to extracting design/technology details of an IC using imaging techniques. It involves several steps that include: depackaging an IC, delayering and imaging individual layers, and analyzing the collected images to identify design/IP details [36]. Companies such as Chipworks provide reverse engineering as a service [9].

Apart from use in benign applications, reverse engineering can also be an enabler for attacks such as cloning or insertion of hardware Trojans (see Sect. 20.1.1); reverse engineering can help an attacker in identifying stealthy locations for inserting Trojans. Reverse engineering can be used along with probing methods to extract secret information such as cryptographic keys [12].

IP Piracy

IP piracy refers to the illegal or unlicensed use of the IP (e.g., netlist or GDS-II files). Similar to the case of software piracy, an attacker can steal valuable IP cores and sell them as genuine. The yearly losses to the semiconductor industry due to IP infringement exceed $4 billion [31].

Overbuilding

IC piracy manifests itself in the form of overbuilding at an untrusted foundry. A rogue foundry can overproduce the ICs and sell the excess ICs illegally. Building extra ICs from the same set of masks increases the costs for the foundry only marginally; the foundry can sell those ICs at a price cheaper than that offered by the original IC company [28].

Hardware Trojans

Malicious agents in the design house or the foundry can insert small, hard-to-detect circuitry in an IC that can reveal secret information or cause service disruption during the IC operation [17]. As the footprint for a Trojan may be very small, it becomes hard to verify/test for Trojans in a circuit, especially when there is no golden (Trojan-free) reference available to compare against. The suspected sources of Trojans are 3PIP used in the design or mask alteration during fabrication [3].

Counterfeiting

Counterfeit ICs are replicas of the genuine ICs that are fraudulently made to appear almost identical to the genuine ICs [12]. Counterfeit ICs include out-of-spec, remarked, and recycled ones, typically salvaged from discarded electronic appliances. More than 5% of all commercial ICs are counterfeits [12, 14]. Counterfeit ICs pose serious reliability and security concerns for the semiconductor industry.

20.1.2 Design-for-Trust Solutions

A number of defense techniques have been proposed to thwart IP piracy, reverse engineering, and other hardware attacks. These techniques include watermarking [16], fingerprinting [6], IC metering [19], IC camouflaging [24], split manufacturing [15], and logic locking [7, 8, 20–22, 26, 26, 29, 30]. The aforementioned techniques differ mainly in the threat model and security objectives. The threat model specifies the entities that are trusted (or untrusted) and the assets each entity has access to.

In watermarking, the designer's signature such as secret design constraints is embedded into the design, whereas in fingerprinting the end user's signature is embedded along with the designer's signature. Watermarking and fingerprinting are *passive* techniques that only help detect piracy but cannot protect against it [6, 16]. IC metering is a set of protocols and techniques, either passive or active, which assign a unique ID to each manufactured IC and help in tracking of individual ICs post-manufacturing [19]. IC camouflaging camouflages selected gates in the design; the camouflaged gates look alike but can implement one of many functions. A reverse engineering attacker faces the additional problem of resolving the functionality of camouflaged gates [24]. In split manufacturing, the design is split into two parts that are manufactured in separate foundries and then stacked together [15].

Logic locking techniques, which are *active*, provide protection against IP/IC piracy, overbuilding, or reverse engineering. The techniques insert additional hardware elements, which can be either combinational or sequential, into a design with the objective of locking the design functionality and implementation [7, 8, 20–22, 26, 26, 29, 30]. While metering assigns a unique ID/key to each IC, locking uses the same key for all the fabricated ICs [28]. A comprehensive review of IP piracy protection techniques is provided in [10], a detailed description of various design-for-trust solutions is presented in [25], and a detailed review of metering techniques can be found in [19]. The scope of this chapter is logic locking.

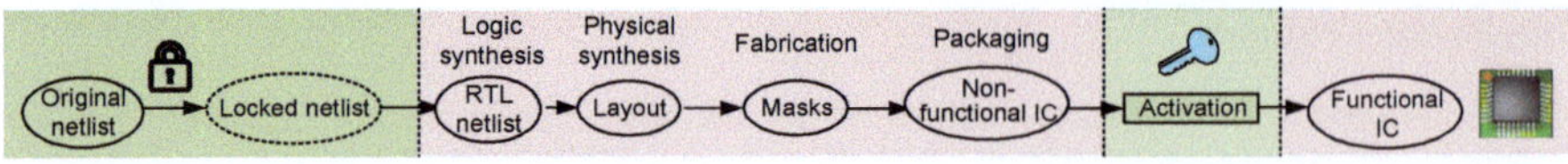

Fig. 20.2 Logic locking in the context of the IC design flow. The design is in the locked form in the untrusted design regime (shaded in red)

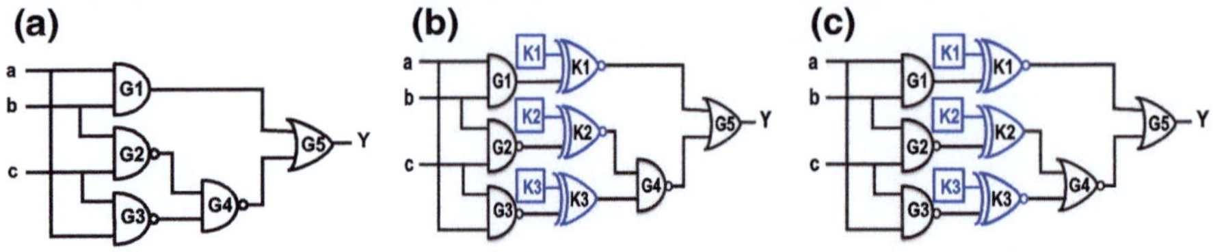

Fig. 20.3 Logic locking using XOR/XNOR gates [29]. **a** An example circuit: majority of three inputs. **b** Circuit locked using XOR/XNOR key gates. The correct key value is 110. **c** Locked circuit with inverters absorbed by the key gates. The correct key value is still 110

20.2 Logic Locking

20.2.1 An Overview

Logic locking [21, 22, 26, 29][1] is a technique that aims at defending against IP piracy, overbuilding, and reverse engineering attacks by locking a design with a secret key. To enable chip-locking features, additional logic, e.g., a set of XOR/XNORs gates referred to as *key gates*, is added to the original netlist to obtain a locked netlist. The locked netlist passes through the untrusted design/fabrication stages as shown in Fig. 20.2. The locked netlist is not useful without the knowledge of the secret key.

Example. Figure 20.3a shows an example (majority circuit) netlist, and Fig. 20.3b shows its locked version with three XOR/XNOR key gates. One of the inputs of each key gate is driven by a wire in the original design, while the other input, referred to as *key input*, is driven by a key bit stored in a tamper-proof memory. The inverters in the netlist can be moved around to increase the obfuscation complexity as shown in Fig. 20.3c. A locked IC (or a locked netlist) will not generate correct output unless it is activated using the correct key. Consider the locked circuit in Fig. 20.3d. When the correct key 110 is applied, the correct output $Y = 0$ is obtained for the input pattern 000. For the same input pattern, however, an incorrect output $Y = 1$ is produced for any incorrect key value.

[1] Researchers have also used the terms "logic obfuscation" [23] and "logic encryption" [22, 26] for this purpose.

Threat Model

The logic locking threat model assumes that:

- The designer is trusted; i.e., the personnel and the tools used in the design house are trustworthy.
- The foundry and the end user are untrusted.
- The attacker has the knowledge of the logic locking algorithm, as well as the location of the key gates. The only unknown is the secret key.

How Logic Locking Addresses Hardware Trust Issues?

- **IP piracy and reverse engineering**: Even if an adversary steals the locked netlist or obtains it by reverse engineering an IC, the netlist will be of no use without the knowledge of the correct key. In XOR-based logic locking, key gates replace either functional buffers or inverters (key values are determined accordingly), introducing ambiguity in design reconstruction without the knowledge of the key values.
- **Overbuilding**: Although a foundry can overproduce the ICs with the intent of selling them illegally, the foundry cannot unlock those ICs without the secret keys.
- **Hardware Trojans**: Logic locking prevents the insertion of Trojans in a netlist by making it harder for the attacker to identify safe locations for inserting Trojans. The key gates alter the transition probabilities of the signals in a manner unknown to the attacker [11].
- **Counterfeiting**: Logic locking is not inherently meant to protect against all forms of counterfeiting. However, it can hinder certain forms of counterfeiting such as cloning that require reverse engineering [14].

20.2.2 Types of Logic Locking

Logic locking can be classified based on:

- The type of security primitives used. Based on these criteria, logic locking techniques can be classified as either sequential or combinational. More details on these techniques are presented in Sect. 20.2.2.
- The algorithm used for selecting the key gate locations. Based on these criteria, logic locking techniques can be classified as random [29], fault analysis-based [26], and key-interference-based [43]. These techniques are described in Sect. 20.4.
- The type of function used for logic locking. Recently, logic locking techniques used a dedicated protection block. The protection block is constructed using Boolean/cryptographic functions that have desirable security properties. Based on the protection function used, logic locking can be classified as a one-way function-based logic locking [43] and complementary functions-based logic locking [37].

Sequential Logic Locking

In sequential logic locking, additional logic (black) states are introduced in the state transition graph [7, 8]. The state transition graph is modified in such a way that the design reaches a valid state only on the application of a correct sequence of key bits. If the key is withdrawn, the design, once again, ends up in a black state and becomes non-functional.

Another sequential locking approach is to withhold a part of the design and replace it with programmable logic/lookup tables (LUTs). This way, the IP owner hides a part of the design from exposure to the rogue elements during manufacturing stages [1]. The withheld design is later configured into the programmable logic. The circuit will function correctly only when these elements are configured/programmed correctly. However, the introduction of programmable memory elements into the circuit incurs significant performance overhead.

Combinational Logic Locking

In combinational logic locking, also referred to as logic encryption or logic obfuscation in the literature, different combinational logic elements are inserted in a circuit to conceal the functionality of a design. These elements can be XOR/XNOR gates [22, 26, 29, 30], AND/OR gates [11], multiplexers [21, 26], or a combination of these elements [20]. The choice of the combinational elements depends upon security objective(s), metric(s) used for logic locking, and the performance overhead.

20.2.3 Metrics for Security Evaluation

Over the years, researchers have developed multiple metrics to quantify the security of logic locking techniques. These include:

- **Output corruptibility**. This metric quantifies protection against black-box usage of an IC. Even if an IC is locked, it may produce correct output for a subset of input patterns, for certain incorrect keys. Output corruptibility computes the Hamming distance between the correct output and the output obtained on supplying a random incorrect key. A percentage Hamming distance of 50% is desirable as it maximizes the ambiguity for an attacker [26].
- **Clique size**. Clique size quantifies the interference among the key gates and represents the resilience against the sensitization attack (see Sect. 20.3.1). A clique is a set of key gates that interfere with one another such that an individual key bit cannot be sensitized to primary outputs without knowing the values of other key bits in the set. The number of key gates that form the largest clique in a circuit is referred to as clique size [43].

- **Number of distinguishing input patterns (# of DIPs)**. Distinguishing input patterns are special input patterns that guide the key-pruning attacks [35]. They are used to differentiate between the design outputs when different key values are applied. The complexity of these attacks can be represented in terms of the number of distinguishing input patterns required for the attack to succeed.
- **Percentage of key bits recovered**. At times, certain attacks may succeed only partially and extract the values for only a subset of the key bits [21, 35, 40]. The success rate of these attacks is quantified in terms of percentage of key bit values recovered correctly.
- **Execution time**. The execution time of an attack can also be used to demonstrate the resilience of a logic locking technique against an attack [35].

20.3 Attacks on Logic Locking

In this section, we describe various attacks against logic locking techniques. A summary of the attacks is presented in Table 20.1.

20.3.1 Sensitization Attack

Sensitization attack makes use of the VLSI test principle of sensitization [23]. Sensitization of a net w to an output o implies that the value of w can be propagated to o, in either true or complementary form. The attacker analyzes the locked netlist and computes attack patterns that can sensitize individual key bit values to primary outputs. These attack patterns are then applied to the functional IC; by analyzing the observed IC responses, an attacker can determine the value of key bits.

Example: Consider the key bit K1 in Fig. 20.4. The key will be sensitized to output O1 if the value at the other input of gate G6 is 0, which can be achieved by setting I1 $=$ 1, I2 $=$ 0, and I3 $=$ 0. As the attacker has access to the functional IC, he/she can apply this pattern and determine the value of K1 on O1. For example, if the observed value of O1 is 0 for that input pattern, then K1 $=$ 0; otherwise K1 $=$ 1.

The sensitization attack is highly effective when key gates do not interfere with each other. A countermeasure against the attack that relies on increased key interference is presented in Sect. 20.4.3.

20.3.2 Key-Pruning Attacks

This class of attacks is the strongest variant of key recovery attacks that have been mounted on logic locking; accordingly, we describe this attack in more detail compared to other attacks. A key-pruning attack that broke all combinational logic locking

Table 20.1 A summary/comparison of the threat models of the attacks against logic locking

Attack	Attacker location	Information available to the attacker	Attack method	Proposed defense
Sensitization [23]	Foundry/end user	1. Locked netlist 2. Functional IC	Sensitization of key bits to circuit outputs	Key-interference-based logic locking [43]
Key-pruning [35]	Foundry/end user	1. Locked netlist 2. Functional IC	SAT-based algorithm that rules out incorrect keys iteratively	One-way function [43], complementary function [37]-based logic locking
Signal probability skew [42]	Foundry/end user	1. Locked netlist	Trace the output of Anti-SAT using signal skew and remove Anti-SAT block	–
Hill-climbing [21]	Foundry/test facility	1. Locked netlist 2. Test patterns and responses	Start with a random key CK. Flip the bits in CK based on the Hamming distance HD_O	Test-aware logic locking [21]
Test-data mining [44]	Foundry/test facility	1. Locked netlist 2. Test patterns and responses	Find the key that maximizes fault coverage and produces correct output for the test patterns	Posttest activation [44]
Differential power analysis [40]	Foundry/end user	1. Locked netlist 2. Functional IC	Collect power samples and analyze to generate a differential trace that identifies correct key value.	–

techniques that existed at that time was proposed by Subramanyan et al. [35]. The attack employs Boolean satisfiability-based (SAT) algorithms, and we refer to it as the *SAT attack*. The SAT attack iteratively prunes out incorrect key values by using *distinguishing input patterns (DIPs)*.

Distinguishing input patterns. A distinguishing input pattern X_d is an input value for which at least two different key values, $k1$ and $k2$, produce differing outputs, $o1$ and $o2$, respectively. Since $o1$ and $o2$ are different, at least one of the key values or both of them are incorrect. It is possible for a single DIP to rule out multiple incorrect key values. The DIPs are computed by constructing a miter-like circuit as illustrated in Fig. 20.5. The SAT solver finds a DIP X_d for which *diff* = *1*; i.e., the outputs

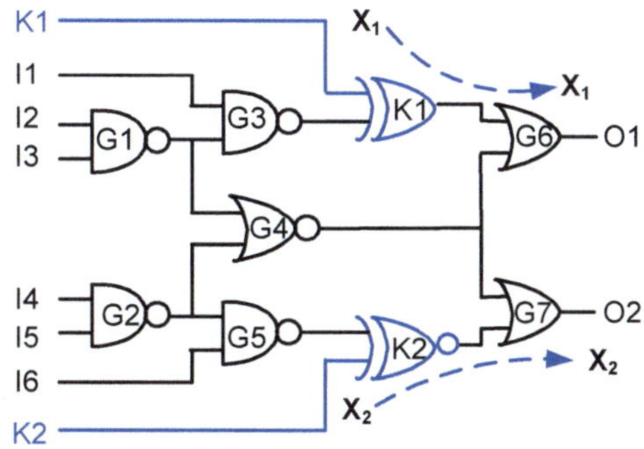

Fig. 20.4 Sensitization attack: By applying the input pattern 100000, an attacker can sensitize key bits K1 and K2 to the outputs O1 and O2, respectively, and observe their values. K1 propagates as is, while K2 is negated

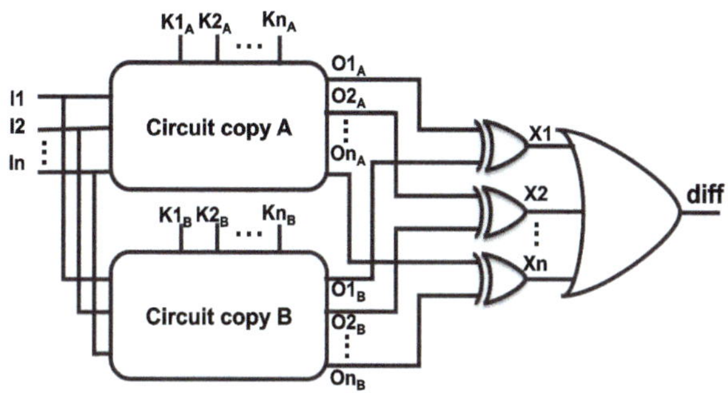

Fig. 20.5 Miter-like circuit to determine distinguishing input patterns [35]. A DIP is found iff diff = 1, implying that there exist at least two key values that produce differing outputs for a DIP. At least one of the two keys is incorrect

of the two circuits are different. X_d is applied to the functional IC, and the correct output I_d is obtained. The input–output pair (X_d, I_d) is used to identify incorrect key values. However, a single pattern may not rule out all incorrect keys. To rule out the remaining incorrect key values, more distinguishing patterns are needed. A new pair (X_d, I_d) is added to the SAT formula in each iteration, and the SAT formula is updated. The attack is successful when no further DIP is found, which implies that all incorrect key values have been pruned.

Example. Let us consider an example SAT attack on the logic-locked circuit shown in Fig. 20.6a. Figure 20.6b represents the output values of the locked circuit for different key input combinations. The values $(k0, \ldots, k7)$ represent all possible values for three key inputs $\{K1, K2, K3\}$. When SAT attack is launched, it takes four DIPs to obtain the correct key. The last column in the table lists the keys eliminated in each iteration. For example, in iteration 4, the pattern 100 is used that eliminates all incorrect keys, and thus identifies $k5$ as the correct key.

Attack complexity. The efficiency of SAT attack depends on the order of choosing the DIPs. The total execution time for SAT attack comprising λ iterations with t_i as the execution time for the ith iteration is $T = \sum_{i=1}^{\lambda} t_i$ [37]. SAT attack can be mitigated if either t_i or λ increases exponentially with the key size. The larger the number of incorrect key values ruled out per DIP, the fewer the patterns needed for the attack, which implies a smaller execution time of the attack algorithm.

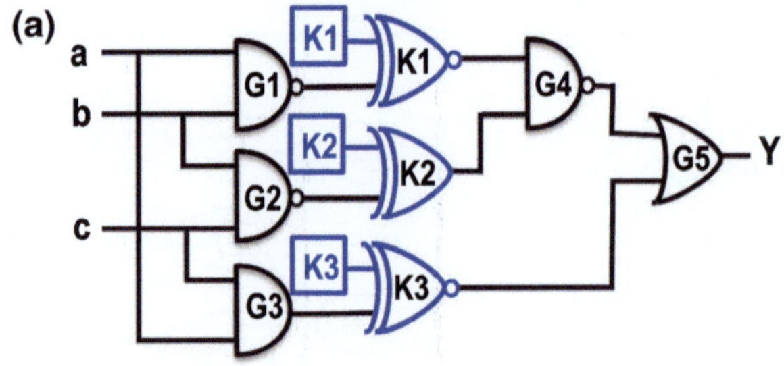

(b)

abc	Y	k0	k1	k2	k3	k4	k5	k6	k7	Incorrect keys detected
000	0	1	1	1	1	1	0	1	1	
001	0	1	1	1	1	1	0	1	1	
010	0	1	1	1	1	1	0	1	1	
011	1	1	1	1	1	0	1	1	1	iter 1: k4
100	0	1	1	1	1	1	0	1	1	iter 4: rest incorrect keys
101	1	1	0	1	1	1	1	1	1	
110	1	1	1	1	1	1	1	1	0	iter 3:k7
111	1	1	1	0	1	1	1	1	1	iter 2: k2

Fig. 20.6 **a** A circuit locked using XOR/XNOR key gates. The correct key $\{K1, K2, K3\}$ value is 101. **b** Analysis of key-pruning attack against logic locking [35]. The key values pruned by a DIP are marked in orange color

Logic locking techniques that thwart the SAT attack include one-way function-based logic locking [43] and complementary function-based logic locking [37] (refer to Sect. 20.4 for details).

20.3.3 Signal Probability Skew Attack

Signal probability skew (SPS) attack exploits structural traces in the netlist to locate and isolate the protection logic [42] in complementary function-based logic locking [37] (see Sect. 20.4.4). The attack uses the notion of signal probability skew (SPS); SPS of a signal x is defined as:

$$SPS(x) = Pr[x = 1] - 0.5 \tag{20.1}$$

where $Pr[x = 1]$ indicates the probability that signal x is 1. For a signal y that is rarely 1, e.g., the output of a large AND tree, $SPS(y) \approx -0.5$; similarly, for a NAND tree the $SPS(y) \approx 0.5$.

The Anti-SAT block delivers maximal key-pruning attack resistance, when functions g and $\overline{g}$ are skewed in opposite directions. This implies that the two inputs of gate Y, which forms the output of the Anti-SAT block, will be skewed in opposite directions. Let us define ADS_Y as the absolute difference of the skew at the inputs of gate Y. As the key size n increases, ADS_Y gets closer to a value of 1, as illustrated in Fig. 20.11. The signal probability skew (SPS) attack [42] exploits the ADS_Y value to locate and remove the logic locking circuitry [42]. The attack simply computes the ADS value for all the gates in a circuit; the gate with the maximum ADS value is the most likely to be the output gate of the block. Despite implementing structural and functional obfuscation, the SPS attack can break Anti-SAT block within minutes. The underlying reason is the structural traces left by the logic locking technique that enable removal attacks.

20.3.4 Test-Data Attacks

While the typical attacks on logic locking rely on input–output observations recorded from a functional IC, test-data attacks [21, 44] extract secret information from the test data. Test data is generated at the design house, sent to the foundry/dedicated test facility, and utilized during manufacturing test to classify ICs as faulty/fault-free. Automatic test pattern generation (ATPG) algorithms are used to generate the test data [5]. Logic-locked ICs may be activated/unlocked prior to the manufacturing test, which is referred to as *pretest activation*. To facilitate pretest activation, the ATPG algorithms require knowledge of the secret key during test generation, producing test data that embeds secret information. While ATPG is run by the trusted designer, the resulting test data may be exploited by the untrusted tester. The two activation methods for logic locking are illustrated in Fig. 20.7.

Hill-Climbing Search Attack

The hill-climbing search-based attack [21] uses test data to guess the secret key for pretest-activated ICs. The attack is based on a hill-climbing search algorithm which uses Hamming distance as a guiding metric.

The attack starts with an initial random key guess K_{guess}. The initial Hamming distance HD_{min} is then computed between the test responses and the locked circuit outputs corresponding to test stimuli. One of the key bits in K_{guess} is then inverted, and Hamming distance HD_{inv} is computed; the inversion is recorded only if HD_{inv} <

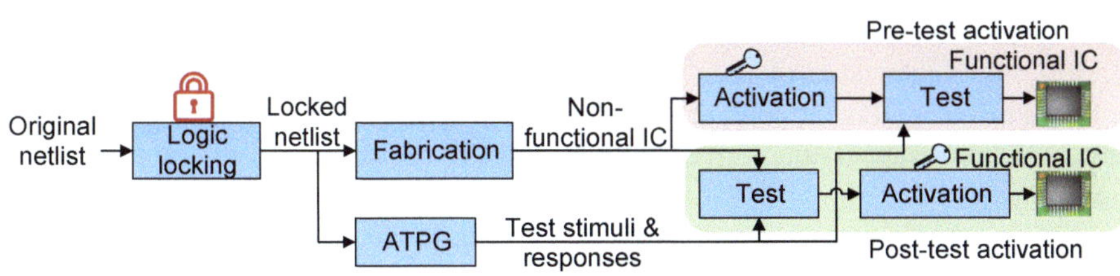

Fig. 20.7 Logic locking and manufacturing test: pretest and posttest activation [44]

HD_{min}; HD_{min} is then updated. The inversion process is repeated for all the bits in K_{guess}. The attack is successful when $HD_{min} = 0$.

While the hill-climbing attack can successfully break random [30] and fault analysis-based logic locking [26], it is not effective against key-interference-based logic locking [43]. Another method to prevent the attack is to insert additional "test-aware" logic into the circuit to produce identical test responses, independent of the key values [21].

Test-Data Mining Attack

Test-data mining attack extracts the secret key value from the test data [44] for pretest-activated ICs. The attack utilizes *fault coverage FC*—the number of faults in a circuit that can be detected with a set of test data—as guiding metric, since the same metric is used by the ATPG algorithms to generate test data [5].

The attack uses test stimuli S and responses Γ as constraints, and determines the key value K_P that maximizes the fault coverage for the locked circuit L_K, as shown in the following optimization:

$$\begin{aligned} &\text{maximize} \quad FC \\ &\text{subject to} \quad \underset{1\le i\le N}{\forall} L_K(K_P, T_i) = \Gamma_i \\ &\text{solve for} \quad K_P \end{aligned} \tag{20.2}$$

where N is the number of test stimuli/responses in the test data.

Compared to the hill-climbing attack [21], test-data mining attack is effective against key-interference-based logic locking as well. The attack can operate successfully using partial test data set as well [44]. Posttest activation with dummy keys used during manufacturing test is an effective countermeasure against the attack [44]. Using dummy keys during test generation does not embed the secret information in the test data and renders test-data attacks ineffective.

20.3.5 Side-Channel Attacks

When a security-critical algorithm is implemented in hardware, its secret information may be leaked through covert channels such as power, electromagnetic radiation, and time, known as side channels [18]. While the previously mentioned attacks exploit algorithmic weaknesses of logic locking, Yasin et al. [40] launched a *differential power analysis (DPA)* attack [18] on logic locking. DPA attack is one of the most powerful side-channel attacks that has been used to break most of the cryptographic algorithms.

DPA attack is launched by collecting power samples and the ciphertext output from the IC under attack corresponding to a large set of plaintext inputs. The collected samples are then analyzed using statistical analysis. The statistical analysis yields a differential trace, which tends to be high for the correct key value and zero for

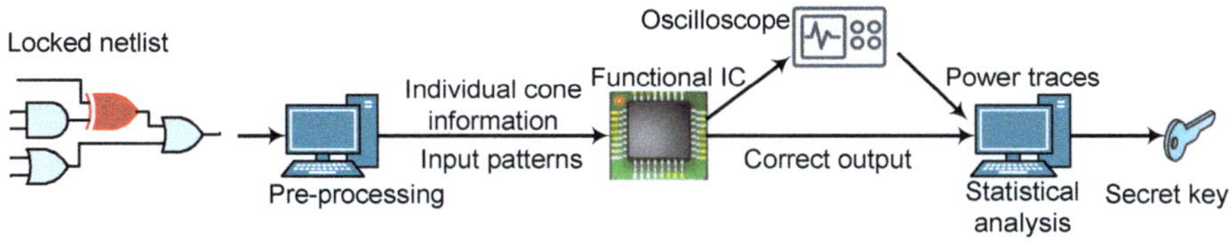

Fig. 20.8 Differential power analysis (DPA) attack against logic locking. The attack is run on individual logic cones to reduce attack complexity. Statistical analysis of the collected power traces yields the secret key value [40]

the incorrect key values [18]. The DPA attack may be launched on individual logic cones in a divide-and-conquer approach. The DPA attack against logic locking is illustrated in Fig. 20.8 [40].

While random logic locking is highly vulnerable to the DPA attack, key-interference-based logic locking exhibits resilience against the DPA attack [40]. As reported in [40], the DPA attack could extract more than 70% of the key bits correctly in 50% of the circuits in case of random logic locking, whereas only about 50% of key bits could be revealed in only 25% of the circuits in case of key-interference-based logic locking (for details of logic locking techniques, refer to Sect. 20.4). Key-interference-based logic locking exhibits strong *key aliasing* which implies that multiple key values generate identical differential traces and reduce the effectiveness of the DPA attack.

20.4 Logic Locking Techniques

20.4.1 Random Logic Locking

The random logic locking technique [30] locks a design by inserting XOR key gates at random locations in a netlist. Figure 20.9a shows an example of a netlist locked with two key gates, K1 and K2, using random logic locking. In random logic locking,

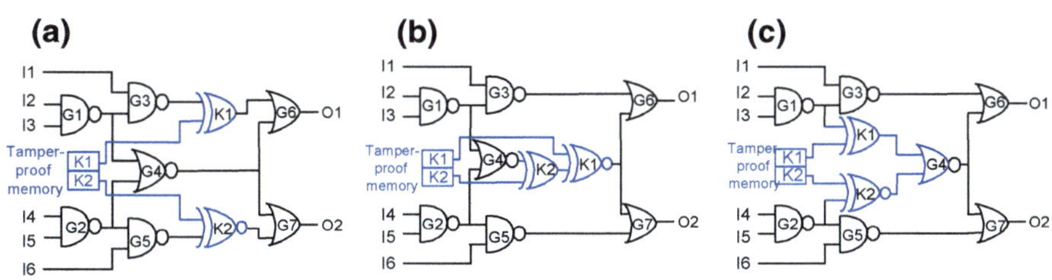

Fig. 20.9 Key gate insertion based on different logic locking techniques. **a** Random [30], **b** fault analysis-based [26], and **c** key-interference-based logic locking [43]

the key gates are spread uniformly in the entire netlist. The interference among the key gates tends to be minimal, rendering random logic locking vulnerable to attacks. Random logic locking is vulnerable to sensitization [23], key-pruning [35], hill-climbing [21], test-data mining [44], and DPA [40] attacks.

20.4.2 Fault Analysis-Based Logic Locking

Fault analysis-based logic locking aims at preventing black-box usage of an IC [26]. In random logic locking, even incorrect keys may lead to correct output for certain input patterns. Fault analysis-based logic locking ensures maximum corruption at the output bits when incorrect keys are used. The output corruption is measured in terms of percentage Hamming distance between the correct output and the incorrect output, obtained upon applying incorrect keys.

In fault analysis-based logic locking, the key gates are inserted at the most *influential* locations in the circuit. The influential locations are those that impact the most number of outputs when incorrect key values are applied. Applying incorrect key values has the same effect as exciting stuck-at faults on the output of key gates. Thus, fault analysis/simulation tools can be used to find the most influential locations in a circuit. Figure 20.9b shows a netlist locked using fault analysis-based logic locking. The key gates tend to be inserted in back-to-back fashion, which undermines the security of logic locking as this creates multiple correct key values [23].

Fault analysis-based logic locking is vulnerable to sensitization [23], hill-climbing [21], test-data mining [44], key-pruning [35], and DPA [40] attacks.

20.4.3 Key-Interference-Based Logic Locking

Key-interference-based logic locking, also referred to as strong logic locking [23], attempts to mitigate the sensitization attack by inserting key gates in a way that maximizes the interference among them while preventing the sensitization of the key bits on an individual basis. With increased interference among key gates, the attacker is forced to brute-force an exponentially increasing number of key combinations [43].

Example. Consider the netlist in Fig. 20.9c. The netlist has two key gates, K1 and K2, that are inserted using the key-interference-based logic locking algorithm. It can be seen that K1 and K2 interfere with each other's path to the primary outputs. It is not possible for an attacker to sensitize either K1 or K2 to a primary output on an individual basis.

While key-interference-based logic locking thwarts sensitization [23], hill-climbing [21], and DPA attacks [40], it is vulnerable to key-pruning attacks [35].

20.4.4 Complementary Functions-Based Logic Locking

Complementary function-based logic locking thwarts key-pruning (SAT-based) attacks by minimizing the number of keys that are pruned by a single DIP [37]. The key-pruning attack requires the maximum number of DIPs to eliminate all incorrect keys when each DIP eliminates at most one incorrect key in an iteration; then, the number of required attack iterations is exponential in the number of key inputs, rendering the attack computationally infeasible for large key inputs.

Complementary function-based logic locking block, also referred to as Anti-SAT, comprises two blocks, $B_1 = g(X, K_{l1})$ and $B_2 = \overline{g(X, K_{l2})}$, as shown in Fig. 20.10a. The two blocks share the same inputs X, but are locked using different key bits K_{l1} and K_{l2}; the two blocks produce complementary outputs when correct keys are applied. The outputs of B_1 and B_2 drive an AND/OR gate to produce the output signal Y. In a functional IC with the correct key in place, Y has a constant value, either 0 or 1.

Example. An instance of Anti-SAT block is shown in Fig. 20.10b [37]. At the inputs of B_1 and B_2, a set of XOR/XNOR key gates are inserted. The number of key inputs is the same as the number of signals tapped from the logic-locked circuit, i.e., $|K_{l1}| = |K_{l2}| = |X| = n$. The resulting key size is thus $2n$. The output Y is implemented as $Y = g(X \oplus K_{l1}) \wedge \overline{g(X \oplus K_{l2})}$. The output Y is 0 in this example when the correct keys K_{l1} and K_{l2} are applied. For incorrect keys, Y may take on the value 1, flipping an internal net in the netlist. Inversions may be added to produce a value of 1 as the correct value of Y.

Attack resilience. The key-pruning attack complexity on Anti-SAT block to decode the $2n$ key bits is defined in terms of the number of input vectors that make the function g equal to 1, i.e., the on-set of g [37]. For an n-bit input vector $\mathbf{L} \in \{0, 1\}^n$, such input vectors are elements of the set,

$$L^T = \{\mathbf{L} | g(\mathbf{L}) = 1\}, |L^T| = p \tag{20.3}$$

For maximum resilience against the key-pruning attack, Anti-SAT constructs g in such a way that the number of such input vectors p is close to either 1 or $2^n - 1$.

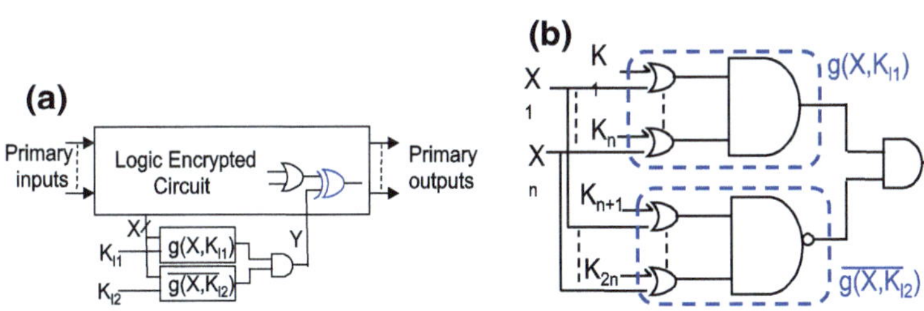

Fig. 20.10 **a** Integration of complementary function-based logic locking with a logic-locked circuit [37]. **b** An instance of Anti-SAT block to resist key-pruning attack [37]

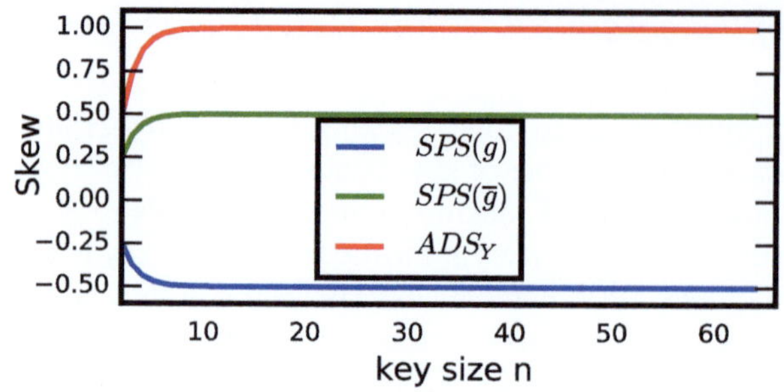

Fig. 20.11 Impact of n on ADS_Y, the absolute difference of skew at the inputs of gate Y in complementary function-based logic locking. $SPS(g)$ and $SPS(\bar{g})$ represent the skew of the AND and NAND tree. The SPS attack uses ADS_Y to locate and remove the circuit that implements complementary function-based logic locking

For the block in Fig. 20.10, $p = 1$. The lower bound on the number of key-pruning attack iterations to recover the $2n$ key bits of Anti-SAT block is $\lambda_l = \frac{2^{2n}-2^n}{p(2^n-p)}$ [37]. For $p \in \{1, 2^n - 1\}$, the number of required iterations λ_l is 2^n, i.e., exponential in half the number of key bits in Anti-SAT block. So, the SAT attack resilience of Anti-SAT hinges on p being very small or very large. As Anti-SAT block provides a provable measure to increase the SAT attack complexity exponentially, the conventional techniques such as fault analysis-based logic locking need to be combined with Anti-SAT block to achieve high-output corruptibility and prevent black-box usage.

A trivial attack could simulate the circuit and find the complementary pair of signal outputs of g and $\bar{g}$ leading to the identification and removal of Anti-SAT block. Key gates can be inserted at the inputs of Anti-SAT block to break complementary relations between signals, thereby providing *functional obfuscation.*

Another simple attack could be in the form of circuit partitioning to identify the isolated Anti-SAT block and remove it from the netlist. To thwart such attacks, *structural obfuscation* based on MUX-based logic locking was proposed to increase the interconnectivity between the logic-locked circuit and Anti-SAT block [37]. The overall obfuscated Anti-SAT block (OA) has $4n$ key gates.

20.4.5 One-Way Function-Based Logic Locking

This logic locking technique integrates one-way random functions (ORFs), such as fixed-key AES [13], with existing techniques to achieve resilience against the SAT attack. A subset of the key inputs in the locked netlist is connected to the outputs of ORF as illustrated in Fig. 20.12. The inputs of the one-way function are the secret keys stored in tamper-proof memory. The overall circuit represents a hard SAT instance that becomes infeasible for a SAT solver when K1 is large. Figure 20.13 presents the execution time of the key-pruning attack on one-way function-based logic locking.

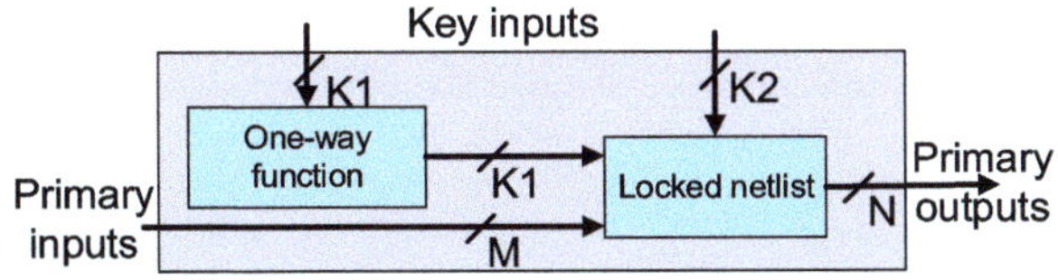

Fig. 20.12 One-way random function (ORF)-based countermeasure against key-pruning attacks. K1 key inputs in the locked netlist are connected to the ORF circuit

Fig. 20.13 Time required by the key-pruning attack [35] to break the fixed-key AES+c7552 circuit for different numbers of key inputs connected to the fixed-key AES circuit (K1). The blue line represents the best-fitting exponential curve to the experimental data

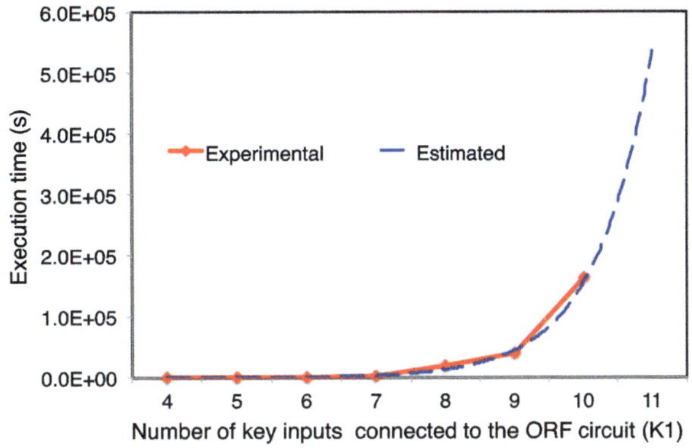

The attack execution time grows exponentially as more key inputs are integrated with the ORF circuit.

AES with a fixed key performs a pseudorandom function [13]. The two circuits (one-way function circuit and the locked netlist) must be synthesized together to prevent removal attack [43]. Another property of the fixed-key AES is that it is computationally infeasible to determine the inputs of AES from its outputs when the key is unknown [13]. Thus, it is infeasible to backtrace from the outputs of the design and determine the inputs to the AES.

20.4.6 Attack/Defense Matrix and Future Guidelines

In this section, we provide a broader view of attacks and defenses presented so far, and identify a set of guidelines for developing secure logic locking countermeasures. Table 20.2 presents the resiliency of various logic locking countermeasures against each of the aforementioned attacks.

Key-pruning attack [35] breaks all logic locking techniques that had been developed prior to its introduction. Thus, any incremental improvement over these defenses, such as achieving better Hamming distance or improved key-interference at lower overhead, will be susceptible to the key-pruning attack. Thus, resilience against the key-pruning attack must be explicitly considered when developing any future countermeasures.

Table 20.2 Representing attack resiliency of logic locking techniques against the existing attacks. ✓ denotes that the technique is vulnerable to the attack, ✗ denotes that the technique is resilient, and - denotes "no data available." It must be noted that the threat models and the information required vary among the attacks

		Logic locking techniques				
		Random [30]	Fault [26]	Key-Inter. [43]	Compl. func. [37]	One-Way func. [43]
Attacks	Sensitization [23]	✓	✓	✗	✗	✗
	Key-pruning [35]	✓	✓	✓	✗	✗
	SPS attack [42]	✗	✗	✗	✓	✗
	Hill climbing [21]	✓	✓	✗	-	-
	Test-data mining [44]	✓	✓	✓	-	-
	DPA [40]	✓	✓	✓	-	-

The key-pruning attack resilient approaches also have their limitations. One-way function-based logic locking incurs prohibitive area and power overheads, and is suitable for only very large designs. Complementary function-based logic locking thwarts the key-pruning attacks, while it is highly vulnerable to simple removal (SPS) attack that essentially isolates and removes the logic locking circuitry. Thus, attack resiliency against multiple attacks must be taken into account, and it must be ensured that no secrets are hard-coded in logic gates. DPA attack is only one example of side-channel attacks; the future countermeasures must also take into account other side-channel attacks. Each countermeasure must be evaluated based on the metrics presented in Sect. 20.2.3.

We would also like to direct the interested to reader to the recent research efforts in logic locking that appeared after the acceptance but prior to the final publication of this chapter. The most notable defense techniques, all of which focus on thwarting the SAT attack, include SARLock [41], TTLock [47], cyclic logic locking [32], and stripped-functionality logic locking (SFLL) [46]. The new classes of attacks that have emerged in the meantime are: (1) removal attacks [45] and approximate attacks [33, 34]. A brief review of the latest logic locking research can be found in [39].

20.5 Conclusions

Globalization of the IC supply chain has given rise to multiple challenges for IC designers. However, the designers can leverage the DfTr solutions such as logic locking to protect their designs from piracy, reverse engineering, and overbuilding attack. Logic locking research has gained significant traction over the last few years. Stronger defenses as well as attacks have been developed over the years, with the SAT attack still being the most powerful attack. Whereas most of the existing countermeasures focus on thwarting individual attacks, it is imperative to develop holistic solutions that offer cost-effective security against generic classes of attacks.

Acknowledgements This work has been supported by the Semiconductor Research Corporation (SRC) under the Abu Dhabi SRC Center of Excellence on Energy-Efficient Electronic Systems (ACE^4S), Contract 2013 HJ2440, with funding from the Mubadala Development Company, Abu Dhabi, UAE.

References

1. A. Baumgarten, A. Tyagi, J. Zambreno, Preventing IC piracy using reconfigurable logic barriers. IEEE Des. Test Comput. **27**(1), 66–75 (2010)
2. M. Berry, G. John, Outsourcing Test—What are the most valuable engagement periods? (2014), http://www.amkor.com/go/outsourcing-test. Accessed 16 May 2016
3. S. Bhunia, M.S. Hsiao, M. Banga, S. Narasimhan, Hardware trojan attacks: threat analysis and countermeasures. Proc. IEEE **102**(8), 1229–1247 (2014)
4. D.S. Board, Defense Science Board (DSB) study on High Performance Microchip Supply (2005), www.acq.osd.mil/dsb/reports/ADA435563.pdf. Accessed 16 March 2015
5. M.L. Bushnell, V.D. Agrawal, *Essentials of Electronic Testing for Digital, Memory, and Mixed-Signal VLSI Circuits* (Kluwer Academic Publishers, Boston, 2000). ISBN 9780792379911
6. A.E. Caldwell, H.J. Choi, A.B. Kahng, S. Mantik, M. Potkonjak, G. Qu, J.L. Wong, Effective iterative techniques for fingerprinting design IP, in *Proceedings of the IEEE/ACM Design Automation Conference*, 1999, pp. 843–848
7. R.S. Chakraborty, S. Bhunia, HARPOON: an obfuscation-based SoC design methodology for hardware protection. IEEE Trans. Comput-Aided Design Integr. Circuits Syst. **28**(10), 1493–1502 (2009)
8. R.S. Chakraborty, S. Bhunia, Security against Hardware Trojan through a novel application of design obfuscation, in *Proceedings of the IEEE/ACM International Conference on Computer-Aided Design*, 2009, pp. 113–116
9. Chipworks, Reverse engineering software (2016), http://www.chipworks.com/en/technical-competitive-analysis/resources/reerse-engineering-software
10. B. Colombier, L. Bossuet, Survey of hardware protection of design data for integrated circuits and intellectual properties. IET Comput. Digit. Techniq. **8**(6), 274–287 (2014)
11. S. Dupuis, P. Ba, G.D. Natale, M. Flottes, B. Rouzeyre, A novel hardware logic encryption technique for thwarting illegal overproduction and hardware Trojans, in *Proceedings of the IEEE International On-Line Testing Symposium*, 2014, pp. 49–54
12. K.M. Goertzel, B. Hamilton, Integrated circuit security threats and hardware assurance countermeasures. CrossTalk (2013), p. 33
13. O. Goldreich, *Foundations of Cryptography*. Basic Tools, vol. 1 (Cambridge University Press, 2001). ISBN 978-0521035361

14. U. Guin, K. Huang, D. DiMase, J.M. Carulli, M. Tehranipoor, Y. Makris, Counterfeit integrated circuits: a rising threat in the global semiconductor supply chain. Proc. IEEE **102**(8), 1207–1228 (2014)
15. R.W. Jarvis, M.G. McIntyre, Split Manufacturing Method for Advanced Semiconductor Circuits. US Patent 7,195,931 (2007)
16. A.B. Kahng, J. Lach, W.H. Mangione-Smith, S. Mantik, I.L. Markov, M. Potkonjak, P. Tucker, H. Wang, G. Wolfe, Watermarking techniques for intellectual property protection, in *Proceedings of the IEEE/ACM Design Automation Conference*, 1998, pp 776–781
17. R. Karri, J. Rajendran, K. Rosenfeld, M. Tehranipoor, Trustworthy hardware: identifying and classifying hardware Trojans. Computer **43**(10), 39–46 (2010)
18. P. Kocher, J. Jaffe, B. Jun, Differential power analysis, in *Advances in Cryptology* (Springer, 1999), pp. 388–397
19. F. Koushanfar, Integrated circuits metering for piracy protection and digital rights management: an overview, in *Proceedings of the Great Lakes Symposium on VLSI*, 2011, pp. 449–454
20. Y.W. Lee, N. Touba, Improving logic obfuscation via logic cone analysis, in *Proceedings of the Latin-American Test Symposium*, 2015, pp. 1–6
21. S.M. Plaza, I.L. Markov, Solving the third-shift problem in IC piracy with test-aware logic locking. IEEE Trans. Comput.-Aided Design Integr. Circuits Syst. **34**(6), 961–971 (2015)
22. J. Rajendran, Y. Pino, O. Sinanoglu, R. Karri, Logic encryption: a fault analysis perspective, *Proceedings Design, Automation and Test in Europe*, 2012, pp. 953–958
23. J. Rajendran, Y. Pino, O. Sinanoglu, R. Karri, Security analysis of logic obfuscation, in *Proceedings of the IEEE/ACM Design Automation Conference*, 2012, pp. 83–89
24. J. Rajendran, M. Sam, O. Sinanoglu, R. Karri, Security Analysis of integrated circuit camouflaging, in *Proceedings of the ACM/SIGSAC Conference on Computer & Communications Security*, 2013, pp. 709–720
25. J. Rajendran, O. Sinanoglu, R. Karri, Regaining trust in VLSI design: design-for-trust techniques. Proc. IEEE **102**(8), 1266–1282 (2014)
26. J. Rajendran, H. Zhang, C. Zhang, G. Rose, Y. Pino, O. Sinanoglu, R. Karri, Fault analysis-based logic encryption. IEEE Trans. Comput. **64**(2), 410–424 (2015)
27. A.P. Room, Silicon Laboratories and ASE announce milestone shipment of 10 million tested integrated circuits (2014), http://www.aseglobal.com/en/News/PressRoomDetail.aspx?ID=45. Accessed 16 May 2016
28. M. Rostami, F. Koushanfar, R. Karri, A primer on hardware security: models, methods, and metrics. Proc. IEEE **102**(8), 1283–1295 (2014)
29. J. Roy, F. Koushanfar, I.L. Markov, EPIC: ending piracy of integrated circuits, in *Proceedings Design, Automation and Test in Europe*, 2008, pp. 1069–1074
30. J.A. Roy, F. Koushanfar, I.L. Markov, Ending piracy of integrated circuits. Computer **43**(10), 30–38 (2010)
31. SEMI, Innovation is at Risk Losses of up to $4 Billion Annually due to IP Infringement (2008), www.semi.org/en/Issues/IntellectualProperty/ssLINK/P043785. Accessed 10 June 2015
32. K. Shamsi, M. Li, T. Meade, Z. Zhao, D.Z. Pan, Y. Jin, Cyclic obfuscation for creating sat-unresolvable circuits, in *Proceedings of the on Great Lakes Symposium on VLSI 2017*, 2017, ACM, pp. 173–178
33. K. Shamsi, M. Li, T. Meade, Z. Zhao, D.Z. Pan, Y. Jin, AppSAT: approximately deobfuscating integrated circuits, in *IEEE International Symposium on Hardware Oriented Security and Trust*, 2017, pp. 95–100
34. Y. Shen, H. Zhou, Double dip: Re-evaluating security of logic encryption algorithms. Cryptology ePrint Archive, Report 2017/290, (2017), http://eprint.iacr.org/2017/290
35. P. Subramanyan, S. Ray, S. Malik, Evaluating the security of logic encryption algorithms, in *Proceedings of the IEEE International Symposium on Hardware Oriented Security and Trust*, pp. 137–143
36. R. Torrance, D. James, The State-of-the-Art in semiconductor reverse engineering, in *Proceedings of the IEEE/ACM Design Automation Conference*, 2011, pp. 333–338

37. Y. Xie, A. Srivastava, Mitigating SAT attack on logic locking. IACR Cryptology ePrint Archive **2016**, 590 (2016)
38. X. Xu, B. Shakya, M. Tehranipoor, D. Forte, Novel bypass attack and BDD-based tradeoff analysis against all known logic locking attacks. Cryptology ePrint Archive, Report 2017/621, 2017, http://eprint.iacr.org/2017/621
39. M. Yasin, O. Sinanoglu, Evolution of logic locking, in *2017 IFIP/IEEE International Conference on Very Large Scale Integration (VLSI-SoC)*, 2017, pp. 1–6
40. M. Yasin, B. Mazumdar, S.S. Ali, O. Sinanoglu, Security analysis of logic encryption against the most effective side-channel attack: DPA, in *Proceedings of the IEEE International Symposium on Defect and Fault Tolerance in VLSI and Nanotechnology Systems*, 2015, pp. 97–102
41. M. Yasin, B. Mazumdar, J. Rajendran, O. Sinanoglu, SARLock: SAT attack resistant logic locking, in *IEEE International Symposium on Hardware Oriented Security and Trust*, 2016, pp. 236–241
42. M. Yasin, B. Mazumdar, O. Sinanoglu, J. Rajendran, Security analysis of anti-SAT. Cryptology ePrint Archive, Report 2016/896, (2016), http://eprint.iacr.org/2016/896
43. M. Yasin, J. Rajendran, O. Sinanoglu, R. Karri, On Improving the security of logic locking. IEEE Trans. CAD of Integr. Circuits Syst. (2016)
44. M. Yasin, S.M. Saeed, J. Rajendran, O. Sinanoglu, Activation of logic encrypted chips: pre-test or post-test?, in *Proceedings Design, Automation Test in Europe*, 2016, pp. 139–144
45. M. Yasin, B. Mazumdar, O. Sinanoglu, J. Rajendran, Removal attacks on logic locking and camouflaging techniques. IEEE Trans. Emerg. Topics Comput. **99**(0) (2017)
46. M. Yasin, A. Sengupta, M.T. Nabeel, M. Ashraf, J.J. Rajendran, O. Sinanoglu, Provably-secure logic locking: from theory to practice, in *Proceedings of the ACM SIGSAC Conference on Computer and Communications Security*, 2017, pp. 1601–1618
47. M. Yasin, A. Sengupta, B.C. Schafer, Y. Makris, O. Sinanoglu, J.J. Rajendran, What to lock?: functional and parametric locking, in *Great Lakes Symposium on VLSI*, 2017, pp. 351–356

Epilog

أنا في أول الطريق والنهاية بعيدة, وأنا جزء من شعب الإمارات الذي هو من أسعد الشعوب في العالم وأكثرها تفاؤلاً في المستقبل. وأشعر الآن بالسعادة لإنجاز هذا الكتاب، وأرى كل شيء من حولي جميلاً، فقد أهدت إليّ الحياة كل نعمها وهداياها حتى الذي يزعجني بدأ يسعدني سعادة الخالدين.

د. عبد الخالق عبد الله، اعترافات أكاديمي متقاعد

I am at the beginning of the road, and the destination is far away. And I belong to the Emerati people - one of the happiest peoples in the world with the brightest outlook on the future. And I am feeling happy that I have completed this book, and I see everything around me beautiful. For life has given me all its blessings and gifts. Even what used to bother me is now giving me everlasting happiness.

Dr. Abdul Khaleq Abdullah, Confessions of a Retired Academic

Ibrahim (Abe) M. Elfadel and M. Ismail (eds.), *The IoT Physical Layer*,
https://doi.org/10.1007/978-3-319-93100-5

Index

Ibrahim (Abe) M. Elfadel and M. Ismail (eds.), *The IoT Physical Layer*,
https://doi.org/10.1007/978-3-319-93100-5

R

S

T

U

V

MIX
Papier aus verantwortungsvollen Quellen
Paper from responsible sources
FSC® C105338

If you have any concerns about our products,
you can contact us on
ProductSafety@springernature.com

In case Publisher is established outside the EU,
the EU authorized representative is:
Springer Nature Customer Service Center GmbH
Europaplatz 3, 69115 Heidelberg, Germany

Printed by Libri Plureos GmbH
in Hamburg, Germany